SALMON AT THE EDGE

SALMON AT THE EDGE

Edited by

Derek Mills
MSc, PhD, FIFM, FLS
Atlantic Salmon Trust

From a symposium organized by

Published with the financial support of

FISHMONGERS' COMPANY   ≡Scottish and Southern Energy plc

by

Blackwell
Science

© 2003 by Blackwell Science Ltd,
a Blackwell Publishing company

Editorial offices:
Blackwell Science Ltd, 9600 Garsington Road,
Oxford OX4 2DQ, UK
 Tel: +44 (0)1865 776868
Iowa State Press, a Blackwell Publishing Company,
2121 State Avenue, Ames, Iowa 50014-8300, USA
 Tel: +1 515 292 0140
Blackwell Science Asia Pty, 550 Swanston Street,
Carlton, Victoria 3053, Australia
 Tel: +61 (0)3 8359 1011

The right of the Author to be identified as the
Author of this Work has been asserted in
accordance with the Copyright, Designs and
Patents Act 1988.

All rights reserved. No part of this publication may
be reproduced, stored in a retrieval system, or
transmitted, in any form or by any means,
electronic, mechanical, photocopying, recording or
otherwise, except as permitted by the UK
Copyright, Designs and Patents Act 1988, without
the prior permission of the publisher.

First published 2003

Library of Congress
Cataloging-in-Publication Data
Salmon at the edge/edited by Derek Mills.
  p. cm.
'From a conference organized by the Atlantic
Salmon Trust, the Atlantic Salmon Federation.'
Includes bibliographical references and index.
ISBN 0-632-06457-9
1. Atlantic salmon–Congresses. 2. Fishery
conservation–Congresses. I. Mills, Derek
Henry. II. Atlantic Salmon Trust.
III. Atlantic Salmon Federation.

QL638.S2S233 2003
639.9'7756–dc21
   2003044332

ISBN 0-632-06457-9

A catalogue record for this title is available from the
British Library

Set in 10/13pt Times
By DP Photosetting, Aylesbury, Bucks
Printed and bound in Great Britain using acid-free
paper by MPG Books Ltd, Bodmin, Cornwall

For further information on Blackwell Publishing,
visit our website:
www.blackwellpublishing.com

Contents

Preface	ix
Message from His Royal Highness The Prince of Wales	xi

1 Opening Address
   Eroding the Edge – The Future of Coastal Waters   1
   Professor Sir F. Holliday

INTO THE UNKNOWN – THE FIRST WEEKS AT SEA   5

2 Migration and Distribution of Atlantic Salmon Post-Smolts in the
   North Sea and North-East Atlantic   7
   *M. Holm, J.C. Holst, L.P. Hansen, J.A. Jacobsen, N. O'Maoiléidigh
   and A. Moore*

3 Smolt Tracking at Sea – A Search Rewarded   24
   G.L. Lacroix

4 The Ecology of Post-Smolts of Atlantic Salmon   25
   L.P. Hansen, M. Holm, J.C. Holst and J.A. Jacobsen

RUNNING THE GAUNTLET – INSHORE AND COASTAL HAZARDS   41

5 The Significance of Marine Mammal Predation on Salmon and
   Sea Trout   43
   S.J. Middlemas, J.D. Armstrong and P.M. Thompson

6 Predation on Post-Smolt Atlantic Salmon by Gannets: Research
   Implications and Opportunities   61
   W.A. Montevecchi and D.K. Cairns

7 Progress in Ending Mixed-Stock Interceptory Fisheries: United Kingdom   78
   A. Whitehead

8 Closing the North American Mixed-Stock Commercial Fishery for
   Wild Atlantic Salmon   84
   S. Chase

9 Assessing and Managing the Impacts of Marine Salmon Farms on
   Wild Atlantic Salmon in Western Scotland: Identifying Priority Rivers
   for Conservation   93
   J.R.A. Butler and J. Watt

vi   *Contents*

10	Relationship Between Sea Lice Infestation, Sea Lice Production and Sea Trout Survival in Ireland, 1992–2001 *P.G. Gargan, O. Tully and W.R. Poole*	119
11	Mortality of Seaward-Migrating Post-Smolts of Atlantic Salmon Due to Salmon Lice Infection in Norwegian Salmon Stocks *J.C. Holst, P. Jakobsen, F. Nilsen, M. Holm, L. Asplin and J. Aure*	136
12	A Two-Generation Experiment Comparing the Fitness and Life History Traits of Native, Ranched, Non-Native, Farmed and 'Hybrid' Atlantic Salmon Under Natural Conditions *P. McGinnity, A. Ferguson, N. Baker, D. Cotter, T. Cross, D. Cooke, R. Hynes, B. O'Hea, N. O'Maoiléidigh, P. Prodöhl and G. Rogan*	138
13	Finding Resolution to Farmed Salmon Issues in Eastern North America *A. Goode and F. Whoriskey*	144
14	Delivering the Solutions – The Salmon Farmer's Point of View *Lord Lindsay and G. Rae*	159

BEACONS OF HOPE   173

Success Stories

15	The Return of Salmon to Cleaner Rivers – A Scottish Perspective *R. Doughty and R. Gardiner*	175
16	The Return of Salmon to Cleaner Rivers – England and Wales *G.W. Mawle and N.J. Milner*	186
17	Opening Up New Habitat: Atlantic Salmon (*Salmo salar* L.) Enhancement in Newfoundland *C.C. Mullins, C.E. Bourgeois and T.R. Porter*	200
18	Optimizing Wild Salmon Production *F.G. Whoriskey*	222

Pointers for the future   233

19	Stream Restoration for Anadromous Salmonids by the Addition of Habitat and Nutrients *B.R. Ward, D.J.F. McCubbing and P.A. Slaney*	235
20	Prospects for Improved Oceanic Conditions *S. Hughes and W.R. Turrell*	255

PUTTING IT TOGETHER — 269

Catchment management

21 The European Water Framework Directive and its Implications for Catchment Management — 271
    J. Solbé

Reports and Recommendations — 281

   The Outlook for Post-Smolts — 283
   D. Mills

   Overcoming Estuarial and Coastal Hazards — 287
   J. Anderson

   Resolving Conflicts Between Aquaculture and Wild Salmon — 289
   M. Windsor

   Can We See a Brighter Future? — 291
   D. Clarke

THE WAY AHEAD — 295

   Conclusions, Discussion and Resolutions — 297
   R. Shelton

Index — 301

Colour plates 20.1–20.3 can be found following page 260.

Preface

The First International Atlantic Salmon Symposium, arranged by the two organizations, based respectively in the United Kingdom and North America, who became the Atlantic Salmon Trust and the Atlantic Salmon Federation, was held in 1972. This Symposium, held in St. Andrews, New Brunswick, has been repeated at intervals of 5 or 6 years, with subsequent Symposia being held in Edinburgh, Biarritz, St. Andrews and Galway. All the Proceedings have been published. The Edinburgh Symposium was of particular importance as it called for an international approach to conservation and led to the signing of the North Atlantic Salmon Convention.

This Sixth Symposium, held over 3 days in July 2002, marked a return to Edinburgh. His Royal Highness Prince Charles, who had opened the previous Edinburgh Symposium, graciously sent the message of encouragement which is reproduced in this volume. The Symposium title, *Salmon at the Edge*, refers to the principal theme of problems faced by wild Atlantic salmon and sea trout in estuaries and the coastal zone, and in the early weeks at sea. More generally, it also reflects the current precarious state of many migratory salmonid populations and the need for new approaches to a number of threats; some successes in the restoration and enhancement of populations were signalled during the programme. The meeting was noteworthy for the constructive way in which over 180 participants from 11 countries examined subjects as diverse as the impact of salmon farming; the behaviour of post-smolts in their early migration and their vulnerability to by-catch in near-surface pelagic trawl fisheries; and the use of nutrient enrichment and habitat enhancement to increase production of juvenile salmonids. Conclusions on these were reached after much discussion in the final session – The Way Ahead – and four Resolutions encapsulating the need for concerted action on specific issues were agreed.

Jeremy Read

ST. JAMES'S PALACE

When the International Atlantic Salmon Symposium was first held in Edinburgh, some twenty four years ago, I gave the opening address. I spoke then of the increasing pressures on wild salmon and, in drawing particular attention to the way in which the survival of salmon at sea appeared to be failing, I emphasised the need for action. All those words could, I fear, be repeated today.

Many hazards face salmon and sea trout as they make their extraordinary transition between a fresh water existence and life at sea. Rightly, then, you will be considering the whole range of problems, including predation in estuaries and an unfriendly marine climate, that confront fish in the coastal zone, in their early days at sea and during their return to their native rivers.

Since that occasion in 1978, some things have changed. In those days salmon farming was just developing. It was hailed as the potential saviour of wild stocks because it would reduce the pressure of commercial fishing. No-one then foresaw the damaging impacts that aquaculture might have on wild fish. These are now recognised and, although there is progress in overcoming them, there is a long way to go. So it is indeed fitting that this Symposium should also concern itself with the subject of salmon farming.

Once again, those who share concern for the well-being of the wild salmon have gathered in Edinburgh under the auspices of The Atlantic Salmon Trust and the Atlantic Salmon Federation. At that earlier Symposium, I expressed the hope that the meeting would be successful and fruitful. It was: the resolution that it passed unanimously, which called for the banning of fishing outside territorial waters, resulted in the signature of the North Atlantic Salmon Convention and in the foundation of NASCO. I can only trust that your work this week will be as well rewarded.

HRH The Prince of Wales

Chapter 1
Eroding the Edge – The Future of Coastal Waters

Professor Sir F. Holliday

This Symposium is about the future of the Atlantic salmon. There is an implication that its future is not simply uncertain, but perhaps even in doubt. The salmon is not unique in being considered in danger. Even if one confines the debate to bony fish, I have seen scientific papers and popular articles expressing concerns about the future of many species, and almost all are the subject of commercial harvesting. Since first I went to the Marine Laboratory, Aberdeen, in 1954 I have watched as governments have factored scientific data into and out of political decisions; I have seen how those data, which seemed of monumental importance to me and my science colleagues, appeared to shrink in stature and consequence as other factors such as economics and international bargaining overshadowed them.

So what can we say about the future of the salmon? Let me talk a little about the salmon and a little about the future. Perhaps we should first remind ourselves about the salmon's past. I do not know exactly when *Salmo salar* first appeared in the waters of the earth, but it probably pre-dates *Homo sapiens* by some time. It has been subject to the forces of natural selection for millennia. It has survived and been shaped by coping with the global changes of those many years. The salmon, as you know, is 'tuned' to an early life in fresh water, a period of life in sea water and a return to fresh water to spawn. Each stage of its life history is based upon physiological systems that allow it to progress to the next stage, provided that predation permits.

Having in the long past studied the ionic and osmotic regulatory systems of developing eggs and larvae of salmon, I know just how specifically each stage of development – blastula, late gastrula, fry, parr, smolt, and so on – brings its own set of systems for survival. Some of the systems last only for days, some for weeks, others for a lifetime. The salmon is, in a sense, not one fish but a series of fish. Shakespeare had some words for it (he usually does!) when he described the ages of man. For our fish to survive from one stage to the next and to the ultimate stage of a spawning adult many physiological systems must come into play and interact with the environment.

Let me now change the subject and talk about railways. A train, belonging to a company for which I chair the Board, went through a red light just outside Paddington Station. It collided with a high-speed train. When we came to analyse that dreadful event we had the help of Professor Jim Reason of Manchester University, a risk analysis expert. He used a simple model to explain to me the basis of that accident. In imaginary terms it looks like this: imagine 20 slices of Swiss cheese, the sort with holes, placed one behind the other, between Paddington Station and the approaching high-speed train. The hole in each slice represents a hazard and a potential accident; solid cheese prevents an accident.

What happened on that fateful morning was that all the holes lined up; the train ran through them all, encountered no solid cheese, and collided head on with the approaching express. Any one of the slices could have prevented it; on 99.999% of days the holes do *not* line up and no accident occurs. Frighteningly, in that Swiss cheese model, the holes can move about. A solid slice one day can have a hole in it the next, for example, sunshine on signals producing 'phantom lights'. Hold that image of slices of Swiss cheese in your mind as we go back to the salmon. Let's turn the train/cheese model on its head. We need thousands of cheese slices between fertilized egg and adult, and this time we *want* all the holes to line up so that the salmon can run through. Between the newly fertilized egg and the potential spawning adult that develops from it are those hazardous cheese slices. The salmon needs the holes. Sadly, for most individuals the holes won't line up; indeed, some slices for some individuals have no holes. For a few, the holes do line up and the fish has a clear run. So the great Swiss cheese race is run afresh each year, and so long as, for enough individuals, the holes are there *and* line up, all is well.

In the 18th–19th centuries we fouled our rivers and we closed many holes. Nets blocked other holes. Seals and other predators were stationed at yet other holes, and so on. We have been unblocking some holes – the Tyne is cleaner, the Tees is cleaner. I am pleased, at Northumbrian Water, to have had a part in that clean-up of rivers and coastal waters.

We do not know how *many* cheese slices there are and we do not know enough about each slice or its hole. What we *do* know is that too few salmon are getting through today. Will the future be different?

Now that's pretty much all I am going to say about the salmon itself, and I want to think now about the future, the future that we shall shape for ourselves and all other species. It is a future that will be determined by many forces but two in particular are my concern, and now I cease to be a biologist and 'modeller' and my expertise is only that of experience in the worlds of water, energy and finance.

I must be rather dogmatic about those two forces: first, I accept the phenomenon of global warming and climate change. Second, I believe that the present inequalities between standards of living around the world will be corrected by globalization. The world in, say, the next 50 years, will be subject to these two powerful and inexorable forces – global warming and globalization. What will these two forces do to our coastal environments, our oceans and rivers and the creatures that inhabit them?

This is where we must turn to social and political matters to help us form a conclusion. I have been struck by what Lord Blake called 'The Tyranny of Democracy' and Lord Hailsham named 'Elective Dictatorship'. It is not easy to be a minority when the majority always gets its way, hence the creation of the Human Rights Convention to protect minorities. Human rights are one thing, but what rights does the salmon have? Or a fragile coastline? I shall return to these matters in a moment, after I have blended into the mix the views of an eminent historian, Professor John

Roberts. He argued convincingly in his book *The Triumph of the West* that the ideologies and lifestyles of the West would, by globalization, be adopted worldwide.

When one combines the powers of democracy and the 'westernization' process, a future is foreshadowed that we and the salmon should enter with caution. For example, the world will need high and constant supplies of energy. A sustainable future for all cannot be built on low or unreliable energy supplies. Thus, the potential for continued global warming is great.

Globalization will be built upon world trade because there is a global mismatch between resources such as space, food and water and the distribution and needs of populations. We must move *either* the resources or the people – one (resources) is relatively easy to move, the other (migration of people) is not easy. As millions are displaced by rising sea levels then migration will become a critical issue. But whether we move the people or the resources we will need copious energy and space in order to provide up to nine billion people with the standard of living that we require our elected representatives to provide for us. The future will put global democracy under great strain; governments need votes and voters are, in the main, motivated by self-interest.

Where will that leave habitats and species? I said earlier that they have no intrinsic rights. How could they? 'Rights' are a product of the human mind. Humans confer rights. I have been greatly influenced by the work of Christopher Stone, an American who developed his thoughts in a book published in 1972 entitled *Should Trees have Standing? Towards Legal Rights for Natural Objects*. His ideas, inspired by the work of John Muir, may need revisiting as globalization and global warming exert their pincer power on the earth.

It is the coastal zones that I see as being subject to the greatest pressures. Sea levels will rise; those deepened and extended coastal waters will be energized by more wind. There will be inundation of low-lying areas. You may expect governments to be forced at some stage to 'barrage out' the sea. The coastal zone will also be the preferred site for wind farms, wave and tidal generators, container ports, new cities and new airports. To deny coastal protection and coastal development may appear to be denying the security and needs of people worldwide. The salmon may be faced with many more obstacles, and it will not be alone.

How do we deal with these challenges?

Science will, of course, have its contribution to make. But I doubt if science will determine policy. Policy will be driven by electoral and political self-interest. Science is socially constructed and used, i.e. it is shaped largely by the priorities of funding bodies and governments.

We might say, 'very well, coastal waters are threatened but surely the oceans will fare better?' But will they? Global warming is predicted to at least weaken the Gulf Stream and perhaps to displace it. All marine organisms including the salmon will need systems to cope with the associated temperature and salinity changes. They've done it before, they have found the holes in the cheese slices; can they find them again?

The challenge with the salmon is part of a greater challenge. How, in future, do we manage the species of the changing world as they come into conflict with humans and with themselves. Today, in the UK, we argue about seals, cormorants, goosanders, hen-harriers, peregrines, badgers, even hedgehogs; tomorrow we may have even longer lists. Worldwide the lists will be long indeed. We are short of sound and generally agreed philosophical principles on which to base future legislation. Politicians are likely to find themselves encircled by vested interests all urging them to 'stand by me', all prepared to invoke the tyranny of democracy.

So what do we say, in conclusion? Change is coming; change in climate, in lifestyles and in ecosystems. We *can* factor conservation thinking into political decisions if we can persuade electorates and governments that it is necessary and right to do so. It will not be easy, but recently I was with the climatologist John Houghton, and he reminded me of the famous words of Edmunde Burke: 'No man made a greater mistake than he who did nothing because he could only do a little'.

INTO THE UNKNOWN
THE FIRST WEEKS AT SEA

Chapter 2
Migration and Distribution of Atlantic Salmon Post-Smolts in the North Sea and North-East Atlantic

M. Holm[1], J.C. Holst[1], L.P. Hansen[2], J.A. Jacobsen[3], Niall O'Maoiléidigh[4] and A. Moore[5]

[1] *Institute of Marine Research, PO Box 1870, Nordnes, N-5817 Bergen, Norway*
[2] *Norwegian Institute for Nature Research, PO Box 736, Sentrum, N-0105 Oslo, Norway*
[3] *Fiskirannsóknarstovan, Nóatún, PO Box 3051, FO-110 Tórshavn, Faroe Islands*
[4] *Marine Institute, Abbotstown, Castleknock, Dublin 15, Ireland*
[5] *CEFAS, Lowestoft Laboratory, Pakefield Road, Lowestoft, Suffolk, NR33 0HT, UK*

Abstract: Earlier records of Atlantic salmon (*Salmo salar* L.) distribution and migration were almost exclusively based on information gathered from commercial fisheries. However, efforts to study salmon in the marine environment during the last 20 years have yielded scientific information which, although still sparse, has thrown light on the marine distribution of the fish in the north-east Atlantic. Knowledge about the marine migrations and distribution in time and space of the salmon is of vital importance to our understanding of both the general ecology of the species and the possible external influences that may lead to changes in growth and survival.

The paper summarizes our current knowledge on the spatial and temporal distribution of the post-smolt salmon in the north-east Atlantic. A special emphasis is placed on data gathered in Norwegian and English coastal tracking studies, Faeroese, Norwegian and Irish tagging experiments, and recent experiments with post-smolt trawling in coastal waters and the open ocean. The relationship between the marine environment and fish distribution is discussed.

Introduction

Atlantic salmon, *Salmo salar* L., are distributed in rivers over large areas in the north Atlantic and in the Baltic and, after more than 100 years of research, the distribution and ecology of Atlantic salmon in fresh water are well documented. Dutil & Cotu (1988) investigated post-smolts in the Gulf of St. Lawrence, and the relatively scarce knowledge of the early marine phase of salmon was reviewed by Thorpe (1994), but there was still much less knowledge about salmon in the oceanic phase (e.g. Friedland, 1998; Friedland et al., 1998; Hansen & Quinn, 1998; Jacobsen, 2000). The most obvious explanation for this apparent lag is that the marine investigations until recently have been suffering from a lack of effective post-smolt sampling methods not justifying the costs connected with sampling over very large areas, opposed to the relative facility of sampling salmon in rivers and coastal waters.

The development of coastal salmon fisheries more than 100 years ago led to an accumulation of information on migration and distribution of the salmon along the coast and in the fjords. However, until the rapid development of oceanic fisheries for salmon in the 1960s at west Greenland and in the northern Norwegian Sea (Fig. 2.1), our knowledge on this important life stage was very poor. To assess these fisheries and provide more information from oceanic areas, The International Council for the Exploration of the Sea (ICES) took the initiative to carry out joint investigations [Parrish & Horsted (1980); ICES Working Group Reports (Anon., 1967, 1981, 1982, 1984)]. Nevertheless, outside the fishing areas (Fig. 2.1) and the fishing seasons the distribution in the north-east Atlantic remained unknown. Jacobsen (2000) notes that in light of the importance of the high-seas fishery for salmon, it is surprising that there was such a scarcity of information on the salmon life history at sea. Although mark–recapture studies gave some information on migrations (e.g. Hansen & Jonsson, 1989; Jonsson et al., 1993) of the younger, non-fishable stages, it was not until the late 1970s to the early 1980s when the first experiments with salmon tracking started that details on the early marine migratory phase of the species were uncovered.

The mortality of salmon in the sea has been anticipated to take place during the

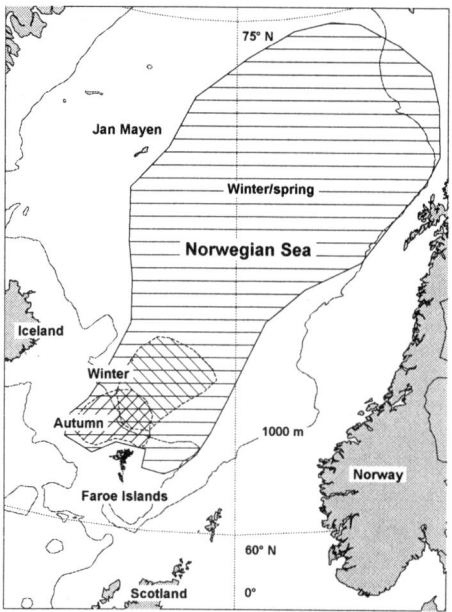

Fig. 2.1 Salmon fishing areas in the north-east Atlantic since 1965. The horizontally hatched area in the Norwegian Sea indicates the multinational fishing area prior to 1984 when the international fishery was closed through the North Atlantic Salmon Conservation Organization (NASCO) convention. The fishery in the northern part of the Norwegian Sea was carried out during January–May. Since 1984 the fishery has been confined inside the Faeroese 200-nm fishing limit (diagonally hatched areas), with the autumn fishery (mid October–December) usually being carried out closer to the isles (area with hatching slanting right) than the winter fishery (January–March; left slanting). (From Jacobsen, 2000.)

first months after the smolts have left fresh water. Reddin & Short (1991) carried out the first systematic efforts to sample post-smolts by drift-net fishing in the Labrador Sea. In the north-east Atlantic, knowledge about the migration and distribution of post-smolts in the open ocean made slow progress until the early 1990s when new trawl technology was developed at the Institute of Marine Research in Bergen, Norway. This buoyed mid-water trawl (Valdemarsen & Misund, 1995) opened up methods of investigating post-smolts at sea (Holst et al., 1993; Holm et al., 2000). Although focusing on larger fish, valuable knowledge on post-smolts approaching their first winter was also gained from an oceanic line fishing and salmon tagging programme at the Faeroes (Jacobsen, 2000).

The aim of this paper is, in light of results from recent studies, to review some aspects of the marine distribution and migration of Atlantic salmon post-smolts in the north-east Atlantic, discuss possible environmental factors explaining the distribution of the species and identify gaps in our knowledge. The migration and distribution of salmon in the north-west Atlantic are presented in other chapters of this volume, and will therefore only be included here as elements of the discussion.

Materials and methods

The background data for the present paper have essentially been collected through:

(1) *Post-smolt tracking experiments performed in the estuaries of south English rivers and in fjords and coastal areas in south-west and mid Norway.*
The movements of Atlantic salmon smolts have been studied in a number of UK river systems as they emigrate from fresh water and enter the sea (Moore et al., 2000). The early migration patterns of smolts trapped and tagged in the Rivers Conwy, Tawe, Test, Tees and Avon have been described in the immediate coastal zone. In general, smolts were trapped at night in fresh water as they migrated seawards and implanted with miniature 300-kHz acoustic transmitters (Moore et al., 1990). The movement of the smolts within the river and estuary was monitored using an array of 300-kHz acoustic signal relay buoys (Moore et al., 1995, 1996, 1998). Migration in the open sea was studied by actively tracking individual fish from a small research vessel, fitted with a 300-kHz acoustic hydrophone, immediately they left the estuary. The Norwegian tracking experiments were performed with hatchery-reared 1-year smolts, first generation offspring of wild parents originating from the River Ims in south-west Norway (1981–85) or from the River Orkla in mid Norway (1996–98). The Ims fish were externally or intragastrically tagged, while the Orkla fish were intraperitoneally tagged. In the Ims study acoustic depth-sensitive transmitters were used (Sintef, 1984), while the transmitters used in mid Norway were pingers (Vemco Ltd., 86–99 kHz), giving horizontal information only. Altogether 60 fish were tracked for some time in these experiments. The fish were individually tracked from boats equipped with hydrophones connected to receiving equipment (Holm et al., 1982, 1984, 2001, unpublished).

(2) *Surface trawling for post-smolts in the Norwegian Sea and adjacent areas.*
The Institute of Marine Research (IMR), Norway, introduced a new pelagic research trawl in 1990 (Åkra trawl; Valdemarsen & Misund, 1995). Post-smolts were captured during the development of the trawl, during a pair trawling experiment for herring, and lastly in an experiment with post-smolt sampling in 1993, all in the Norwegian Sea. Since 1995, surface trawling for post-smolts has been performed regularly during the IMR summer surveys on pelagic species in the Norwegian Sea and adjacent areas. The surface trawl in current use for salmon investigations in the ocean is wider and shallower than the original Åkra trawl, and has an opening during towing of approx 14 × 40 m. It is operated with a fish lift for live fish capture (Holst & McDonald, 2000), and a strictly on-surface operation is maintained by a large number of floats on the head rope and additional large floats on the trawl wings. The method is described in greater detail in Holm et al. (2000). Continuous temperature depth (CTD) profiles were taken before or after each haul. Since 1990, a total of close to 2500 surface tows have been performed in the North, the Norwegian and the Barents Seas from mid May–early September (Fig. 2.2). The current data material (1990–2001) consists of more than 3000 post-smolts (Holst et al., 1993, 1996; Holm et al., 2000; Holm, unpublished). The floating trawl technology was also used at post-smolt investigations in the Faeroes–Shetland Trench in 1996–97 and in 1998 outside Cromarty Firth and in an area 50 miles north-north-east of Montrose, eastern Scotland (Shelton et al., 1997; Shelton, 1999, unpublished).

(3) *Long-line fishing and salmon tagging experiments around the Faeroes.*
In the salmon fishing seasons (November–March) of 1992/93, 1993/94 and 1994/95 the Faeroes Fisheries Research Institute carried out tagging in the areas north and north-east of the Faeroes from a commercial salmon long-line fishing boat chartered for the investigations. The line was rigged and set as a normal salmon long-line (cf. Jacobsen, 2000) and consisted of around 2000 hooks. CTD profiles were taken before and after each line setting. 3811 wild salmon and 1637 fish of hatchery origin were tagged with Lea's tags and released, of which 87 and 19 fish respectively were recaptured in sea or river fisheries over large areas of the north Atlantic. In addition, 121 Carlin tags and 74 micro tags were retrieved during this experimental fishery (Hansen & Jacobsen, in press).

(4) *Various tagging and release experiments in River Ims, south-west Norway, and a number of Irish rivers.*
In the River Ims, different genetic groups of hatchery smolts or groups receiving various pre-release treatments have been tagged and released. In the order of 20 000– 40 000 fish have been released annually since 1979, and wild smolts from the river captured in a trap in the river have also been tagged and released (Hansen & Jonsson, 1989; Jonsson et al., 1990a, b, 1993, 2002). Extensive release programmes with groups of microtagged smolts of predominantly hatchery origin of different age and size groups, sometimes subjected to various pre-release treatments have been performed

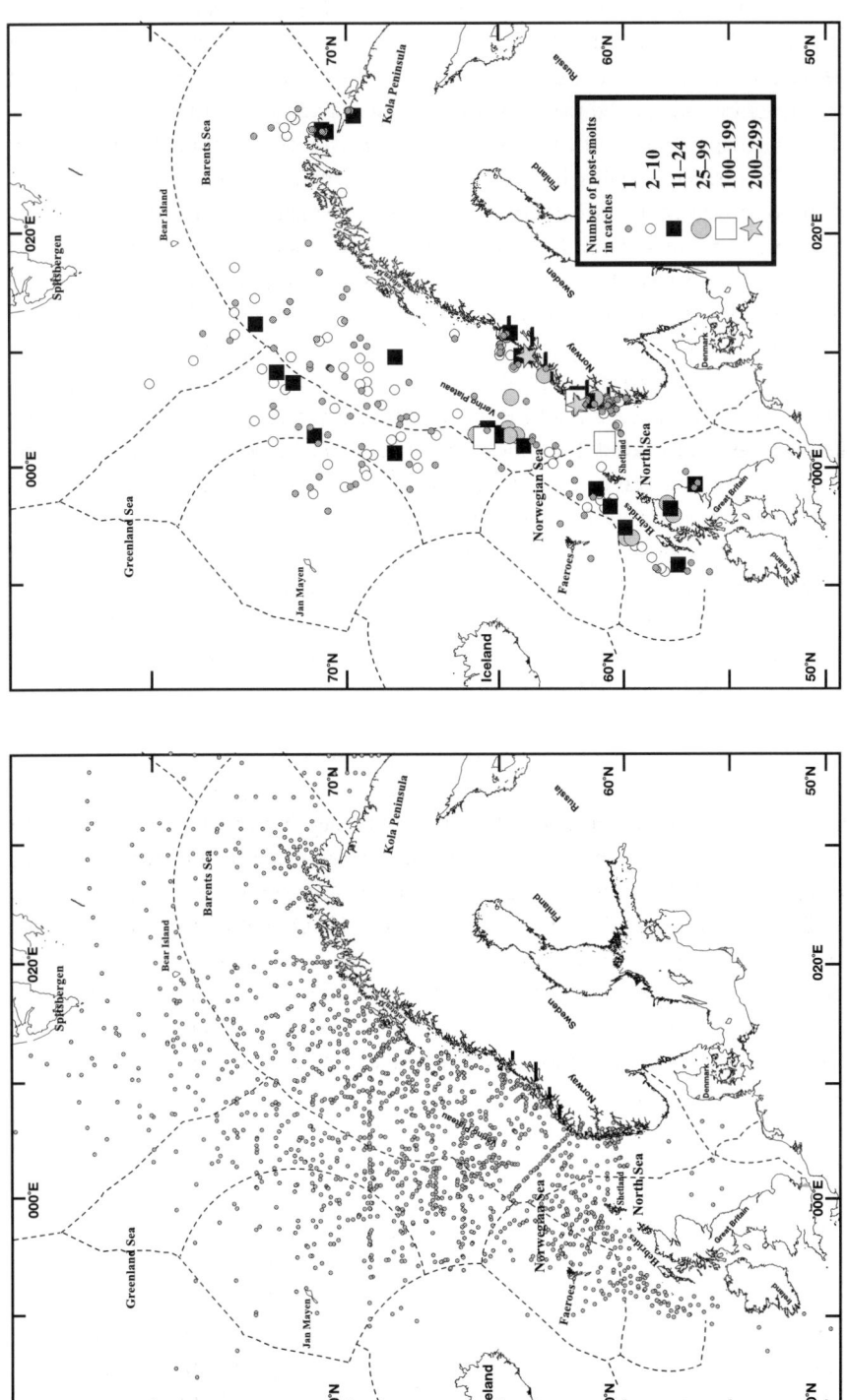

Fig. 2.2 Left panel shows the starting points (dots) of surface trawl hauls carried out from mid May–August/September in 1990–2001. Hyphenations delineate the Exclusive Economic Zones of the respective countries. Right panel presents the distribution of surface trawl hauls containing post-smolts during the same time frame. Size of symbol (dots or stars) indicates number of post-smolts in each catch.

by the Marine Institute all over Ireland during the last two decades (O'Maoiléidigh, pers. comm., and various reports from the ICES compilation of microtags, finclips and external tag releases).

Migration

After onset of smoltification when the young salmon start their seaward migration, their displacement in the rivers is largely nocturnal and affected by factors influencing the water speed. The downstream progress is considered rather as the result of passive transportation with the currents than active migration (e.g. Fried et al., 1978; Tytler et al., 1978; Hansen & Jonsson, 1989; Moore et al., 1990; Lacroix & McCurdy, 1996).

Whether the seaward migration is the result of passive transportation or active swimming was debated for a long time. As opposed to the relative uniformity of the riverine environment, in the estuaries, fjords and coastal waters, the post-smolts will be encountering a complexity of environmental conditions, where the tides or winds will influence the speed and directions of the surface currents as well as the distribution of different water layers and any fronts that may evolve between these waters. Thus the fish may face strategic choices for their seaward displacement: to use energy in efforts to cross counter currents that would divert them from heading towards the sea, or to save energy by staying outside currents moving in a 'wrong' direction, or travel along with them, but then prolonging the dwelling in estuaries/ fjord areas where the risk of predation (Hvidsten & Møkkelgjerd, 1987), parasite infestation (Heuch, 1995) and scarcity of food may have fatal consequences.

Experiments with hydroacoustic tagging and tracking of post-smolts have made a considerable contribution to our knowledge of the early marine migration of the salmon. Studies performed in south and south-west England (Moore et al., 1995), conclude that the movement of smolts through the upper river estuaries was indicative of a nocturnal selective ebb-tide transport pattern of migration, with a marked change of behaviour when reaching the lower parts. The significant change from moving passively with the current to actively swimming seawards could have been initiated when the fish encountered a particular salinity threshold (Moore et al., 1995). There was no apparent period of acclimation required when moving from fresh to salt water, and a physiological requirement to move to a saline environment may therefore be an important cue in initiating smolt migration. Similar results have also been observed in smolt trackings in the River Ims estuary in south-west Norway (Holm et al., 1982, 1984) and on the west side of the Atlantic (LaBar et al., 1978).

Migration within coastal waters, again, occurred during both day and night. The studies in the coastal zones adjacent to the Rivers Conwy, Tawe, Test, Tees and Avon have shown that the post-smolts migrated rapidly (ground speeds of $\sim$ 50 cm/s), close to the surface, and there was a strong tidal current component to the speed and direction of movement (exemplified by tracks from the River Avon, Fig. 2.3). High correlations between the displacement of the fish and the surface current directions have been found when tracking post-smolts released in the estuaries of the Rivers

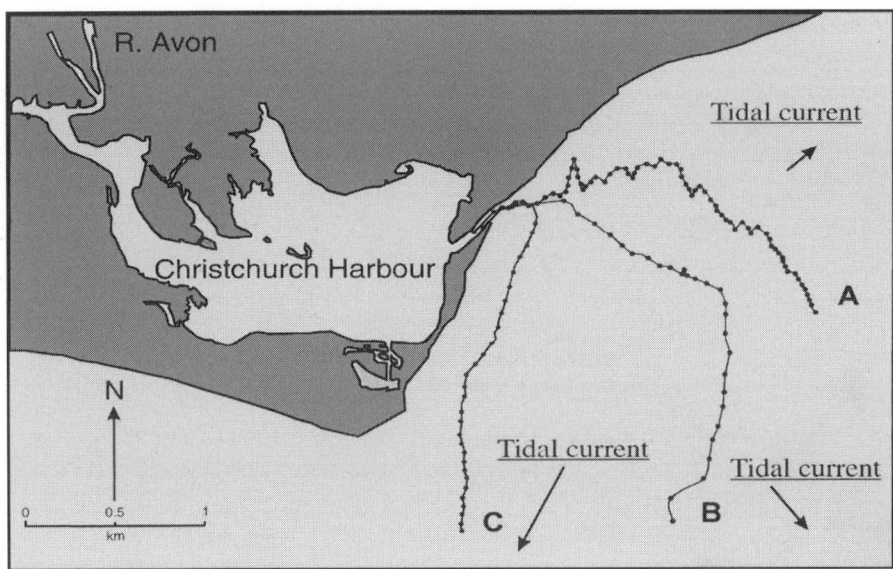

Fig. 2.3 Influence of tidal currents on post-smolt displacement along the coast outside the River Avon estuary in southern England. Movements of three Atlantic salmon smolts in the sea. The fish (A: 15.6 cm, B: 17.4 cm, C: 16.5 cm in length) had been trapped and tagged in fresh water 3–7 days previously using miniature acoustic transmitters. The speeds over the ground of the smolts during the tracks were 18, 23 and 45 cm/s, respectively. Lines with solid dots denote the GPS position of the tracking vessel every 5 min.

Ingdal and Ims, mid- and south-west Norway respectively (Holm, unpublished). In the latter study, the driving forces of the surface waters were mostly wind induced, as the tidal amplitude in this area is only 35–60 cm. In all three studies, however, clear indications of active and directed swimming were observed.

Like in the south England tracks, the relationship between tides, winds and fish displacement can be observed in Norwegian studies in various fjord systems, here exemplified by Fig. 2.4 which shows four tracks from a study in the Trondheim fjord in mid Norway 1996–98. (Holm et al., 2001). These fish were F_1 hatchery-reared progeny of wild parents from the River Orkla 20 km away in the same fjord system. Fish encountering rising tide or winds stronger than force 2 on the Beaufort scale when leaving the Ingdal Bay generally moved south-eastwards and eastwards (inwards, track 9603 and 9801) but could turn and go northwards if the conditions changed.

On the contrary, fish moving out of Ingdal on southerly or south-easterly winds or falling tide moved rapidly outwards (track 9603 and 9802), but would slow down when the tides changed. In these experiments the fishes' speed over ground varied from ~ 10 to > 50 cm/s and mostly surpassed the registered surface current speed. Table 2.1 shows data from selected tracks in the Trondheim fjord. It is worth noting that in some cases there was a considerable displacement inwards the fjord, but at present we have no possibility of concluding whether this behaviour resulted from

14 *Salmon at the Edge*

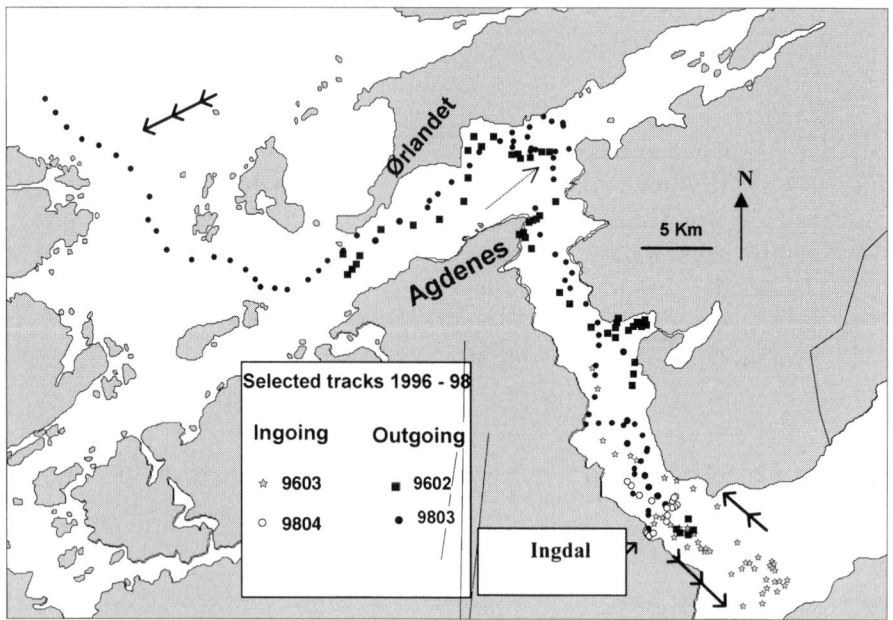

Fig. 2.4 Map of the Trondheim fjord with selected tracks of acoustically tagged post-smolts tracked in 1996 and 1998. The fish were released in the estuary of the small Ingdal River at about 150 m from the river mouth. Each symbol indicates the GPS position of the tracking boat every 30 min. The boat was kept at a certain distance from the fish (around 100 m). Solid arrows indicate wind-force and direction. Three-headed arrows indicate winds of force 4, while two heads represent winds of force 2.5–3. Dotted thin arrow indicates flooding water at a part of the track of fishes 9602 and 9803. The tidal amplitude in this fjord is 1.6–1.9 m.

Table 2.1 Track data for post-smolts released in River Ingdal estuary, mid Norway (cf. Fig. 2.4).

Year–Fish no.	Track time (h)	Track length (km)	Net seaward transport (km)	Speed over ground (cm/s)
96–02	57.4	51.1	26.7	24.7
96–03	40.9	45.5	12.1	30.9
98–03	40.8	82.8	49.8	56.4
98–04	5.5	8.0	4.2	41.0
97–01	48.2	60.0	−22.0	34.6
97–05	25.8	22.5	−3.5	24.2
97–09	22.8	53.4	31.0	65.2
97–10	9.6	10.3	−5.7	29.7
97–11	6.7	16.4	16.4	68.5
97–13	14.8	16.8	−13.1	31.6
97–17	20.3	38.0	25.0	51.9

tagging and handling malaise of the fish making them less willing to search for the outgoing currents, or whether they were actively making feeding excursions inwards the fjord. The consequences of such prolonged stay in the fjordic environment may be increased risk of predation and infestation by salmon lice. Certain meteorological conditions, e.g. on the west coast of Norway, strong and long-lasting westerly winds during spring at the time of the seaward migrations of young salmon, might lead to the surface water being pressed inwards in the fjords, with a subsequent accumulation of post-smolts in these waters for a longer time, which might lead to greater mortality of the cohorts migrating in such conditions, or due to salmon lice infestation (cf. Heuch, 1995; Holst et al., this vol.). Likewise, the cohorts leaving the rivers in conditions favourable to rapid seaward transportation may have a chance of higher survival.

Distribution of post-smolts and salmon in the north-east Atlantic

Spatial distribution

Figure 2.2 shows the total distribution of surface trawl hauls and the distribution of hauls with post-smolt captures during mid May–early September 1990–2001. As can be seen, the station net covers large parts of the northernmost North Sea and the central and north-eastern parts of the Norwegian Sea fairly well, but the captures are not evenly distributed over the surveyed area. The pattern observed up to 1998 (Holm et al., 2000) is still prevailing even with the addition of new data. Thus the relationship anticipated by Holm et al. (1996) and Shelton et al. (1997) of post-smolt distribution with the distribution of the current patterns of the North Atlantic Current (NAC) as presented by Huthnance & Gould (1989) and Poulain et al. (1996) seems confirmed and further strengthened. Some particularly high-density capture areas can be distinguished, especially west of the Vøring plateau in the Norwegian Sea, and west and north of the Hebrides and Shetland. The former area has been the target of particular interest in recent years, confirming the hypothesis that this would be a good sampling place where many post-smolts from the southern areas would be passing, due to the topography of the plateau forcing the NAC to deflect westward and creating a strong current before it spreads out in a 'fanlike' formation with several weaker branches and gyres out of the main core. The Hebrides–Shetlands areas have not been surveyed since 1997, but there is little reason to believe that the strong slope current in the Faeroes–Shetland Channel should not be the main 'carrier' for the Irish and British post-smolts towards the feeding grounds in the north as suggested by Shelton et al. (1997) and Holm et al. (2000). Shelton (1999, unpublished) reports smolt captures up to 50 n.m. from shore during a survey of relatively short duration in the Cromarty Firth and east-south-east of Montrose in east Scotland (Fig. 2.2). Unfortunately these surveys were discontinued, so it is not possible to establish a distribution and migration pattern for post-smolts emigrating from some of the most important salmon rivers

16 *Salmon at the Edge*

in Scotland. However, it is reasonable to assume that the fish leaving the east Scottish rivers will enter the Dooley current branch, taking them across the North Sea where they could enter the Norwegian coastal current, taking them northwards along the Norwegian coast as proposed by Holst et al. (2000a).

Distribution and migration of post-smolts of different origins

Forty-seven tags, 94% of which were microtags, have been retrieved from the marine post-smolt surveys up to 2000. The site of release, distance from site of origin and distribution of these tagged fish can be seen in Fig. 2.5 (redrawn from Holm et al., 1996; Shelton et al., 1997; Holst et al., 2000a; ICES, 2001). The fish originated from a

Fig. 2.5 Migration and distribution of post-smolts of different origins. Corresponding symbols (squares, filled dots or stars) on land and in the sea indicate sites of release and recapture of microtagged or Carlin-tagged (Norway) salmon smolts. [Redrawn from data in Holst et al. (1996, 2000a) and Holm et al. (2000).]

variety of rivers: Great Britain, Ireland, Norway and Spain were represented. Irish rivers dominate the captures, as could be expected since the Marine Institute tagged around 290 000 (mostly hatchery reared) fish in 2000 for release in a large number of rivers for enhancement or experimental purposes. The Irish microtaggings account for around 45% of the total microtaggings in the north-east Atlantic (cf. ICES, 2001). The number of tags from Ireland and Great Britain retrieved in the Norwegian Sea supports the results from scale readings of the fish captured in these areas where a preponderance of post-smolts of low age at entry at sea was observed (Holm et al., 1998; Holst et al., 2000a). Smolts leaving the rivers at 1–2 years are predominantly of southern European origin, while the northern stocks produce smolts where the higher river ages dominate (cf. e.g. Balignière, 1976; Metcalfe & Thorpe, 1990). The calculated minimum speed of migration of the tagged post-smolts from the site of release to the recapture point was estimated to lie between 7 and 30 cm/s (Holst et al., 1996, 2000a; Shelton et al., 1997), which is well below the speed of the core of the slope current (Huthnance & Gould, 1989; Poulain et al., 1996) and considerably lower than the speeds observed in the fjordic and coastal trackings by Moore et al. (1995, 1998) and Holm et al. (2001) (Table 2.1). The difference may, in the absence of concrete evidence, be hypothesized as a behavioural adaptation of the newly emigrated smolts to rapidly head for areas where the food density is higher and the predator density lower. Once arrived in the relative safety of a northward carrying current the fish could concentrate on feeding excursions or they may get entrapped in large current eddies, which both might delay their northward progress as suggested by Shelton et al. (1997).

Distribution in relation to salinity and temperature

The close association between temperatures (9–11°C; > 35 ppt salinity) characteristic of the warm and saline Atlantic Current and the number and size of post-smolt captures in the open ocean has been documented by Shelton et al. (1997) and Holm et al. (2000). Figure 2.6 exemplifies this relation. The distribution of post-smolt captures in three south-west and mid Norwegian fjords and at the coast clearly differs from the oceanic areas, and in the light of the relatively low number of 'northern-type' river ages (3 years and higher) in the scale readings from the oceanic areas, one might discuss whether there might be an adaptation to cooler environment by the northern smolts that keeps the stocks separated at the early marine life stage, or whether the separation is purely an effect of a shorter time for migration of the northern stocks that leave the rivers 1–3 months later than the southern ones.

Vertical distribution of post-smolts

The distribution of the post-smolts in the water column may have implications for sampling effectiveness and for the interception by other pelagic fisheries. In a study of food and feeding of fish released for sea ranching purposes in west Iceland,

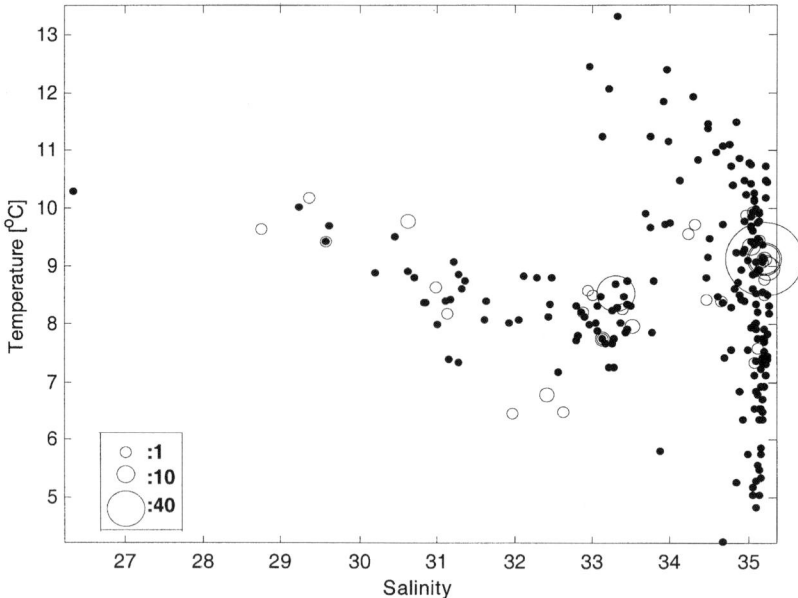

Fig. 2.6 Distribution of salmon captures in 244 surface trawl hauls in 2001 in relation to temperature and salinity in 5-m depth. Post-smolt captures are represented by open circles, the size of which are relative to the number of fish captured (see figure). Black dots signify no salmon captured. Hauls in lower than 34-ppt salinity are from the fjords and coastal areas of Norway, while hauls in > 34 ppt are from the open ocean.

Sturlaugsson & Thorisson (1995) recorded surface-near locations in drift nets of most fish captured in the coastal areas outside the release area. Similar surface-near locations were registered by Reddin & Short (1991) in drift-net sampling of post-smolts in the north-west Atlantic. The vertical tracks of post-smolts attached with depth-sensitive transmitters in the River Ims, fjordic and coastal areas in south-west Norway show a surface-near location most of the time, with deeper dives and increased diving activity in areas with less linear current systems and hydrographical fronts (Holm et al., 1982, 1984, unpublished). The question of diurnal distribution of post-smolts has been discussed, and, for example, Shelton et al. (1997) suggested that post-smolts might seek a deeper nocturnal location, as night-time trawling in the Shetland–Hebrides area gave no smolt captures. However, when data from the 2001 trawl surveys in the Norwegian Sea were grouped according to the registered position of the head-line and the catch per unit of effort or incidence of hauls with smolt captures (prevalence), a very clear association with the uppermost water layers occurs (Fig. 2.7). Post-smolts were captured regardless of time of day, but it should be pointed out that in the northern area the sun is above or in the very vicinity of the horizon (depending on latitude and time-lapse from the summer equinox) even at night.

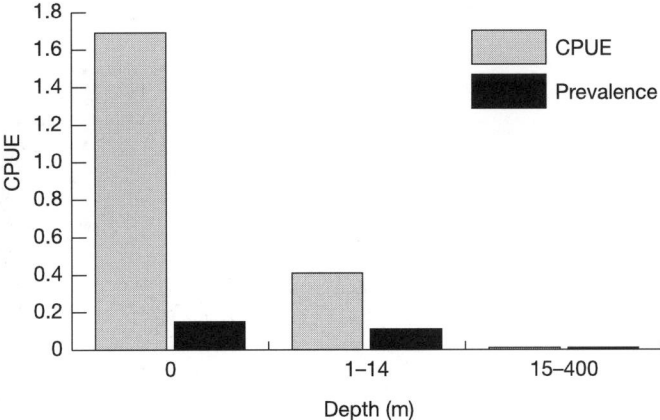

Fig. 2.7 Catch per unit of effort (CPUE, trawl hours) and prevalence of post-smolts in trawl hauls carried out with the float line at 0, < 15 and 15–400 m depth respectively during Institute of Marine Research cruises to the Norwegian Sea and adjacent areas during mid May–late August 2001. Number of hauls = 350.

Distribution overlap with other species in the North and Norwegian Seas

The salmon share the feeding areas, at least in parts of the north-east Atlantic, with, among others, large commercially exploited fish stocks, such as Atlantic herring and mackerel. This fact has been receiving attention over recent years, and growing concern has been expressed about the possibility for and the possible magnitude of interception by these fisheries. Post-smolt and herring seem to overlap spatially mostly in July–early August in the areas north of 68°N (Holst et al., 1998; Hansen et al., unpublished).

Information in Belikov et al. (1998) on the migratory pattern and the distribution of older year classes of mackerel shows a great similarity to that of the post-smolts. The mackerel start a northward summer migration from the waters north and west of Ireland in May, and some stock components may migrate as far into the Norwegian Sea as 74°N in July before they turn back to their southerly winter-feeding areas. Catch distribution diagrams for 1989–97 for mackerel indicate for some years a strong overlap with the inferred northward migratory routes of post-smolts in June and July in the Shetland–Faeroes Channel, the North Sea and the Norwegian Sea (Belikov et al., 1998).

Mackerel distribution, as expressed by the distribution of commercial catches in second and third quarters in 1977–2000 (ICES, 2002), is superimposed upon the distribution of post-smolt captures 1990–2001 in Fig. 2.8. It should be pointed out that the mackerel distribution might vary considerably from year to year due to environmental factors and stock size. For example, during an aerial survey in the Norwegian Sea in 2001, the major aggregations of feeding mackerel were distributed more easterly than in previous years (ICES, 2002).

However, there is little doubt that there is a substantial overlap in the mackerel and

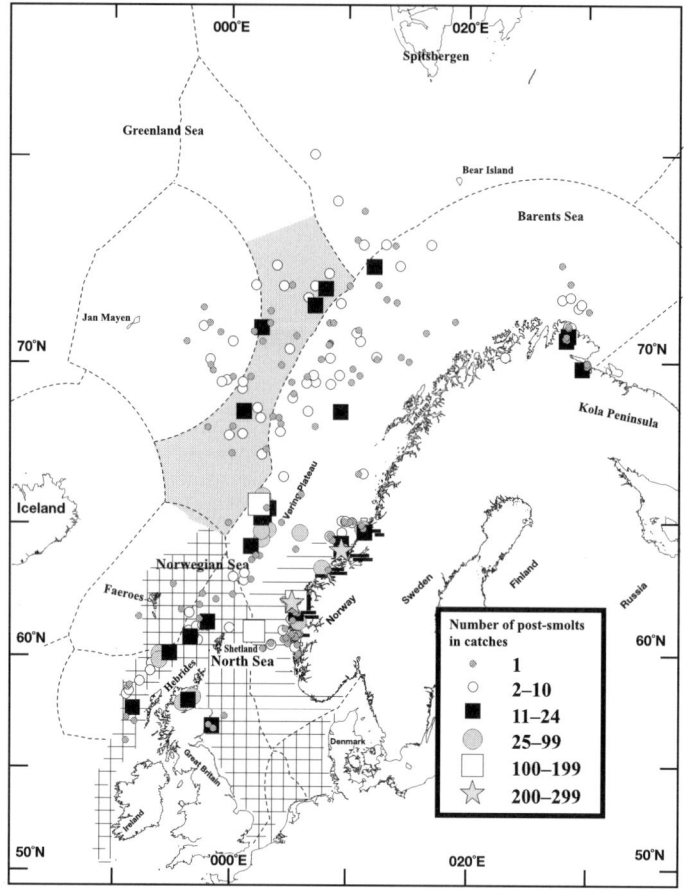

Fig. 2.8 Spatial overlap between post-smolts and mackerel in the North and the Norwegian Sea in second and third quarter. Post-smolt legends in figure. Mackerel trawl and purse seine fishing areas 1977–2000 are superimposed on post-smolt distributions 1990–2001. Shaded area indicates predominantly trawl fisheries. In the cross-ruled area both purse seine and trawl fishery (pairtrawling and trawling) are carried out. Horizontal hatching indicates purse seine fishing areas. [Redrawn from ICES (2002).]

salmon post-smolt habitats during summer, but it is less likely that the post-smolt area and mackerel habitat area are overlapping in the same magnitude throughout the whole season.

The purse seine fishery for herring takes place in the areas west of Iceland up to Jan Mayan Island as early as April and June and is therefore not likely to intercept young salmon. With reference to Norwegian post-smolt research surveys in 2000–02 showing frequent and large co-occurrences of salmon and mackerel in the surface trawl captures (ICES, 2002, unpublished), both inside the Norwegian Exclusive Economic Zone (EEZ) and in the international zone, the fishery with the greatest potential for catching post-smolts is identified to be the trawl fishery for mackerel in the Faeroes EEZ and the international area of the Norwegian Sea (ICES, 1999; Holst et al., 2000b).

Conclusions

Although some gaps in our knowledge on the distribution and migration of post-smolts have been filled in recently, there are still areas where a 'black hole' remains, notably the distribution of the northernmost salmon stocks and the total overview of the temporal and spatial distribution of young salmon at sea. Considering both the size of the potential distribution area (roughly 2 million miles^2) and the relatively good success in retrieving tags in sampling programmes at sea, a coordinated international effort in tagging and sampling both on shore and at sea could help fill in these gaps in knowledge. Implementation of the rapidly developing electronic tag technology, as a supplement to conventional methods, should also be evaluated in conjunction with such a programme.

With regard to the interception of salmon in pelagic fisheries, the screening of post-smolt by-catches on board commercial mackerel fishing vessels may prove difficult due to the resemblance both in coloration and sometimes size between the species (Holm, pers. observ.) and the requirement to move large quantities of fish very quickly, but may eventually be possible to arrange at landing sites.

Data on the methods applied and the timing of the mackerel trawl fishery in relevant areas of the EU's EEZ (cf. Fig. 2.8) are at present too insufficient to allow conclusions on the potential danger of interception, but should be further addressed also.

References

Anon. (1967) *Report of the ICES/ICNAF Joint Working Party on North Atlantic Salmon, 1996.* ICES, Cooperative Research Services A, no. 8.

Anon. (1981) *Report of the ICES Working Group on North Atlantic Salmon, 1979 and 1980.* ICES, Cooperative Research Report no. 104.

Anon. (1982) *Report of the ICES Working Group on North Atlantic Salmon, 1981.* ICES, CM 1982/Assess: 19.

Anon. (1984) *Report of the ICES Working Group on North Atlantic Salmon, 1983.* ICES, CM 1982/Assess: 16.

Baglinière, J.-L. (1976) Etude des populations de saumon Atlantique (*Salmo salar* L., 1766) en Bretagne – Basse Normandie. II – Activité de devalaison des smolts sur l'Elle. *Annales Hydrobiologiques*, 7, 159–77.

Belikov, S.V., Jàkupstovu, S.H.i., Shamrai, E. & Thomsen, B. (1998) *Migration of mackerel in the Norwegian Sea.* ICES CM 1998/AA:8.

Dutil, J.D. & Cotu, J.M. (1988) Early marine life of Atlantic salmon, *Salmo salar*, post-smolts in the northern Gulf of St. Lawrence. *Fisheries Bulletin*, 86, 197–212.

Fried, S.M., McCleave, J.D. & LaBar, G.W. (1978) Seaward migration of hatchery reared Atlantic salmon, *Salmo salar*, smolts in the Penobscot River estuary, Maine: riverine movements. *Journal of the Fisheries Research Board of Canada*, 35, 76–87.

Friedland, K.D. (1998) Ocean climate influences on critical Atlantic salmon (*Salmo salar*) life history events. *Canadian Journal of Fisheries and Aquatic Sciences*, 55 (Suppl. 1), 119–30.

Friedland, K.D., Hansen, L.P. & Dunkley, D.A. (1998) Marine temperatures experienced by post-smolts and the survival of Atlantic salmon, *Salmo salar* L., in the North Sea area. *Fisheries Oceanography*, 7, 22–34.

Hansen, L.P. & Jacobsen, J.A. (in press) Origin and migration of wild and escaped farmed Atlantic salmon, *Salmo salar* L., in oceanic areas north of the Faroe Islands. *ICES Journal of Marine Science*, **00**, 000–000.

Hansen, L.P., & Jonsson, B. (1989) Salmon ranching experiments in the River Imsa: effect of timing of Atlantic salmon (*Salmo salar*) smolt migration on survival to adults. *Salmonid smoltification III*. Proceedings of Workshop sponsored by the Directorate for Nature Management, Norwegian Fisheries Research Council, Norwegian Smolt Producers Association and Statkraft, University of Trondheim, Norway, 27 June–1 July 1988. *Aquaculture*, **82**, 367–73.

Hansen, L.P. & Quinn, T.P. (1998) The marine phase of the Atlantic salmon life cycle, with comparisons to Pacific salmon. *Canadian Journal of Fisheries and Aquatic Science*, **55** (Suppl. 1), 104–18.

Heuch, P.A. (1995) Experimental evidence for aggregation of salmon louse copepodids (*Lepeophtheirus salmonis*) in step salinity gradients. *Journal of the Marine Biological Association of the United Kingdom*, **75**, 927–39.

Holm, M., Huse, I., Waatevik, E., Døving, K.B. & Aure, J. (1982) *Behaviour of Atlantic salmon smolts during seaward migration. I. Preliminary report on ultrasonic tracking in a Norwegian fjord system.* ICES CM 1982/M:7, 10 pp.

Holm, M., Huse, I., Waatevik, E., Aure og, J., Døving, K.B. (1984) Tracking seaward migrating post-smolts (in Norwegian). In: *Behaviour of Marine Animals* (eds M. Holm, A. Fernø & J.W. og Voldemarsen) (in Norwegian), Proceedings from Symposium, 9–10 February 1983, pp. 8–14.

Holm, M., Holst, J.C. & Hansen, L.P. (1996) *Sampling Atlantic salmon in the NE Atlantic during summer: methods of capture and distribution of catches.* ICES CM 1996/M:12, 7 pp.

Holm, M., Holst, J.C. & Hansen, L.P. (1998) *Spatial and temporal distribution of Atlantic salmon post-smolts in the Norwegian Sea and adjacent areas – origin of fish, age structure and relation to hydrographical conditions in the sea.* ICES CM 1998/N:15, 8 pp.

Holm, M., Holst, J.C. & Hansen, L.P (2000) Spatial and temporal distribution of post-smolts of Atlantic salmon (*Salmo salar* L.) in the Norwegian Sea and adjacent areas. *ICES Journal of Marine Science*, **57**, 955–64.

Holm, M., Axelsen, B.E., Sturlaugsson, J., Hvidsten, N.A., Johnsen, B.O. & Ikonen, E. (2001) *Seaward migration of Atlantic salmon post-smolts (*Salmo salar *L.) in the Trondheim fjord, mid-Norway.* Poster presented at the 4th Conference on Fish Telemetry in Europe, Trondheim, 26–30 June.

Holst, J.C. & McDonald, A. (2000) FISH-LIFT: a device for sampling live fish with trawls. *Fisheries Research*, **48**, 87–91.

Holst, J.C., Nilsen, F., Hodneland, K. & Nylund, A. (1993) Observations of the biology and parasites of post-smolt Atlantic salmon, *Salmo salar*, from the Norwegian Sea. *Journal of Fish Biology*, **42**, 962–6.

Holst, J.C., Hansen, L.P. & Holm, M. (1996) *Preliminary observations of abundance, stock composition, body size and food of post-smolts of Atlantic salmon caught with pelagic trawls in the NE Atlantic in the summers 1991 and 1995.* ICES CM 1996/M:4, 8 pp.

Holst, J.C., Arrhenius, F., Hammer, C., et al. (1998) *Report on surveys of the distribution, abundance and migrations of the Norwegian spring-spawning herring, other pelagic fish and the environment of the Norwegian sea and adjacent waters in late winter, spring and summer of 1998.* ICES CM/D:3.

Holst, J.C., Shelton, R., Holm, M. & Hansen, L.P. (2000a) Distribution and possible migration routes of post-smolt Atlantic salmon in the north-east Atlantic. In: *The Ocean Life of Atlantic Salmon: Environmental and Biological Factors Influencing Survival* (ed. D. Mills). Fishing News Books, Oxford, pp. 65–74.

Holst, J.C., Couperus, B., Hammer, C., et al. (2000b) *Report on surveys of the distribution, abundance and migrations of the Norwegian spring-spawning herring, other pelagic fish and the environment of the Norwegian Sea and adjacent waters in late winter, spring and summer of 2000.* ICES CM 2000/D:03. Ref. ACFM.

Huthnance, J.M. & Gould, W.J. (1989) On the northeast Atlantic slope current. In: *Poleward Flows Along Eastern Ocean Boundaries*, Vol. 1 (eds S.J. Neshyba, C.N.K. Mooers, R.L. Smith & R.T. Barger). Springer, New York, pp. 76–81.

Hvidsten, N.A. & Møkkelgjerd, P.I. (1987) Predation on salmon smolts (*Salmo salar* L.) in the estuary of the River Surna, Norway. *Journal of Fish Biology*, **30**, 273–80.

ICES (1999) *Report of the Working Group on North Atlantic Salmon.* ICES CM 1999/ACFM:14. Quebec City, Canada, 12–22 April 1999, 288 pp.

ICES (2001) *Report of the Working Group on North Atlantic Salmon.* ICES CM 2001/ACFM:15. Aberdeen, 2–11 April 2001, 290 pp.

ICES (2002) *Report of the Working Group on the Assessment of Mackerel, Horse Mackerel, Sardine and Anchovy.* ICES CM 2002/ACFM:06.

Jacobsen, J.A. (2000) *Aspects of the marine ecology of Atlantic salmon (*Salmo salar *L.).* PhD Thesis, University of Bergen, Norway, 51 pp + appendices.

Jonsson, B., Jonsson, N. & Hansen, L.P. (1990a) Does juvenile experience affect migration and spawning of adult Atlantic salmon? *Behavioral Ecology and Sociobiology*, **26**, 225–30.

Jonsson, B., Jonsson, N. & Hansen, L.P. (1990b) Partial segregation in the timing of migration of Atlantic salmon of different ages. *Animal Behaviour*, **40**, 313–21.

Jonsson, N., Hansen, L.P. & Jonsson, B. (1993) Migratory behaviour and growth of hatchery-reared post-smolt Atlantic salmon *Salmo salar. Journal of Fish Biology*, **42**, 435–43.

Jonsson, N., Jonsson, B. & Hansen, L.P. (2003) Marine survival and growth of sea ranched and wild Atlantic salmon. *Journal of Animal Ecology* (in press).

LaBar, G.W., McCleave, J.D. & Fried, S.M. (1978) Seaward migration of hatchery-reared Atlantic salmon (*Salmo salar*) smolts in the Penobscot River estuary, Maine: open-water movements. *Journal de Conseil International pour l'Exploration de la Mer*, **38** (2), 257–69.

Lacroix, G.L. & McCurdy, P. (1996) Migratory behaviour of post-smolt Atlantic salmon during initial stages of seaward migration. *Journal of Fish Biology*, **49**, 1086–101.

Metcalfe, N.B. & Thorpe, E.J. (1990) Determinants of geographical variation in the age of seaward-migrating salmon, *Salmo salar. Journal of Animal Ecology*, **59**, 135–45.

Moore, A., Russell, I.C. & Potter, E.C.E. (1990) The effects of intraperitoneally implanted dummy acoustic transmitters on the behaviour and physiology of juvenile Atlantic salmon, *Salmo salar* L. *Journal of Fish Biology*, **37**, 713–21.

Moore, A., Potter, E.C.E., Milner, N.J. & Bamber, S. (1995) The migratory behaviour of wild Atlantic salmon (*Salmo salar* L.) smolts in the estuary of the River Conwy, north Wales. *Canadian Journal of Fisheries and Aquatic Sciences*, **52** (9), 1923–35.

Moore, A., Stonehewer, R., Kell, L.T., et al. (1996) The movements of emigrating salmonid smolts in relation to the Tawe barrage, Swansea. In: *Barrages: Engineering Design and Environmental Impacts* (eds N. Burt & J. Watts). Wiley, Chichester, pp. 409–17.

Moore, A., Ives, S.I., Mead, T.A. & Talks, L. (1998) The migratory behaviour of wild Atlantic salmon (*Salmo salar* L.) smolts in the River Test and Southampton Water. *Hydrobiology*, **371/372**, 295–304.

Moore, A., Lacroix, G.L. & Sturlaugsson, J. (2000) Tracking Atlantic salmon post-smolts in the sea. In: *The Ocean Life of the Atlantic Salmon: Environmental and Biological Factors Influencing Survival* (ed. D. Mills). Fishing News Books, Oxford, pp. 49–64.

Parrish, B.B. & Horsted, S.A. (1980) Ed. ICES/ICNAF joint investigations on North Atlantic salmon. *Rapports et Procés-Verbaux des Réunion du Conseil International pour l'Exploration de la Mer*, **176**, 1–146.

Poulain, P.-M., Warn-Varnas, A. & Niiler, P.P. (1996) Near-surface circulation of the Nordic Seas as measured by Lagrangian drifters. *Journal of Geophysical Research*, **101** (C8), 18.237–58.

Reddin, D.G. & Short, P.B. (1991) Post-smolt Atlantic salmon (*Salmo salar*) in the Labrador Sea. *Canadian Journal of Fisheries and Aquatic Sciences*, **48**, 2–6.

Shelton, R.G.J. (1999) *Post-smolt sampling by R/V 'Clupea' – cruise report 1998.* ICES Working Group on North Atlantic Salmon 1999/WP:17, 3 pp.

Shelton, R.G.J., Turrell, W.R., Macdonald, A., McLaren, I.S. & Nicoll, N.T. (1997) Records of post-smolt Atlantic salmon, *Salmo salar* L., in the Faroe–Shetland Channel in June 1996. *Fisheries Research*, **31**, 159–62.

Sturlaugsson, J. & Thorisson, K. (1995) *Post-smolts of ranched Atlantic salmon (*Salmo salar*) in Iceland: II. The first days of the sea migration.* ICES CM 1995/M:15.

Thorpe, J.E. (1994) Salmonid fishes and the estuary environment. *Estuaries*, **17** (1A), 76–93.

Tytler, P., Thorpe, J.E. & Shearer, W.M. (1978) Ultrasonic tracking of the movements of Atlantic salmon smolts (*Salmo salar* L.) in two Scottish rivers. *Journal of Fish Biology*, **12**, 575–86.

Valdemarsen, J.W. & Misund, O.A. (1995) Trawl designs and techniques used by Norwegian research vessels to sample fish in the pelagic zone. In: *Precision and relevance of pre-recruit studies for fishery management related to fish stocks in the Barents Sea and adjacent waters* (ed. A. Hylen). Proceedings of the 6th IMR-PINRO Symposium, Bergen, 14–17 June 1994. Institute of Marine Research, Bergen, Norway, pp. 129–44.

Chapter 3
Smolt Tracking at Sea – A Search Rewarded

G.L. Lacroix

Department of Fisheries and Oceans, St. Andrews Biological Station, 531 Brandy Cove Road, St. Andrews, New Brunswick, E5B 2L9, Canada

Developments in acoustic telemetry over the past decade now make it possible to monitor the movements of wild Atlantic salmon smolts during the first summer at sea. A case study reveals how acoustic telemetry coupled with new tagging and receiver deployment methodologies were used to passively track and monitor post-smolts throughout the Bay of Fundy on the east coast of Canada. Smolts were captured and tagged near the mouth of rivers, and their movements were monitored from May to October. The tracking method provided estimates of migratory success to various points: departure from the river, from the coastal zone near the river mouth, from the inner sector of the Bay of Fundy and from the outer sector of the bay into the open waters of the Gulf of Maine. Movements to and from a sector of the Bay of Fundy where salmon aquaculture is centred were also monitored. The estimates of success helped identify the areas of potentially high early marine mortality and to focus on the potential causal factors. The routes and timing of migration were compared for smolts originating from rivers of the inner and outer Bay of Fundy to test the hypothesis that these genetically different stocks show different behaviour and use different habitats at sea, which could account for the differences in marine mortality observed. Salmon stocks from the inner Bay of Fundy rivers have crashed and they were declared 'endangered' under the Canadian Species at Risk Act.

This is a summary of the complete paper upon which the author's presentation at the Symposium was based. The author plans to publish the material later, possibly modified by additional field data obtained in 2002.

Chapter 4
The Ecology of Post-Smolts of Atlantic Salmon

L.P. Hansen[1], M. Holm[2], J.C. Holst[2] and J.A. Jacobsen[3]

[1] *Norwegian Institute for Nature Research, PO Box 736, Sentrum, N-0105 Oslo, Norway*
[2] *Institute of Marine Research, PO Box 1870, Nordnes, N-5817 Bergen, Norway*
[3] *Fiskirannsóknarstovan, Nóatún, PO Box 3051, FO-110 Tórshavn, Faeroe Islands*

Abstract: Anadromous Atlantic salmon move from fresh water to the ocean and gain weight. The fish move over large areas and are exposed to different environments and ecological communities during the marine phase of the life cycle. Although the biomass of salmon in the ocean is very small compared with the biomass of other pelagic marine species, Atlantic salmon are still part of the marine ecosystem and subjected to fluctuations and changes in that system.

The distribution of salmon in the sea is probably dependent on several environmental factors like surface temperature and surface currents. Furthermore, the oceanic distribution of salmon may probably also be determined by genetic components, for example that salmon have developed navigation systems that bring the fish to the right place at the right time, and thus maximize fitness. It is also reasonable to assume that the availability of suitable food organisms affects the distribution of salmon in the sea, as growth and survival are important fitness characters as well.

Salmon spend most of their time in the ocean close to the water surface, and prey on different pelagic animals such as crustaceans, fish and squid. Several authors have suggested that Atlantic salmon are opportunistic feeders, but selective feeding has also been described.

The abundance of Atlantic salmon in the north Atlantic has declined considerably in recent years. The decline is evident from most of the salmon distribution area, and is most pronounced for large salmon, but there appears to be no significant density-dependent mortality of salmon in the sea as there is in fresh water. The abundance of salmon is dependent on several factors such as smolt production, natural and man-made mortality. It has been shown that marine mortality accounts for a significant proportion of the decline, and is associated with a decline in the temperature to which post-smolts are exposed during the first months at sea.

There are many factors thought to influence marine mortality of salmon. Although it has been suggested that the highest mortality takes place at the post-smolt stage, there are also indications that heavy mortality may occur later in the life cycle. Important sources of post-smolt mortality are predation, infestations of parasites and diseases, influences from freshwater life, as well as synergistic effects between these and a number of other factors. Furthermore, salmon may respond to changes in the ecosystem by altering their life histories, which may be seen by, for example, changes in growth rates, sea age at maturity and seasonal return pattern.

Salmon are harvested in the ocean, but exploitation in commercial fisheries is now close to zero. On the other hand, recent information from surface trawling experiments in the Norwegian Sea has raised concern about the significance of by-catches of salmon post-smolts in pelagic fisheries for marine species.

Introduction

Atlantic salmon, *Salmo salar* L., undertake long migrations on their way from fresh water to the feeding areas in the ocean and on return to fresh water. There is substantial information on Atlantic salmon in fresh water, but in contrast much less is known about their sea life (e.g. Friedland, 1998; Hansen & Quinn, 1998).

When coastal fisheries developed more than 100 years ago some information became available on salmon in the local environment, but distribution and migration of salmon in oceanic areas was a 'black box' until oceanic fisheries developed in the 1960s at west Greenland and later in the northern Norwegian Sea. Research on material sampled from these fisheries was undertaken (e.g. Parrish & Horsted, 1980; Hansen & Pethon, 1985; Jacobsen, 2000), and new knowledge was gained. However, these investigations did not address the distribution and ecology of fish outside the fishing seasons and areas and the post-smolt biology.

It is well known that there are large variations in survival of salmon among different smolt year classes. Although there was a lack of knowledge, it was suggested that the heaviest mortality of salmon in the sea took place during the first months after the smolts left fresh water (e.g. Doubleday et al., 1979; Ritter, 1989).

Knowledge of salmonids in the early marine phase is relatively scarce and has been reviewed by Thorpe (1994). Many factors affect smolt survival and return of adult salmon, although many of them are poorly known. Research on this issue has been strongly recommended by a number of organizations, but it was only recently that there were systematic efforts to sample post-smolts at sea (Reddin & Short, 1991; Holst et al., 1993; Shelton et al., 1997; Holm et al., 2000). Results from these studies together with the development of new techniques to analyse life history signals from scales, bones and tissue of salmon have improved our understanding of the biology of salmon post-smolts, but there are still major gaps.

Several parasites and diseases may affect salmon post-smolts. Some of them like the copepod sea lice are of great concern, and can cause mortality of salmon post-smolts (Holst et al., this volume). The disease infectious salmon anaemia is also a matter of concern. The increased salmon farming activity has contributed significantly to these problems.

The aim of this paper is to review some aspects of the ecology of Atlantic salmon post-smolts, and particularly to focus on abiotic and biotic factors that affect salmon in the early marine phase.

Freshwater influences

The timing of smolt migration seems to be crucial for survival of Atlantic salmon (Hansen, 1987), and the development of the 'smolt window' is presumably an adaptation to the area's prevailing environmental conditions. This may be related to interaction between environmental factors and physiological processes (e.g. McCormick et al., 1998). For example, a temperature-related smolt window indicates that delays in migration can have negative consequences that will be greater in warmer, southern rivers.

In addition to the change in physiological state of smolts, it is likely that there are other mechanisms behind the observations that there is an optimal time for smolt migration. These are less well understood, but predators, diseases, parasites and food conditions clearly contribute to smolt mortality. Atlantic salmon smolts are subjected to heavy predation and important predators include birds such as gulls, cormorants, herons and mergansers (Anthony, 1994), and small smolts may be more vulnerable than large ones. There may be an interaction between migration timing, physiological readiness and predation. Under laboratory conditions the risk of predation is greater during acclimation of smolts to sea water, and osmotic imbalance during seawater acclimation is greater in the presence of predators (Järvi, 1990; Handeland et al., 1997). Thus, even if smolts with reduced salinity tolerance (such as those migrating too early or too late in their development) survive when they enter sea water, they are less likely to escape predation due to both their greater susceptibility and the increased time they need to spend in an estuary for gradual acclimation.

Several studies strongly suggest the existence of a survival window which is open for a few weeks during spring and gradually closes during the summer (Larsson, 1977; Hansen & Jonsson, 1989). This effect is likely to be driven endogenously, but it may be affected by extrinsic factors in the river, estuary, or at sea. In Norwegian rivers draining areas with high snow fall, the current velocity and turbidity increase considerably during snow melt. Survival of hatchery-reared smolts in the Gaula and Surna rivers was greater when they were released at high water discharge within the normal period of migration (Hvidsten & Hansen, 1988), consistent with results in Maine (Hosmer et al., 1979). Survival may have been improved because the smolts move close to the water surface (Holm et al., 1982) and descend more quickly at high current velocity (Youngson et al., 1989), and they may be less vulnerable to predation owing to the high turbidity.

One important example is the effect of acidity and aluminium on smolt development. It has been known for some time that acid (and acid plus aluminium) exposures that have little effect on survival and growth of parr or smolts in fresh water can completely eliminate the development of gill Na^+, K^+-ATPase activity and salinity tolerance (Saunders et al., 1983). Exposures as short as 12 hours to relatively mild acidity with high aluminium levels can compromise salinity tolerance (Staurnes et al., 1996a). The timing of exposure is also important; acid conditions that have little effect on smolts in April can greatly reduce salinity tolerance in May. Acid exposures at the smolt stage have devastating effects on adult return rates. Fish released in an acidic river in Norway often have no adult returns and have an average of one-tenth the returns of fish released at the mouth of the river or in a neighbouring limed river (Staurnes et al., 1996b). These return rates were strongly correlated with the effect of acidity on gill Na^+, K^+-ATPase activity and osmotic balance of fish held in cages at the different release sites.

Distribution of post-smolts in the sea

Smolting and the subsequent seaward migration of Atlantic salmon smolts usually occur during one month in the spring, earlier in the southern range of distribution than in the north. The estuarine residence time is relatively brief (Tytler et al., 1978). In estuaries post-smolts are displaced with the surface current, and movement is influenced by the tide and the direction of the water flow (e.g. Holm et al., 1982; Thorpe, 1988; Pethon & Hansen, 1990). Several authors have concluded that smolts in the upper estuary move by selective ebb tide transport (Fried et al., 1978; Moore et al., 1995) but display active swimming in the lower estuary or bays (Moore et al., 1995; Lacroix & McCurdy, 1996).

Most of the records available suggest that post-smolts move relatively quickly into the ocean, swimming as fast as 2 body lengths/s (Lacroix & McCurdy, 1996), after a period of more passive movements with tidal currents. Further evidence, albeit indirect, of rapid migration comes from the fact that very few post-smolts are recorded in fjords and coastal waters during summer and autumn, although they are already present in oceanic areas in the east Atlantic (Holm et al., 2000) and west Atlantic (Reddin & Short, 1991) at this time of the year. However, Dutil & Coutu (1988) caught many post-smolts in a nearshore zone of the northern Gulf of St. Lawrence, suggesting that this trait may vary among populations or areas. Friedland et al. (1998a) also found evidence that post-smolts in some years used the Gulf of St. Lawrence as a nursery area prior to seaward migrations.

In Atlantic salmon there is evidence that the marine distribution of this species is dependent on temperature (Reddin & Shearer, 1987), but whether distribution of food is an important factor is still an open question. The biomass of Atlantic salmon in the ocean relative to other pelagic oceanic fish species is extremely small, and salmon during its marine phase is thought to be an opportunistic pelagic and midwater predator, supporting rapid growth rate by exploiting a wide range of invertebrates and fish prey. The wide variety of food in different areas and periods suggests that salmon abundance is unlikely to be very sensitive to annual changes in the availability of any particular prey (Jacobsen & Hansen, 2001).

A number of post-smolts have been caught in oceanic areas of the north-east Atlantic in recent years during pelagic trawl surveys in the Norwegian Sea in July and August (Holm et al., 2000), and north of Scotland in May and June (Holm et al., 1996; Holst et al., 1996; Shelton et al., 1997), and the distribution of those in the first summer at sea is shown in Fig. 4.1. Based on the distribution of catches north of Scotland, the fish appeared to move northwards with the shelf edge current (Shelton et al., 1997). Farther north in the Norwegian Sea post-smolts were caught beyond 70°N in July. Analysis of growth and smolt age distribution strongly suggested that most of the post-smolts originated from rivers in southern

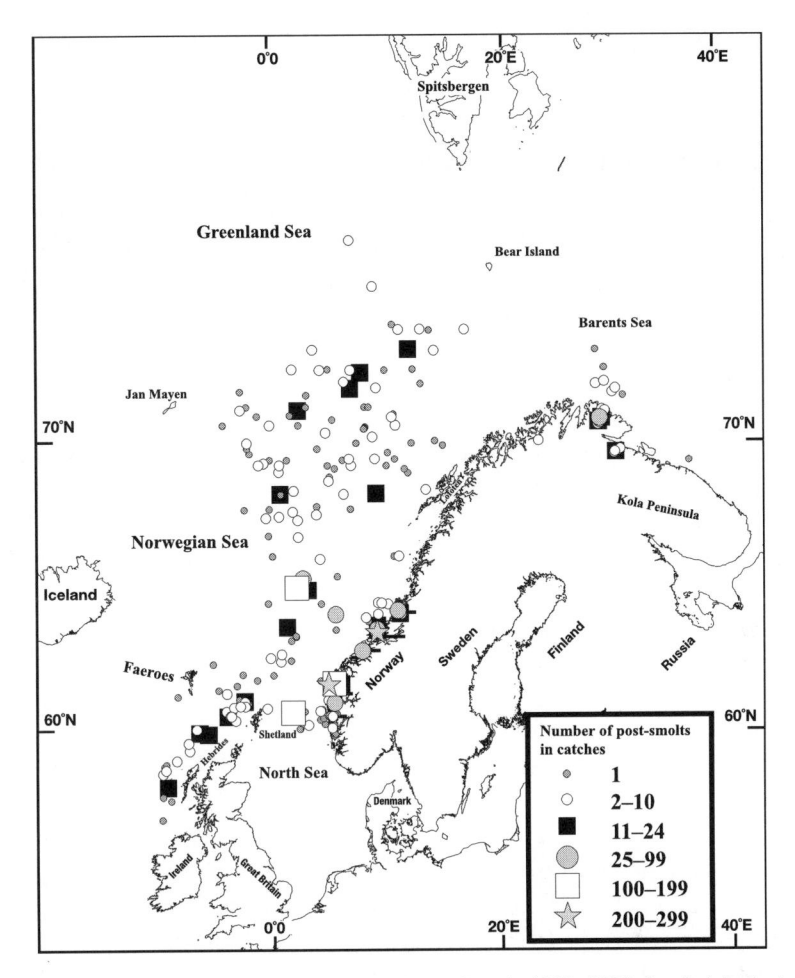

Fig. 4.1 Distribution of post-smolt captures in the Nordic Seas in 1990–2000. Symbols indicate number of post-smolts in the catch. (From Holm et al., 2000. Reproduced with permission of Elsevier Science.)

Europe (Holst et al., 1996). This was supported by the recapture of a salmon that had been tagged in April 1995 in southern England, and recovered about 2000 km farther north 3 months later, demonstrating the capacity for rapid travel by post-smolts (Holst et al., 1996).

In the Labrador Sea, Reddin & Short (1991) caught many post-smolts in the early autumn of 1987 and 1988, using surface drift nets. The highest catch rates occurred between 56° and 58°N, and they also caught many older salmon in the area. Based on age analysis, they concluded that post-smolts originating from Maine to Labrador were present. This was supported by recoveries of two post-smolts that had been tagged in the Penobscot River in Maine, and Western Arm Brook, Newfoundland. In addition, four older fish in the Labrador Sea had tags, three from the Penobscot River and one from the Middle River, Nova Scotia.

Food of post-smolts

Coastal areas

Morgan et al. (1986) reported stomach contents of 21 salmon post-smolts caught shortly after emigration from the rivers into the Firth of Clyde, west Scotland. They found that the diet consisted principally of sand eels. A larger study of post-smolts was undertaken in the northern Gulf of St. Lawrence by Dutil & Coutu (1988) who examined salmon stomachs from July until October. In July the main food items were invertebrates (Insecta, mainly Chironomidae, and Crustacea, mainly Gammaridae). Later in the summer and autumn, the post-smolts had consumed mainly small fish, where *Ammodytes americanus* were most prevalent and occurred in most of the stomachs. Capelin were also relatively abundant. Among the crustaceans, euphausiids were the most frequent.

Levings et al. (1994) caught post-smolts of Atlantic salmon by pair-trawling at different sites in Trondheimsfjord, Norway, in May and beginning of June 1992–94. These fish may have been at sea for a few days to a few weeks. The fjord area was divided into zones, starting in the estuary, followed by the fjord area and ending in the outer coastal area. In the estuary, the stomach contents were dominated by prey taken in the river (Levings et al., 1994), whereas prey occurring in brackish and salt water were found in stomach samples collected farther out. The volume frequency of the main prey groups for the material collected in the fjord area just inside the open coastal zone showed great temporal variations, e.g. insects dominated the first 2 years and crustaceans the third year. There were also spatial variations within a year, e.g. in 1994 euphausiids dominated in the fjord area, whereas farther out the importance of this group was somewhat reduced, as amphipods (Hyperiidae) and fish (herring larvae) became important as well (Table 4.1; Hvidsten et al., 1995). This pattern was further confirmed by Andreassen et al. (2001) and they pointed out that the gradual change in diet suggested that feeding conditions in the early marine phase were important for survival and growth of the post-smolts.

Sturlaugsson (1994, 2000) examined the food of ranched post-smolts of Atlantic salmon in coastal waters off west Iceland. He found that the most important prey of post-smolts were decapods (crab larvae), amphipods, copepods, dipterans (imago/ adult stage), euphausiids and fish larvae. The importance of some of the main prey types was area restricted, e.g. dipterans and epibenthic amphipods (Gammaridea) were important in nearshore areas as opposed to the offshore areas where pelagic amphipods (Hyperiidea) and fish larvae (*Ammodytes* spp.) became important prey.

Oceanic areas

The feeding habits of post-smolts have been analysed in the area west of the British Isles and in the Norwegian Sea (Holst et al., 1993, 1996). In total, stomach contents from 34 post-smolts that were captured in the Norwegian Sea in 1991, 46 off the

Table 4.1 Food content of Atlantic salmon post-smolts collected in the Trondheimsfjord and along the coast in 1994. Percentage frequency of occurrence (%F) and weight (volume) percentages (%V). + indicates less than 1%. (From Hvidsten et al., 1995.)

Prey group	Area					
	Fjord area ($n=84$)		Coastal area ($n=108$)		Total	
	%F	%V	%F	%V	%F	%V
Insects						
Diptera larvae	3.7	3.44	—	—	1.4	0.01
Diptera adults	37.8	0.03	39.8	1.40	37.0	1.68
Coleoptera	2.4	2.40	0.9	0.02	1.4	0.02
Insecta remains	17.1	0.01	17.6	0.65	15.9	0.52
Crustaceans						
Calanus	6.1	0.03	14.8	0.34	10.1	0.17
Euphausiidae	64.6	92.34	38.0	39.78	49.5	63.05
Hyperiidae	3.7	0.11	63.0	8.74	35.1	4.20
Gammaridae	—	—	0.9	0.01	0.5	0.01
Decapoda (Zoea)	8.5	0.06	0.9	0.01	3.8	0.03
Corophium spp.	2.4	—	0.9	0.08	1.4	0.05
Isopoda	1.2	0.03	—	—	0.5	0.01
Crustacea remains	9.8	0.03	38.9	23.68	27.4	14.10
Fish						
Herring	+	0.92	7.4	15.77	4.3	7.93
Fish remains	3.7	+	24.1	9.42	15.9	6.40

Hebrides in 1995 and 62 in the Norwegian Sea in 1995 were analysed. The proportion of empty stomachs was relatively small (3–15%), which might indicate favourable feeding regimes in the sampled areas (Holst et al., 1996). In the Norwegian Sea, the most common prey observed in post-smolts were hyperiid amphipods (*Themisto* spp.), euphausiids (Euphausiidae), 0-group herring and redfish (*Sebastes* spp.) larvae, whereas off the Hebrides blue whiting (*Micromesistius poutassou*) larvae/0-group was the only food item recorded (Holst et al., 1996).

The total dominance of 0-group/larval blue whiting eaten by the post-smolts off the Hebrides in May–June 1995 coincided with the very strong 1995 year-class of blue whiting, which was spawned earlier in the year mainly on the Porcupine Bank area and along the slope west of Ireland and the Hebrides (Holst et al., 1998).

Survival of post-smolts may also be connected with abundance of suitable prey, particularly during the period immediately after they leave fresh water. When the smolts enter salt water their expenditure of energy increases (Farmer et al., 1978), and scarce food resources may result in increased mortality. Lack of food would also reduce growth rate and increase their vulnerability to size-selective predation.

Jacobsen & Hansen (2001) found differences in the feeding habits of salmon with different sea ages. The smaller 1 sea-winter (1 SW) salmon had mainly eaten amphipods and lanternfishes, and some barracudinas, while the larger 3+ SW salmon had mainly eaten large fish, barracudinas and lanternfishes, and the 2 SW salmon had eaten the various prey groups in intermediate proportions compared to the 1 SW and 3+ SW salmon (Fig. 4.2). Thus, with increasing age (size) of salmon, the larger fish prey was preferred to smaller mesopelagic fish and crustaceans.

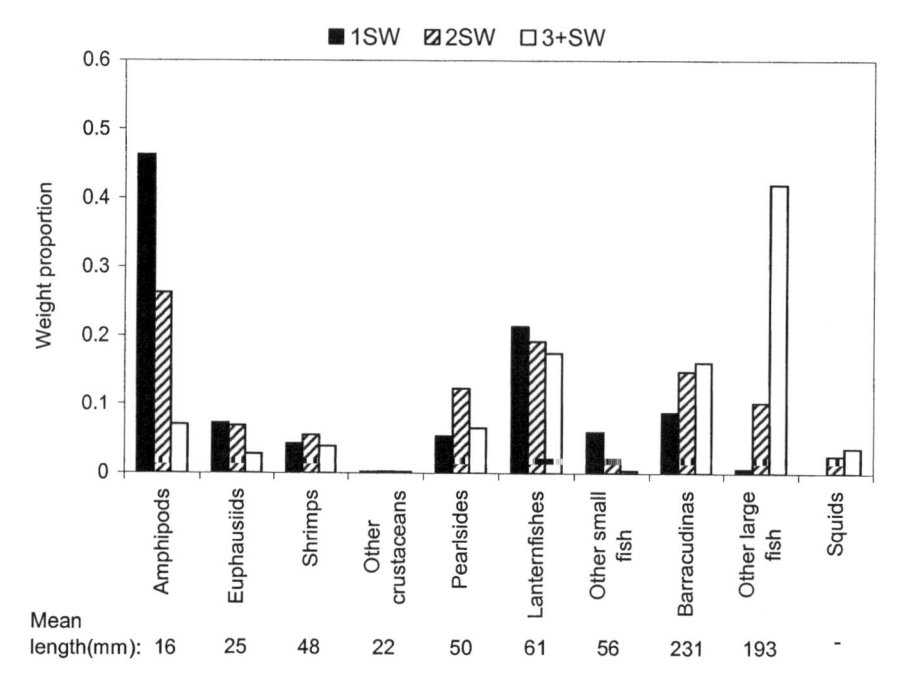

Fig. 4.2 Diet of 1 (filled), 2 (hatched) and 3+ SW (sea-winter) (open) Atlantic salmon (proportions by weight) for ten prey groups observed in salmon stomachs at the Faeroes. Mean prey length (mm) is indicated below each prey group. Squid were not measured, but were juveniles. (From Jacobsen & Hansen, 2001. Reproduced with permission of Elsevier Science.)

Competition

In some areas post-smolts probably compete with other fish species for food as they pass through various coastal and offshore areas on their way to the high seas. One such example could be from the North Sea, where post-smolts from the west coast of Sweden, southern Norway, Denmark, south-east England and east Scotland are anticipated to travel through the northern North Sea in spring and summer before entering the Norwegian Sea. The often abundant resource of Norway pout (*Trisopterus esmarkii*) in the North Sea, a fish the same size as post-smolts (juvenile 8–13 cm and adult 14–21 cm), take food of the same class as post-smolts are expected to prey on, i.e. mainly pelagic zooplankton (Raitt & Adams, 1965), including calanoid copepods, euphausiids, decapod larvae and fish larvae. Furthermore, both juvenile whiting and haddock are feeding in the same areas. However, Raitt & Adams (1965) considered only whiting to feed on the same type of food as Norway pout. Probably the phase of the post-smolts stage in coastal areas and on the continental shelves could be considered food limited due to competition from other pelagic predators.

In the Norwegian Sea, spring spawning herring (NSSH) and mackerel are also present at the same time as post-smolts (Holst et al., 1996). Herring larvae may be important food for post-smolts in coastal areas, but adult herring and mackerel may be competitors in the ocean. The biomass of herring has increased considerably in recent years, and thus it is reasonable to ask if this may affect growth and survival of post-smolt salmon. We have made a first approach to examine this, by comparing time series of biomass of herring, catches of salmon in the east Atlantic and marine survival of salmon from the River Figgjo (Hansen et al., 1999). The results from the preliminary analysis support the hypothesis that the presence of large numbers of NSSH in the Norwegian Sea may contribute to increased mortality of salmon in the ocean. This is indicated by an observed negative relationship between herring abundance and salmon catch, as well as between herring abundance and marine survival of smolts from the River Figgjo. However, the quality of the time series is variable, and in some cases estimates have been used without taking consideration of the fact that there may be considerable variation among the variables.

Growth

Salmon grow relatively fast in the sea, but the plasticity of salmon growth makes the discrimination between the extent of genetic and environmental influences difficult. Differences in growth rates could, for example, be due to variable food supply, or more general changes in oceanographic processes and structural features in the water column (Levings et al., 1994). These changes might affect some populations more than others, depending on their marine distribution. Differences in growth rates may also be shaped by patterns of natural selection in individual rivers. Jonsson et al. (1991) observed that Atlantic salmon in rivers with annual average flows of less than $40 \text{ m}^3/\text{s}$ were smaller than salmon in larger rivers, and suggested that selection in small rivers may act against large salmon because low water flows make successful ascent and breeding by large salmon risky and difficult. Mortality of salmon at sea and fish size have been hypothesized to be inversely related (Doubleday et al., 1979). Friedland et al. (2000) found a significant relationship between sea surface temperatures and survival observed, and that the back-calculated growth in the first sea year varied between 32 and 43 cm in the period 1965–93. High growth rates were associated with high temperatures, which in turn resulted in higher survival rates.

Predation

Survival of Atlantic salmon smolts to adulthood varies greatly among years and stocks. Marine survival rates of salmon have been estimated at 0.3–0.4 (30–40%) for wild salmon from the River Bush, Northern Ireland (Crozier & Kennedy, 1993). Similar figures have also been observed in salmon from the River North Esk in

Scotland (Shearer, 1992). These observations have been made in years when survival rates were high. On the other hand, marine survival rates of salmon can be less than 0.05. Several factors may affect the survival of salmon in the sea, including fish disease, predation and food resources. Predation is probably the most important source of mortality, though diseased fish are more likely to be eaten before they perish. A comprehensive review by Anthony (1994) discussed predatory birds, fish and mammals as sources of marine mortality in salmon. The main mortality is believed to take place at the post-smolt stage, the first months after the smolts leave their home rivers. This is based on the assumptions that small fish are exposed to a larger range of predator species than large fish, and that there are more predators inside the continental shelf than in oceanic areas.

Data on predation in the ocean are scarce, but it has been speculated that sharks and skates may eat salmon, and there is some documentation of consumption by seals and whales (Pierce et al., 1991; Shearer, 1992; Hislop & Shelton, 1993).

Predation on smolts and post-smolts may be most severe in estuaries and fjords, just after the smolts have left fresh water. In the estuaries of the Rivers Surna and Orkla, Norway, heavy predation by fish, mainly cod, has been observed on both hatchery-reared and wild smolts (Hvidsten & Møkkelgjerd, 1987; Hvidsten & Lund, 1988). Higher survival of hatchery-reared smolts was obtained when they were transported by well-boat and released in the ocean outside the River Surna (Gunnerød et al., 1988), but straying of the returning adults increased considerably. In another Norwegian estuary, Reitan et al. (1987) showed that many migrating hatchery-reared smolts were eaten by sea birds.

Fishing mortality

Post-smolts have been recorded as a by-catch in other marine fisheries, e.g. herring and mackerel nets in fjords, purse seines and surface trawls. However, there is hardly any quantitative information available to assess the effects of this. There have been approximate estimates of the potential risk of salmon post-smolts being taken in commercial fisheries for pelagic fish, such as floating trawl fisheries for mackerel.

In June 2001 the estimates were improved when large catches of mackerel together with a varying number of post-smolts were registered by the Institute of Marine Research in Bergen during a special post-smolt survey west of the Vøring Plateau in the Norwegian Sea. This survey was carried out at approximately the same time as the commercial mackerel fishery starts in these areas. The simultaneous occurrence of salmon and post-smolts in areas where the commercial fleet is known to operate, for the first time gave the opportunity to examine the possible magnitude of the by-catches of post-smolts of salmon in the commercial fishery. The survey took place between 13 and 17 June 2001 between 64.3–67.9°N and 1–3°E (west of the Vøring Plateau). In total, 14 hauls with a mean tow-duration of 1.8 h were conducted, and 198 post-smolts and 5 salmon were captured [Holm et al. reported in ICES (2002)]. Simultaneously, a total of 7959 kg mackerel was taken. This corresponds to a catch of

0.025 post-smolts per kg mackerel caught. This suggests a significant potential for by-catch of post-smolts in this fishery. It was also interesting to note that the smolt age of post-smolts captured in this area is relatively low and thus many may originate from southern and mid areas of Europe. However, more research should be carried out to confirm these estimates.

Environment

The decreased Atlantic salmon abundance in recent years has been most pronounced in North America, although there has also been a significant decrease in Europe, particularly in southern areas. There are several factors causing this, e.g. freshwater habitat degradation, pollution, parasites like *Gyrodactylus salaris* and effects of overexploitation. However, the decline is more pronounced for multi-sea-winter (MSW) salmon than for grilse, suggesting that changes in the marine environment may affect the survival of salmon in the sea (e.g. Ritter, 1989). Friedland et al. (1993) showed that interannual variation in the area of potential post-smolt habitat at sea, defined as the area combining their optimal temperature and full marine salinity, is significantly correlated with catches of salmon from this area. Further studies, using time series of smolt survival rates, seem to confirm this.

If the driving force behind marine mortality is temperature-related, one might predict that survival of salmon stocks migrating as post-smolts into the same area would be correlated. This hypothesis was tested using time series of survival estimates of wild salmon from the Rivers Figgjo, Norway, and the North Esk, Scotland (Friedland et al., 1998b). These rivers enter at opposite sides of the North Sea at about the same latitude and salmon from both rivers have been observed in the Faroes fisheries at the same time. Hence it was assumed that both groups are subjected to the same marine environment for at least the first months at sea (both stocks migrate to sea in early May). Return rates of 1 SW and 2 SW salmon from Figgjo were significantly correlated, as were 1 SW and 2 SW salmon from the North Esk, supporting the hypothesis that most mortality takes place in the first year at sea. Furthermore, there were significant correlations in return rates between salmon from both rivers, suggesting that survival was driven by common marine factors. A positive correlation was found between the area of temperature 8–10°C in the Norwegian Sea north of the outlet of the respective rivers in May and the survival of salmon, which supports the hypothesis of temperature-related mortality in the sea.

Changes in marine environments may change the life history of salmon. A changed climate may, for example, change the timing of smolt migration. There is an apparent optimal sea temperature at post-smolt entry. If temperatures in the freshwater and marine environments are not synchronized, this may affect the mortality of post-smolts. For example, if temperature increases in fresh water, the smolts may be triggered to migrate to sea earlier. But if the sea temperature has not changed, or has decreased, this would result in the smolts moving into an environment that is not optimal for them. A changed marine climate may also affect growth rate and thus sea age at maturity. Thus a faster growth in the marine environment may produce salmon with younger sea age.

There are numerous factors that affect survival of salmon in the marine environment, abiotic as well as biotic (Fig. 4.3), and the interactions between these factors are poorly known. Despite the small biomass of salmon present in the ocean, salmon are a component in the marine ecosystem and are affected by long- and short-term fluctuations in this ecosystem.

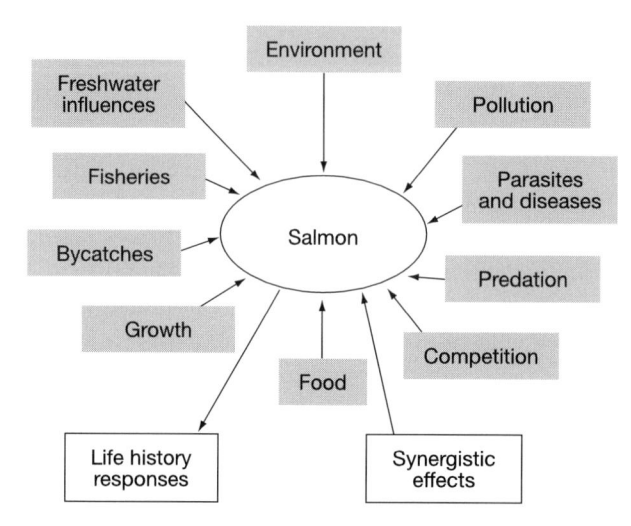

Fig. 4.3 Factors influencing salmon at sea.

References

Andreassen, P.M.R., Martinussen, M.B., Hvidsten, N.A. & Stefansson, S.O. (2001) Feeding and prey selection of wild Atlantic salmon post-smolts. *Journal of Fish Biology*, **58**, 1667–79.

Anthony, V.C. (1994) The significance of predation on Atlantic salmon. In: *A hard look at some tough issues* (eds S. Calabi & A. Stout). Proceedings of New England Salmon Management Conference. Silver Quill, Camden, Maine, pp. 240–88.

Crozier, W.W. & Kennedy, G.J.A. (1993) Marine survival of wild and hatchery-reared Atlantic salmon (*Salmo salar* L.) from the River Bush, Northern Ireland. In: *Salmon in the Sea and New Enhancement Strategies*. (ed. D. Mills). Fishing News Books, Oxford, pp. 139–62.

Doubleday, W.G., Rivard, D.R., Ritter, J.A. & Vickers, K.U. (1979) *Natural mortality rate estimates for north Atlantic salmon in the sea*. ICES CM 1979/M:26.

Dutil, J.-D. & Coutu, J.-M. (1988) Early marine life of Atlantic salmon, *Salmo salar*, postsmolts in the northern Gulf of St. Lawrence. *Fisheries Bulletin*, **86**, 197–212.

Farmer, G.J., Ritter, J.A. & Ashfield, D. (1978) Seawater adaptation and parr-smolt transformation of juvenile Atlantic salmon, *Salmo salar*. *Journal of the Fisheries Research Board of Canada*, **35**, 93–100.

Fried, S.M., McCleave, J.D. & LaBar, G.W. (1978) Seaward migration of hatchery-reared Atlantic salmon, *Salmo salar*, smolts in the Penobscot River estuary, Maine: riverine movements. *Journal of the Fisheries Research Board of Canada*, **35**, 76–87.

Friedland, K.D. (1998) Ocean climate influences on critical Atlantic salmon (*Salmo salar*) life history events. *Canadian Journal of Fisheries and Aquatic Sciences*, **55** (Suppl. 1), 119–30.

Friedland, K.D., Reddin, D.G. & Kocik, J.F. (1993) Marine survival of North American and European Atlantic salmon: effects of growth and environment. *ICES Journal of Marine Science*, **50**, 481–92.

Friedland, K.D., Dutil, J.-D. & Sadusky, T. (1998a) Growth patterns in postsmolts and the nature of the marine juvenile nursery for Atlantic salmon, *Salmo salar*. *Fisheries Bulletin*, **97**, 472–81.

Friedland, K.D., Hansen, L.P. & Dunkley, D.A. (1998b) Marine temperatures experienced by postsmolts and the survival of Atlantic salmon, *Salmo salar* L., in the North Sea area. *Fisheries and Oceanography*, **7**, 22–34.

Friedland, K.D., Hansen, L.P., Dunkley, D.A. & MacLean, J.C. (2000) Linkage between ocean climate, post-smolt growth and survival of Atlantic salmon (*Salmo salar* L.) in the North Sea area. *ICES Journal of Marine Science*, **57**, 419–29.

Gunnerød, T.B., Hvidsten, N.A. & Heggberget, T.B. (1988) Open sea releases of Atlantic salmon smolts, *Salmo salar*, in central Norway, 1973–83. *Canadian Journal of Fisheries and Aquatic Sciences*, **45**, 1340–45.

Handeland, S.O., Järvi, T., Fernö, A. & Stefansson, S.O. (1997) Osmotic stress, antipredator behaviour, and mortality of Atlantic salmon (*Salmo salar*) smolts. *Canadian Journal of Fisheries and Aquatic Sciences*, **53**, 2673–80.

Hansen, L.P. (1987) Growth, migration and survival of lake reared juvenile anadromous Atlantic salmon, *Salmo salar*. *Fauna Norvegica Series A*, **8**, 29–34.

Hansen, L.P. & Jonsson, B. (1989) Salmon ranching experiments in the River Imsa: effect of timing of Atlantic salmon (*Salmo salar*) smolt migration on survival to adults. *Aquaculture*, **82**, 367–73.

Hansen, L.P. & Pethon, P. (1985) The food of Atlantic salmon, *Salmo salar* L. caught by long-line in northern Norwegian waters. *Journal of Fish Biology*, **26**, 553–62.

Hansen, L.P. & Quinn, T.P. (1998) The marine phase of the Atlantic salmon life cycle, with comparisons to Pacific salmon. *Canadian Journal of Fisheries and Aquatic Sciences*, **55** (Suppl. 1), 104–18.

Hansen, L.P., Holst, J.C., Jensen, A.J. & Johnsen, B.O. (1999) *Norwegian spring spawning herring and Atlantic salmon: do they interact?* ICES North Atlantic Salmon Working Group, 13 pp.

Hislop, J.R.G. & Shelton, R.G.J. (1993) Marine predators and prey of Atlantic salmon (*Salmo salar* L.). In: *Salmon in the Sea and New Enhancement Strategies* (ed. D. Mills). Fishing News Books, Oxford, pp. 104–18.

Holm, M., Huse, I., Waatevik, E., Døving, K.B. & Aure, J. (1982) *Behaviour of Atlantic salmon smolts during seaward migration. I. Preliminary report on ultrasonic tracking in a Norwegian fjord system.* ICES CM 1982/M:7, 17 pp.

Holm, M., Holst, J.C. & Hansen, L.P. (1996) *Sampling Atlantic salmon in the NE Atlantic during summer: methods of capture and distribution of catches.* ICES CM 1996/M:12, 7 pp.

Holm, M., Holst, J.C. & Hansen, L.P. (2000) Spatial and temporal distribution of post-smolts of Atlantic salmon (*Salmo salar* L.) in the Norwegian Sea and adjacent areas. *ICES Journal of Marine Science*, **57**, 955–64.

Holst, J.C., Nilsen, F., Hodneland, K. & Nylund, A. (1993) Observations of the biology and parasites of postsmolt Atlantic salmon, *Salmo salar*, from the Norwegian Sea. *Journal of Fish Biology*, **42**, 962–6.

Holst, J.C., Hansen, L.P. & Holm, M. (1996) *Preliminary observations of abundance, stock composition, body size and food of postsmolts of Atlantic salmon caught with pelagic trawls in the NE Atlantic in the summers 1991 and 1995.* ICES CM 1996/M:4, 8 pp.

Holst, J.C., Arrhenius, F., Hammer, C., et al. (1998) *Report on surveys of the distribution, abundance and migrations of the Norwegian spring-spawning herring, other pelagic fish and the environment of the Norwegian Sea and adjacent waters in late winter, spring and summer of 1998.* ICES CM/D:3.

Hosmer, M.J., Stanley, J.G. & Hatch, R.W. (1979) Effects of hatchery procedures on later return of Atlantic salmon to rivers of Maine. *Progressive Fish Culturist*, **41**, 115–19.

Hvidsten, N.A. & Hansen, L.P. (1988) Increased recapture rate of adult Atlantic salmon, *Salmo salar* L., stocked at high water discharge. *Journal of Fish Biology*, **32**, 153–4.

Hvidsten, N.A. & Lund, R.A. (1988) Predation on hatchery-reared and wild smolts of Atlantic salmon, *Salmo salar* L., in the estuary of River Orkla, Norway. *Journal of Fish Biology*, **33**, 121–6.

Hvidsten, N.A. & Møkkelgjerd, P.I. (1987) Predation on salmon smolts (*Salmo salar* L.) in the estuary of the River Surna, Norway. *Journal of Fish Biology*, **30**, 273–80.

Hvidsten, N.A., Johnsen, B.O. & Levings, C.D. (1995) Vandring og ernæring hos laksesmolt i Trondheimsfjorden og på Frohavet. *NINA Oppdragsmelding*, **332**, 1–17.

ICES (2002) *Report of the Working Group on North Atlantic salmon.* ICES CM 2002/ACFM:14, 297 pp.

Jacobsen, J.A. (2000) *Aspects of the marine ecology of Atlantic salmon (*Salmo salar *L.)* PhD Thesis, University of Bergen, Norway.

Jacobsen, J.A. & Hansen, L.P. (2001) Feeding habits of wild and escaped farmed Atlantic salmon, *Salmo salar* L., in the northeast Atlantic. *ICES Journal of Marine Science*, **59**, 916–33.

Järvi, T. (1990) Cumulative acute physiological stress in Atlantic salmon smolts: the effect of osmotic imbalance and the presence of predators. *Aquaculture*, **89**, 337–50.

Jonsson, N., Hansen, L.P. & Jonsson, B. (1991) Variation in age, size and repeat spawning of adult Atlantic salmon in relation to river discharge. *Journal of Animal Ecology*, **60**, 937–47.

Lacroix, G.L. & McCurdy, P. (1996) Migratory behaviour of post-smolt Atlantic salmon during initial stages of seaward migration. *Journal of Fish Biology*, **49**, 1086–101.

Larsson, P.-O. (1977) *The importance of time and place of release of salmon and sea trout on the results of stocking*. ICES CM 1977/M:42, 4 pp.

Levings, C.D., Hvidsten, N.A & Johnsen, B.O. (1994) Feeding of Atlantic salmon (*Salmo salar*) postsmolts in a fjord in central Norway. *Canadian Journal of Zoology*, **72**, 834–9.

McCormick, S., Hansen, L.P., Quinn, T.P. & Saunders, R.L. (1998) Movement, migration and smolting of Atlantic salmon. *Canadian Journal of Fisheries and Aquatic Sciences*, **55** (Suppl. 1), 77–92.

Moore, A., Potter, E.C.E., Milner, N.J. & Bamber, S. (1995) The migratory behaviour of wild Atlantic salmon (*Salmo salar*) smolts in the estuary of the River Conwy, north Wales. *Canadian Journal of Fisheries and Aquatic Sciences*, **52**, 1923–35.

Morgan, R.I.G., Greenstreet, S.P.R. & Thorpe, J.E. (1986) *First observations on distribution, food and fish predators of post-smolt Atlantic salmon*, Salmo salar, *in the outer Firth of Clyde*. ICES CM 1986/M:27, 12 pp.

Parrish, B.B. & Horsted, S.A. (eds) (1980) ICES/ICNAF Joint Investigation on North Atlantic Salmon. *Rapports et Procès-Verbaux des Réunions du Conseil International pour l'Exploration de la Mer*, **176**, 1–146.

Pethon, P. & Hansen, L.P. (1990) Migration of Atlantic salmon smolts *Salmo salar* L. released at different sites in the River Drammenselv, SE Norway. *Fauna Norvegica Series A*, **11**, 17–22.

Pierce, G.J., Thompson, P.M., Miller, A., Diack, J.S.W., Miller, D. & Boyle, P.R. (1991) Seasonal variation in the diet of the common seals (*Phoca vitulina*) in the Moray Firth area of Scotland. *Journal of Zoology, London*, **223**, 641–52.

Raitt, D.F.S. & Adams, J.A. (1965) The food and feeding of *Trisopterus esmarkii* (Nilsson) in the northern North Sea. *Marine Research*, **3**, 1–28.

Reddin, D.G. & Shearer, W.M. (1987) Sea-surface temperature and distribution of Atlantic salmon in the northwest Atlantic Ocean. *American Fisheries Society Symposium*, **1**, 262–75.

Reddin, D.G. & Short, P.B. (1991) Post-smolt Atlantic salmon (*Salmo salar*) in the Labrador Sea. *Canadian Journal of Fisheries and Aquatic Sciences*, **48**, 2–6.

Reitan, O., Hvidsten, N.A. & Hansen, L.P. (1987) Bird predation on hatchery-reared Atlantic salmon smolts, *Salmo salar* L., released in the River Eira, Norway. *Fauna Norvegica Series A*, **8**, 35–8.

Ritter, J.E. (1989) Marine migration and natural mortality of North American Atlantic salmon (*Salmo salar* L.). *Canadian Manuscript Report Fisheries and Aquatic Sciences*, **2041**, 1–136.

Saunders, R.L., Henderson, E.B., Harmon, P.R., Johnston, C.E. & Eales, J.G. (1983) Effects of low environmental pH on smolting Atlantic salmon (*Salmo salar*). *Canadian Journal of Fisheries and Aquatic Sciences*, **40**, 1203–11.

Shearer, W.M. (1992) *The Atlantic Salmon. Natural History, Exploitation and Future Management*. Fishing News Books, Oxford.

Shelton, R.G.J., Turrell, W.R., MacDonald, A., McLaren, I.S. & Nicoll, N.T. (1997) Records of post-smolt Atlantic salmon, *Salmo salar* L., in the Faroe–Shetland Channel in June 1996. *Fisheries Research*, **31**, 159–62.

Staurnes, M., Kroglund, F. & Rosseland, B.O. (1996a) Water quality requirement of Atlantic salmon (*Salmo salar*) in water undergoing acidification or liming in Norway. *Water, Air and Soil Pollution*, **85**, 347–52.

Staurnes, M., Hansen, L.P., Fugelli, K. & Haraldstad, Ø. (1996b) Short term exposure to acid water impairs osmoregulation, seawater tolerance and subsequent marine survival of smolts of Atlantic salmon. *Canadian Journal of Fisheries and Aquatic Sciences*, **53**, 1695–704.

Sturlaugsson, J. (1994) Food of ranched Atlantic salmon (*Salmo salar* L.) post-smolts in coastal waters, west Iceland. *Nordic Journal of Freshwater Research*, **69**, 43–57.

Sturlaugsson, J. (2000) The food and feeding of Atlantic salmon (*Salmo salar* L.) during feeding and spawning migrations in Icelandic coastal waters. In: *The Ocean Life of Atlantic Salmon. Environmental and Biological Factors Influencing Survival* (ed. D. Mills). Fishing News Books, Oxford, pp. 193–210.

Thorpe, J.E. (1988) Salmon migration. *Science Progress Oxford*, **72**, 345–70.

Thorpe, J.E. (1994) Salmonid fishes and the estuary environment. *Estuaries*, **17** (1A), 76–93.

Tytler, P., Thorpe, J.E. & Shearer, W.M. (1978) Ultrasonic tracking of the movements of Atlantic salmon smolts (*Salmo salar* L.) in two Scottish rivers. *Journal of Fish Biology*, **12**, 575–86.

Youngson, A.F., Hansen, L.P., Jonsson, B. & Næsje, T.B. (1989) Effects of exogenous thyroxine or prior exposure to raised water flow on the downstream movement of hatchery-reared Atlantic salmon smolts. *Journal of Fish Biology*, **34**, 791–7.

RUNNING THE GAUNTLET
INSHORE AND COASTAL HAZARDS

Chapter 5
The Significance of Marine Mammal Predation on Salmon and Sea Trout

S.J. Middlemas[1], J.D. Armstrong[1] and P.M. Thompson[2]

[1] *Fisheries Research Services, Freshwater Laboratory, Faskally, Pitlochry, Perthshire, PH16 5LB, UK*
[2] *Lighthouse Field Station, Department of Zoology, University of Aberdeen, Cromarty, IV11 8YJ, UK*

Abstract: There is evidence that stocks of Atlantic salmon (*Salmo salar*) are declining on both sides of the Atlantic. There is also concern over the health of sea trout (*Salmo trutta*) populations, particularly on the west coast of Scotland. The decline in salmon abundance appears to be related to a decrease in marine survival. This decline, together with a concurrent growth of some marine mammal populations, has led to an increased interest in the impacts of marine mammals on salmonids. We review the published information on predation by marine mammals on salmon and sea trout. There is no evidence that marine mammals are the main causal agents for the decline in salmonid abundance. However, in areas where salmonid abundance is already low, they could have substantial local effects on populations. Problems in assessing, and defining, significant predation are discussed. Although to assess the significance of marine mammal predation fully it must be put in an ecosystem context, we are some way from achieving this.

Introduction

Atlantic salmon and marine mammals may be bracketed together as charismatic megafauna, which tend to have special value in the minds of many people. Moreover, these animals support local sporting and tourist industries and each group is a focus for conservation plans in many parts of their ranges. The possibility that marine mammals may seriously adversely affect populations of salmon and sea trout is therefore particularly concerning and potentially difficult to manage.

There is an evident need for good information on the importance of predation by marine mammals on Atlantic salmon and sea trout in the sense of how much damage occurs and what the consequences are for the prey populations and the fisheries. This issue has been notoriously difficult to address, in large part for two main reasons. First, the population biomass of salmonids is limited by the extent of their freshwater rearing habitat with the consequence that, overall, they are rare animals at sea, although they may occur at high densities locally. It has often been calculated that even if the population of grey or harp seals consumed the entire population of Atlantic salmon, they would be rare items in the diet. Extrapolating from measurements of rare diet items to estimate the consumption of a prey species by a population is inevitably associated with very high levels of uncertainty. Second, the technical problems involved in studying the diets of marine mammals are formidable. This issue is a recurring theme throughout this chapter, which concludes on the optimistic note that new methods might soon be available to help improve estimates of the impacts of some species of marine mammal.

The chapter is arranged as a series of topics. The first section describes, in broad terms, some of the types of interactions between fisheries and marine mammals. A brief outline is then provided of the aspects of population dynamics that are particularly important to consider when evaluating impacts of predators on prey. A section then summarizes the biology of the prey and is followed by an account of the available data on consumption of Atlantic salmon and sea trout by marine mammals. Finally, the issue of assessing the significance of impacts is considered and the chapter is concluded with a brief discussion.

Interactions between marine mammals and fisheries

Most species of marine mammal interact with fisheries (Northridge & Hofman, 1999). Such interactions are often classified as operational or biological. Operational interactions occur when the marine mammal becomes directly involved with fishing operations, for example, the by-catch of dolphins in the east tropical Pacific tuna fishery (Beverton, 1985; Gosliner, 1999). Biological interactions tend to be more complex, and their effects on prey densities are less obvious, than those resulting from operational interactions (Northridge & Hofman, 1999). One type of biological interaction occurs when there is competition between marine mammals and man over commercially exploited fish stocks, such as Atlantic salmon (Beverton, 1985).

Biological interactions may have beneficial or detrimental effects on the marine mammal and the fishery (Northridge & Hofman, 1999). Considering effects of fisheries on mammals, it has been suggested that industrial fisheries in the North Sea have reduced sand eel stocks producing negative impacts on predatory fish, sea birds and seals (Naylor et al., 2000). However, it is also possible that declines in the stocks of predatory fish, possibly due to fishing, have allowed sand eel, and hence seal and sea bird, populations to increase (Northridge & Hofman, 1999; Furness, 2002). Predators have the potential to deplete fisheries seriously as indicated by the estimate that Californian sea lions (*Zalophus californianus*) were consuming more than half of the threatened steelhead trout (*Oncorhynchus mykiss*) returning to the Lake Washington watershed (Fraker, unpublished observations). Fishermen often view marine mammals as competitors, and historically their hostility towards seals in the UK has been due in part to the perceived effect they may have on the Atlantic salmon population (Harwood & Greenwood, 1985; Woodward, 2002).

Some principles of predator–prey interactions

Understanding the interactions of predators and prey is key to the study of ecology (Begon et al., 1996). Predators and prey affect each other in a variety of different ways and over wide ranges of spatial and temporal scales. On an evolutionary timescale, predation can produce both morphological and behavioural adaptations in predator and prey species (Krebs & Davies, 1993; Skelton, 1993). Prey tend to evolve ever-increasing abilities to avoid predators, while predators tend to evolve increased efficiency at capturing prey. The impacts of particular predators on their prey may depend to an extent on where they are in this evolutionary arms race and how important the prey are in the predators' overall diet.

Predation is one of the key mechanisms underpinning population and community dynamics (Matson & Hunter, 1992). Consequences of predation can be pronounced and some predators can produce 'top down' control of some ecosystems by regulation of their prey populations (Pace et al., 1999). In most cases, the role of marine mammals in shaping communities is unclear (Bowen, 1997). However, it has been suggested that the presence of harbour seals (*Phoca vitulina*) in Lower Seal Lake, Quebec, has drastically modified the fish community, particularly the population of lake trout (*Salvelinus namaycush*) (Power & Gregoire, 1978), although this work did not account for other potential causal agents.

Salmon and sea trout are subject to predation throughout their life cycles, from a wide variety of different predators (e.g. Mather, 1998). However, the impact of predation probably varies with life stage. Self-thinning is thought to occur in the early stages of freshwater development in some populations of salmonid, implying that an increase in the mean weight of individuals (i.e. growth) can occur only through a reduction in density by emigration or death (Grant & Kramer, 1990; Elliott, 1993; Grant, 1993; Armstrong, 1997). This natural removal of individuals may provide a surplus that can be harvested by predators without them having an effect on the eventual population size (compensatory mortality). It is even possible that predation could have a positive effect by thinning populations of stunted prey to speed the growth of the survivors to a size at which they can reproduce and contribute to harvest. There is little evidence for density-dependent mortality (and hence compensation) beyond the early parr stage, although it may occur in some situations (Armstrong, 1997). It has been considered unlikely that *Salmo* spp. would be affected by density-dependent mortality in the marine environment where they occur at comparatively low densities (Crozier & Kennedy, 1993; Hansen, 1993). Predation on salmonids from the smolt stage onwards may therefore result in a proportional reduction in the number of adults returning to spawn.

The effect of predation on prey populations depends in part on the relationship between the densities of predators and their prey. It has long been recognized that there are two major components to the population dynamics of predators and prey. The functional response is the relationship between the number of prey eaten per predator and prey density (Solomon, 1949). The numerical response is the relationship between the predator density and prey density (Solomon, 1949). A numerical response can occur through changes in predator density due to mortality and reproduction (reproductive response) and immigration and emigration of predators (aggregative response) (Readshaw, 1973). It is important to consider these responses when assessing impacts of predators. For example, regarding the functional response, a prey species may be ignored when it is at low density, but at higher densities a predator may switch to feed almost exclusively on it. At the higher prey densities, the predators may aggregate in the area where the prey species occurs, in a numerical response.

Salmon and sea trout

The genus *Salmo* contains two species, Atlantic salmon (*Salmo salar*) and brown trout (*Salmo trutta*), both of which are commercially important (Crisp, 2000). Atlantic salmon occur on both sides of the Atlantic Ocean and follow an anadromous life history during which many of them leave fresh water to feed in the marine environment before returning to spawn (Crisp, 2000). However, in many populations, a large proportion of the males spawn before migrating to sea, whereas substantial female maturation in fresh water is restricted to a few populations in North America. Populations of brown trout, which is originally a European species, exhibit both resident (entirely freshwater) and anadromous forms of both sexes. The anadromous form of *Salmo trutta* is commonly known as sea trout. Due to human introductions, trout are now distributed around the world (Elliott, 1994).

Although the life histories of both salmon and trout are remarkably plastic, they share a common underlying pattern of development (Crisp, 2000) (Table 5.1). The timing of the various life stages can produce differences in life strategies, denoted by nomenclature. For example, a salmon may spend only one winter at sea and return to the river as a grilse (one-sea-winter fish) or it may stay in the sea for more than one winter (multi-sea-winter fish) and is then known as a salmon (Mills, 1989). Similarly,

Table 5.1 Stages in the life cycles of Atlantic salmon and trout. Italics are used to indicate life stages that occur in sea trout and salmon but not brown trout. [Adapted from Mills (1989) and Elliott (1994).]

Stage	Definition
Alevin	Hatched fish still dependent on the yolk sac for nutrition.
Fry	Short transitional stage where the fish emerge from the redd and start to feed and disperse.
Parr	Stage between full absorption of the yolk sac and smoltification or maturity.
Smolt	Stage when seaward migration occurs.
Post-smolt	Stage from departure from the river (usually in spring) to the end of the first winter in the sea.
Adult	Mature fish which return to the place of their birth to spawn.
Kelt	Adult fish which has lost condition due to spawning.

finnock or whitling are brown trout that migrate to sea and return to fresh water within the same year, whereas sea trout are those that return in different years (Elliott, 1994). These details of the life history are important, because the impact of marine predation on the population depends on how the sea-going fraction of the population interacts with those that remain in fresh water. Furthermore, the ratio of grilse to multi-sea-winter fish may affect the impact of predation depending on where it occurs. If losses are mainly in coastal areas, then salmon experience two periods of high vulnerability, irrespective of time spent at sea, as they leave and return to rivers. However, if predation occurs in the open sea, then multi-sea-winter salmon are most vulnerable because of the overall time they spend in the marine environment. The timing of return to fresh water may also be important because the burst swimming speed of fish depends on the water temperature (which varies seasonally), while that of the warm-blooded seals is more constant. Hence, early-running spring fish may be particularly susceptible to predation in the coastal region.

After entering the marine environment, salmon undertake long migrations to reach their feeding grounds (e.g. Mills, 1989; Reddin & Short, 1991; Hansen, 1993; Shelton et al., 1997; Reddin & Friedland, 1999). For example, salmon in the Faeroes area have been shown to originate from Norway, Sweden, Iceland, Canada, Russia, Scotland, Ireland and Spain (Hansen & Jacobsen, 2000; Jacobsen et al., 2001). Lengths of migration among sea trout are highly variable; some fish travel as much as 600 km, whereas 'slob trout' never leave the estuary (Berg & Berg, 1987; Crisp, 2000). It appears, however, that most sea trout stay in coastal areas, relatively close (within 80 km) to their natal river (Pratten & Shearer, 1983; Berg & Berg, 1987). An estuary may contain a mixture of sea trout from neighbouring rivers (Pratten & Shearer, 1983).

There is evidence that catches of Atlantic salmon, and hence stocks, are in decline on both sides of the Atlantic (Hawkins, 2000) and sea trout are declining on the west coast of Scotland (MacLean & Walker, 2002). The decline in abundance of salmon appears to be caused by a decrease in marine survival. The cause of this decline has been the subject of conjecture and may relate to variations in ocean climate, perhaps through growth-mediated predation (Friedland et al., 2000; Hawkins, 2000).

Marine mammal predators of salmonids

The species for which we have found documentary evidence of predation on Atlantic salmon at the post-smolt and marine adult stages are listed in Table 5.2. In the following section we summarize the published information on five of the known marine mammal predators of salmonids in the Atlantic Ocean. These are selected merely on the basis that they have been the subjects of most study.

Grey seal (Halichoerus grypus)

Grey seals are found throughout the north Atlantic and form three distinct populations: north-west Atlantic, north-east Atlantic and Baltic (Macdonald, 2001). Male

Table 5.2 Mammalian predators that have been documented to have consumed salmonid post-smolts and adults in the Atlantic. The references are given as examples and not as an exhaustive list.

Common name	Latin name	References
Otter	*Lutra lutra*	Carss et al. (1990)
Grey seal	*Halichoerus grypus*	Rae (1960); Pierce et al. (1991)
Harbour seal	*Phoca vitulina*	Rae (1960); Pierce et al. (1991)
Harp seal	*Phoca groenlandica*	Beck et al. (1993)
Harbour porpoise	*Phocoena phocoena*	Fontaine et al. (1994)
Bottlenose dolphin	*Tursiops truncatus*	Santos et al. (2001)
Beluga whale	*Delphinapterus leucas*	Anon. (1999)

grey seals are larger than females; for example, NE Atlantic population females reach 180 cm in length (150 kg) and males 200 cm (200 kg) (Anderson, 1990).

Grey seals are primarily benthic foragers and take a wide variety of prey species (e.g. Thompson et al., 1991; Bowen et al., 1993; Bowen & Harrison, 1994; Hammond et al., 1994; Goulet et al., 2001). Although they are largely pelagic, salmonids have been reported to occur in the diets of grey seals on both sides of the Atlantic (e.g. Pierce et al., 1989; Hammill & Stenson, 2000). Of the 361 grey seal stomachs examined in Scotland between 1958 and 1971 that contained prey items, 97 (27%) contained salmonids (Rae, 1960, 1968, 1973). More recent work based on faecal analysis has failed to find evidence of salmonids in the diets of UK grey seals (e.g. Hammond et al., 1994; Thompson et al., 1996). It has been suggested that salmonid otoliths (the ear bones used to identify many types of prey) are not represented in seal faeces because they are fragile (Boyle et al., 1990) and because of the possibility that seals may not consume the heads of large prey (Pitcher, 1980). It is known that heads are discarded when salmon are available to seals at unusually high densities, for example, in salmon nets and at salmon farms. However, seals are capable of consuming the heads of large fish and there is little or no evidence that heads are discarded routinely in the wild. Preliminary studies have also revealed no tendency for captive seals to discard salmon heads irrespective of prey availability (Middlemas, unpublished observations). The discrepancy between the previous estimates of diet might be explained by the fact that many of the seals in the early studies originated from near salmon nets on the east coast of Scotland (Pierce et al., 1991). This is an important consideration; there seems to be no doubt that seals can use nets as traps and/or as ambush points to catch substantial numbers of salmon. The difficulty in drawing meaningful comparisons between studies is further compounded by the fact that the diet of grey seals is known to vary by area, season and year (e.g. Bowen et al., 1993; Hammond et al., 1994; Mohn & Bowen, 1996).

Some individual grey seals have been observed consuming salmon in rivers, but generally they use rivers less often than harbour seals (e.g. Anderson, 1990; Carter et al., 2001) even though they are the more numerous of the two species in the UK (Anon., 2001).

Harbour seal (Phoca vitulina)

Although known as the common seal in the UK, in the rest of the world *Phoca vitulina* is known as the harbour seal, after its habit of frequenting sheltered waters (Anderson, 1990). Of the five recognized subspecies, three occur within the Atlantic. *P. vitulina vitulina* occurs in coastal areas of the eastern north Atlantic, *P. v. concolor* is found around the coasts of the western north Atlantic and *P. v. mellonae* is resident in freshwater lakes and rivers in the Hudson Bay area of Canada (Rice, 1998). *P. v. vitulina* and *P. v. concolor* and the Pacific harbour seal, *P. v. richardii*, are known to enter rivers and lakes (e.g. Smith et al., 1996; Rice, 1998; Yurk & Trites, 2000; Carter et al., 2001). Harbour seals have been observed consuming salmonids (*Salmo* spp. and *Onchorhynchus* spp.) in fresh water (e.g. Brown & Mate, 1983; Roffe & Mate, 1984; Carter et al., 2001; Yurk & Trites, 2000).

Salmonids have rarely been reported in the diet of *P. v. concolor* (e.g. Selzer et al., 1986; Bowen & Harrison, 1996). In contrast, they have been found in numerous studies of *P. v. vitulina* diet. Of the 117 harbour seal stomachs collected in Scotland between 1956 and 1971 that contained prey items, 21 (18%) contained salmonids (Rae, 1960, 1968, 1973). Subsequent studies have used analyses of faecal samples and have tended to find less evidence of salmonids (Pierce et al., 1990, 1991). For example, annually salmonids occurred in 0–2.86% of the faecal samples collected during the summer in the Inverness Firth (Tollit, 1996). Based on a comparison of the digestive tracts of seals shot near salmon nets and elsewhere, Pierce et al. (1991) concluded that between-study differences in diet may reflect variation in the abundance of different prey.

Detailed observations of the behaviour of harbour seals have been undertaken within the estuaries of the Rivers Don and Dee in the UK (Carter et al., 2001). Seals were observed to consume a variety of fish species, and the minimum number of adult salmonids consumed within the study area was estimated to be an order of magnitude lower than the rod catch. The presence of seals in the rivers was highly seasonal, and appeared to be related to the breeding and moulting behaviour of the seals rather than the abundance of salmonids (Carter et al., 2001). Numbers of seals associated with rivers have been shown to be correlated with numbers of Atlantic salmon smolts leaving the River Lussa, Scotland, (Greenstreet et al., 1993) and the return of adult chum salmon (*Oncorhynchus keta*) to Netarts Bay, Oregon (Brown & Mate, 1983). Caution should be applied to interpretation of such results. For example, Greenstreet et al. (1993) suggested that the increase in the abundance of seals at the mouth of the Lussa may have been related to an increase in marine productivity rather than a response to the increase in smolt numbers.

Harp seal (Phoca groenlandica)

This species is named after the harp-shaped markings on the backs of the mature animals (Macdonald, 2001). They weigh up to 130 kg, grow to a length of 170 cm and live for up to 30 years. Harp seals are found in three breeding populations: north-east

Newfoundland and the Gulf of St. Lawrence, the east coast of Greenland and around Jan Mayen Island, and in the White Sea off the coast of Russia. They are most widespread outwith the breeding season when they are dispersed throughout the north Atlantic. The total population size is estimated to be approximately 7 million (Macdonald, 2001) and there is evidence that the north-west Atlantic population is increasing (Stenson et al., 2002). Analyses of stomach contents have shown that salmon are rare or absent in the diets of harp seals (e.g. Lydersen et al., 1991; Beck et al., 1993; Lawson et al., 1995; Lindstrom et al., 1998; Wathne et al., 2000).

To date, studies have suggested that salmon is either rare (harp and grey) or absent (harbour and hooded) from the diets of seals in Atlantic Canada. Even so, the consumption by seals in 1996 was estimated to be possibly in the order of 3300 tons (Hammill & Stenson, 2000). The majority of this consumption was by harp seals.

Harbour porpoise (Phocoena phocoena)

Harbour porpoises inhabit shallow coastal waters in temperate north Atlantic, north Pacific, Baltic and Black Seas (Rice, 1998; Macdonald, 2001). They are small cetaceans that grow to 140–200 cm and weigh 40–80 kg (Macdonald, 2001).

Predation by porpoise was once thought to have a large controlling influence on the abundance of Atlantic salmon in the Baltic (Svärdson, 1955). However, analysis of their diet and habitat use led Lindroth (1962) to reject this hypothesis. Similarly, there is little evidence of salmon occurring in the diets of harbour porpoises in the Atlantic (e.g. Rae, 1965; Recchia & Read, 1989). Salmonid remains were found in only a single porpoise out of a sample of 138 animals from the Gulf of St. Lawrence (Fontaine et al., 1994). The species of salmonid could not be determined from its remains.

Bottlenose dolphin (Tursiops truncatus)

Bottlenose dolphins are found in the coastal waters of most tropical, subtropical and temperate regions (Reynolds et al., 2000; Macdonald, 2001). Individuals found in warmer tropical regions tend to be smaller than those found in cooler temperate locations and males in the UK can grow up to 410 cm in length (Thompson & Wilson, 1994).

Information on the diet of bottlenose dolphins in UK waters comes from analysis of stomachs of stranded animals. Remains of salmon were present in two of eight dolphins stranded within the Moray Firth region of Scotland, including one animal found entangled in a salmon net (Santos et al., 2001). It is not possible to extrapolate these figures meaningfully to a population or species level due to the small number of dolphins examined (Santos et al., 2001).

Bottlenose dolphins in the Moray Firth are often seen consuming large salmonids (Wilson et al., 1997; Janik, 2000) (Fig. 5.1). It has been suggested that temporal and spatial patterns of habitat use exhibited by dolphins in this area may be linked to the migration of salmonids (Wilson et al., 1997; Hastie, 2000). It has also been suggested

Fig. 5.1 Bottlenose dolphin with salmon in the inner Moray Firth, Scotland. (Photo: Paul Thompson.)

that these dolphins may use specific vocalizations to manipulate the behaviour of salmon to make them easier to catch (Janik, 2000).

Assessing the significance of predation

Modelling and experimentation can be used to estimate the impact of marine mammals on salmonid stocks. Experimentation would involve the manipulation of marine mammal numbers by removal or addition and the monitoring of subsequent changes in the population of fish. This approach has a number of drawbacks. For example, there are logistic and political difficulties associated with the manipulation of numbers of marine mammals; it is difficult or impossible to achieve the necessary replication; natural fluctuations in salmon mortality caused by other factors may mask any effect of predation; due to the long life cycle of salmonids the experiment may have to run for many years; and it is difficult to devise suitable controls. In some circumstances it may be possible to use natural changes in predator populations, such as an increase in mortality caused by disease, in place of experimental manipulations.

In the absence of direct experimental data, modelling may be considered to be a useful tool for developing further understanding of the impacts of marine mammal predators. We suggest that there are four key questions that need to be addressed in order to assess fully the significance of marine mammal predation on stocks of salmonids.

Does the predator eat salmon and sea trout?

Information on the diet of animals can be obtained through direct observations of feeding and this method has been used to study seal predation on salmonids (Roffe & Mate, 1984; Carter et al., 2001). However, this technique is dependent on the availability of sufficient numbers of suitable observation points. Information obtained is limited to the area of observation and biased to large prey items consumed at the water surface (Roffe & Mate, 1984). Indirect methods used to study diet include the use of stable isotopes, fatty acids and the analysis of prey remains found in stomachs or faecal material (e.g. Pierce et al., 1991; Hammond et al., 1994; Hooker et al., 2001).

As mentioned in the previous section, salmonids have been documented in the diets of a number of marine mammals (Table 5.2). It is possible that other species may consume salmonids. For example, although we did not find evidence of killer whales eating salmonids in the Atlantic, they are known to eat Pacific salmon (*Oncorhynchus* spp.) (e.g. Saultis et al., 2000), and would be unlikely to select against Atlantic salmon.

How much of each life stage of salmon and sea trout does the predator consume?

There are several approaches to estimating the amount of salmonids consumed by predators, one of which is the use of bioenergetic models (e.g. Mohn & Bowen, 1996; Hammill & Stenson, 2000; Boyd, 2002). It is possible to use this type of model to incorporate changes in energy requirements associated with season and the ages and sexes of the individuals within populations (Mohn & Bowen, 1996; Boyd, 2002). Using such a bioenergetics model, Hammill & Stenson (2000) estimated that during 1996, grey and harp seals consumed 3229 tonnes of Atlantic salmon. However, it is difficult to evaluate this figure without further information on the stock size and other causes of mortality.

The impact of predation will partly depend on the numbers and the life stages of salmonids consumed. Each individual smolt has a lower chance of contributing to a rod fishery or spawning stock than a returning adult does and this factor has to be taken into account when comparing predation on different life stages. For example, a predator could consume 2.5 kg of food as a 2.5-kg adult or 100 25-g smolts. If each smolt had a 10% chance of returning to the river, then the impact of predation on the smolts would be greatest, as ten of these would have returned as adults had they not been removed.

The consequences of consumption of kelts are difficult to assess. In many systems, kelts have a low chance of returning to spawn again. But does this mean that losses to predators are of little relevance, or are losses to predators the reason for the low return rate? The answer to this question is not known.

How does consumption of salmonids compare with stock sizes and other causes of mortality?

Estimates of consumption can be used to give an impression of the likely extent of interactions between predators and prey (e.g. Hammond, et al., 1994; Brown et al., 2001). For example, the consumption of salmonids by a small number of seals in the mouths of the Rivers Don and Dee in northern Scotland was estimated to be an order of magnitude less than the rod-and-line catch (Carter et al., 2001). In this instance, the impact of seal predation on the salmonid stocks is likely to be small, although disturbance caused by seals entering fishing beats may have a larger impact on the fishery. It is possible, however, that proportionately greater damage occurs during some specific months of the year, for example during spring, when salmon are most highly valued as sport fish. Carter et al., (2001) did not analyse their data by season, presumably because of inadequate sample sizes.

How would changes in the level of predation affect the prey population?

A problem with modelling impact is that usually the populations of predator and prey and the diet of the predator are sampled at one or few points in time. Both the numbers and the proportion of the prey population consumed by each predator are likely to vary depending on the densities of predator and prey. Therefore, models may have very limited value for predicting impacts of predators. As previously mentioned, predators can respond to changes in the abundance of prey species through numerical and functional responses. Mohn & Bowen (1996) compared estimates of grey seal predation on cod assuming two different types of functional response. The proportion of cod in the diet was constant in one model but increased with the availability of cod in the environment in the other model. The estimated consumption of cod was considerably larger when they were assumed to comprise a constant fraction of the seal diet. Therefore, predicting the future consumption of cod depends on the assumptions made regarding the responses of seals to changes in cod abundance. A numerical response may be important when salmon become aggregated at high densities due to a year of high abundance or local conditions, such as drought, impairing up-river migration. Any factor that tends to delay salmon in areas where they are particularly vulnerable to predation is likely to increase the prey death rate. It is probably important that salmon move quickly through coastal waters where they are aggregated and may be most vulnerable to ambush from seals and dolphins.

It is often felt that a reduction in predator numbers will result in an increase in prey abundance, and thereafter fisheries catch. Figure 5.2 illustrates this scenario as a food web (Beverton, 1985). Although seemingly counter-intuitive, it appears that under certain circumstances predation may be beneficial to a fishery. An example of this effect is the interaction between the fishery for deep-water hake (*Merluccius paradoxis*) and the Cape fur seal (*Arctocepahlus pusillus pusillus*) in the Benguela ecosystem in South Africa (Fig. 5.3). In this system, fur seals may be beneficial to the

54 Salmon at the Edge

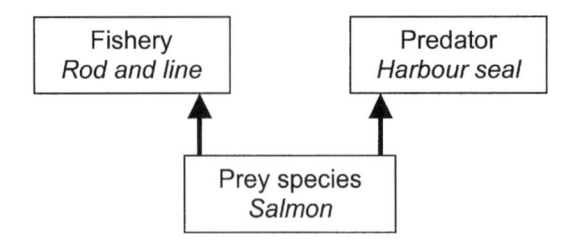

Fig. 5.2 Graphical representation of a simple food web linking a fishery and predator of a single prey species. Direction of arrow indicates the flow of energy from the prey species to the fishery and predator (Beverton, 1985).

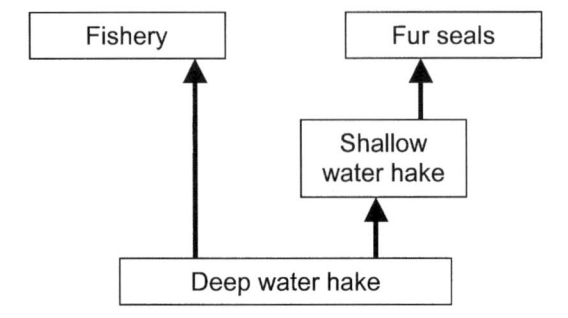

Fig. 5.3 Food web characterizing the interaction between Cape fur seals and the hake fishery in the Benguela ecosystem (Punt & Butterworth, 1995; Yodzis, 2001).

fishery because they consume the shallow-water hake (*Merluccius capensis*) which consumes juvenile deep-water hake (Punt & Butterworth, 1995).

A more complicated situation arises when the marine mammal consumes both a predator of the target species and the target species itself. An example of such a situation may occur in Rogue River, Oregon, where harbour seals are known to consume steelhead (*Oncorhynchus mykiss*) and Pacific lamprey (*Lampetra tridentatus*) (Fig. 5.4; Roffe & Mate, 1984; Beverton, 1985). Any changes in seal predation will have consequences for both the steelhead and the lamprey. If seal numbers decreased

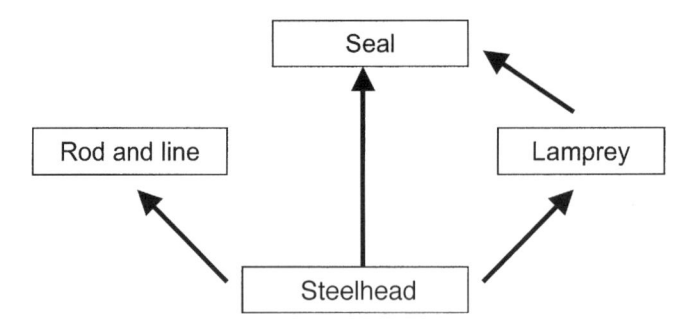

Fig. 5.4 Hypothetical food web showing the possible interactions between steelhead, lamprey, seals and a rod-and-line fishery (Beverton, 1985).

in the short term there should be a subsequent rise in the abundance of steelhead. However, in the longer term there will also be an increase in the lamprey population, which may well result in a decrease in the abundance of steelhead. Depending on the strengths of the interactions involved, the net result of predation by marine mammals may be either beneficial or detrimental to the fishery (Beverton, 1985; Yodzis, 2001).

Such theoretical models are useful in highlighting the highly variable, and occasionally counter-intuitive, effects marine mammals may have on prey populations. A complete food web covering the whole life cycle of salmon and sea trout would necessarily need to be highly complex (Yodzis, 2001). Further complications arise from the need to model interactions explicitly within a spatial and ontogenic framework (Hollowed et al., 2000). For example, because of the density-dependent factors discussed earlier, compensation for predation may vary depending on when and where it occurs.

It is not always necessary to model explicitly every component of an ecosystem to gain insights into predator–fisheries interactions (Bax, 1998; Yodzis, 1998; Livingston & Jurado-Molina, 2000; Hilton et al., 2001; Yodzis, 2001). For example, Bjørge et al. (2002) explored the interactions of harbour seals and fisheries using a combination of energetic and spatial modelling. They were able to identify areas of potential conflict and suggest both negative and positive effects of predation on fisheries.

Discussion

Predators may be considered to have an adverse impact if they cause a reduction in the value of fisheries or affect the conservation status of stocks by reducing the densities of spawning fish below those needed to saturate systems with eggs and maintain genetic diversity. At present there appear to be no cases documented in the scientific literature of marine mammals having been shown to have a large impact on stocks of Atlantic salmon or sea trout, except at farms and netting stations, although there is good evidence that they damage stocks. There have been no large-scale experiments to estimate impacts of marine mammals, and in most cases the second of the four key questions is still being addressed: how much does the predator consume?

Research into the diet and behaviour of seals has resulted in some useful insights into the application of science for managing interactions between marine mammals and salmon. There is clear evidence that some individual seals enter rivers and consume salmon. However, it also appears that damage caused by a few such salmon specialists is low compared to the rod catch (Carter et al., 2001). Simple energy balance equations readily demonstrate that individual seals specializing on adult grilse can potentially remove only small components of the stocks from east coast Scottish fisheries (Middlemas et al., unpublished). However, seals near smaller rivers and where salmonid numbers are low could have larger impacts on stocks. There may be other costs that are more difficult to quantify. In particular, little is known about the impact of seals on smolt runs, and the presence of seals in pools in rivers is widely anecdotally reported to have a negative impact on the chances of anglers catching salmon. Certainly, the presence of seals may deter anglers from a fishery.

Targeted removal, for example by scaring, is likely to have maximum benefit, per animal removed, if it is directed at individual 'salmon-specialist' seals. However, it should be considered whether the costs in time, money and public relations balance what may be rather limited benefits. The cost–benefit analysis may vary depending on whether the damage is primarily from consumption of adult salmon near the estuary or reduction in the catchability of fish further up the river. It is important that it is clearly established whether there are distinct individual seals that specialize in feeding within rivers or a larger sector of the population that uses rivers more occasionally. Furthermore, it is important to establish how rapidly removed specialist river seals (should they exist) are replaced by other individuals.

The remainder of the seal population (non-salmon-specialists), because of their large number, could have a high impact on the salmon stock if they occasionally took salmon at levels that have been too low to detect or quantify. In cases where seals haul-out near rivers, existing methodology using faecal analysis may give an indication of the likely scale of impact (Middlemas et al., unpublished). However, in other cases, new techniques are required to detect traces of salmon in the flesh of marine mammal predators. Analysis of fatty acids may offer some improvement in the scope for identifying the past diets of seals and other mammals, but it is not clear whether rare dietary items, such as salmonids, could be differentiated from common prey.

Major technical problems with faecal analyses include the uncertainty about the proportion of occasions when heads of salmon are discarded by predators and the difficulty of differentiating salmon and sea trout from their otoliths. It is possible that these problems can be circumvented if assays can be developed to identify and distinguish between the DNA of salmon and sea trout in seal faeces.

Marine mammals have the potential to cause serious damage to populations of Atlantic salmon and sea trout. There is direct evidence of local impacts, particularly from harbour seals and bottlenose dolphins, in areas where salmon aggregate near river mouths and in rivers. However, in very few cases have attempts been made, or has it been possible, to quantify the magnitude of these impacts. The estimation of broader-scale impacts has been seriously hampered by the rarity of salmon remains in the diets of marine mammals and the low numbers of samples available.

Acknowledgements

We are grateful for comments from Alison Douglas, David Dunkley and Malcolm Beveridge.

References

Anderson, S.S. (1990) *Seals*. Whittet Books, London.
Anon. (1999) *Report of the working group on marine mammal population dynamics and trophic interactions*. ICES CM 1999/G:3. Ref. ACFM, ACME E.

Anon (2001) *Scientific advice on matters related to the management of seal populations: 2001.* Report to Special Committee on Seals. SCOS 01/02, 9 pp.

Armstrong, J.D. (1997) Self-thinning in juvenile sea trout and other salmonid fishes revisited. *Journal of Animal Ecology*, **66**, 519–26.

Bax, N.J. (1998) The significance and prediction of predation in marine fisheries. *ICES Journal of Marine Science*, **55**, 997–1030.

Beck, G.G., Hammill, M.O. & Smith, T.G. (1993) Seasonal-variation in the diet of harp seals (*Phoca groenlandica*) from the Gulf of St. Lawrence and western Hudson Strait. *Canadian Journal of Fisheries and Aquatic Sciences*, **50**, 1363–71.

Begon, M., Harper, J. & Townsend, C. (1996) *Ecology*, 3rd edn. Blackwell Science, Oxford.

Berg, O.K. & Berg, M. (1987) Migrations of sea trout, *Salmo trutta* L., from the Vardnes River in northern Norway. *Journal of Fish Biology*, **31**, 113–21.

Beverton, R.J.H. (1985) Analysis of marine mammal–fisheries interactions. In: *Interactions Between Marine Mammals and Fisheries* (eds J.R. Beddington, R.J.H. Beverton & D.M. Lavigne). George Allen and Unwin, London, pp. 3–32.

Bjørge, A. Bekkby, T., Bakkestuen, V. & Framstad, E. (2002) Interactions between harbour seals, *Phoca vitulina*, and fisheries in complex coastal waters explored by combined geographic information system (GIS) and energetics modelling. *ICES Journal of Marine Science*, **59**, 29–42.

Bowen, W.D. (1997) Role of marine mammals in aquatic ecosystems. *Marine Ecology Progress Series*, **158**, 267–74.

Bowen, W.D. & Harrison, G.D. (1994) Offshore diet of grey seals *Halichoerus grypus* near Sable Island, Canada. *Marine Ecology Progress Series*, **112**, 1–11.

Bowen, W.D. & Harrison, G.D. (1996) Comparison of harbour seal diets in two inshore habitats of Atlantic Canada. *Canadian Journal of Zoology*, **74**, 125–35.

Bowen, W.D., Lawson, J.W. & Beck, B. (1993) Seasonal and geographic-variation in the species composition and size of prey consumed by gray seals (*Halichoerus grypus*) on the Scotian shelf. *Canadian Journal of Fisheries and Aquatic Sciences*, **50**, 1768–78.

Boyd, I.L. (2002) Estimating food consumption of marine predators: Antarctic fur seals and macaroni penguins. *Journal of Applied Ecology*, **39**, 103–19.

Boyle, P.R., Pierce, G.J. & Diak, J.S.W. (1990) Sources of evidence for salmon in the diet of seals. *Fisheries Research*, **10**, 137–50.

Brown, R.F. & Mate, B.R. (1983) Abundance, movements and feeding habits of harbor seals, *Phoca vitulina*, at Netarts and Tillamook bays, Oregon. *Fisheries Bulletin*, **81**, 291–301.

Brown, E.G., Pierce, G.J., Hislop, J.R.G. & Santos, M.B. (2001) Interannual variation in the summer diets of harbour seals *Phoca vitulina* at Mousa, Shetland (UK). *Journal of the Marine Biological Association UK*, **81**, 325–37.

Carss, D.N., Kruuk, H. & Conroy, J.W.H. (1990) Predation on adult Atlantic salmon, *Salmo salar* L., by otters, *Lutra lutra* (L.), within the River Dee system, Aberdeenshire, Scotland. *Journal of Fish Biology*, **37**, 935–44.

Carter, T.J., Pierce, G.J., Hislop, J.R.G., Houseman, J.A. & Boyle, P.R. (2001) Predation by seals on salmonids in two Scottish estuaries. *Fisheries Management and Ecology*, **8**, 207–25.

Crisp, D.T. (2000) *Trout and Salmon: Ecology, Conservation and Rehabilitation*. Fishing News Books, Oxford.

Crozier, W.W. & Kennedy, G.J.A. (1993) Marine survival of wild and hatchery-reared Atlantic salmon (*Salmo salar* L.) from the River Bush, Northern Island. In: *Salmon in the Sea and New Enhancement Strategies* (ed. D.H. Mills). Fishing News Books, Oxford.

Elliott, J.M. (1993) The self-thinning rule applied to juvenile sea trout, *Salmo trutta*. *Journal of Animal Ecology*, **62**, 371–9.

Elliott, J.M. (1994) *Quantitative Ecology and the Brown Trout*. Oxford University Press, Oxford.

Fontaine, P.-M., Hammill, M.O., Barrette, C. & Kingsley, M.C. (1994) Summer diet of the harbour porpoise (*Phocoena phocoena*) in the estuary and the northern Gulf of St. Lawrence. *Canadian Journal of Fisheries and Aquatic Sciences*, **51**, 172–8.

Friedland, K.D., Hansen, L.P., Dunkley, D.A. & MacLean, J.C. (2000) Linkage between ocean climate, post-smolt growth and survival of Atlantic salmon (*Salmo salar* L.) in the North Sea area. *ICES Journal of Marine Science*, **57**, 419–29.

Furness, R.W. (2002) Management implications of interactions between fisheries and sandeel-dependent seabirds and seals in the North Sea. *ICES Journal of Marine Science*, **59**, 261–9.

Gosliner, M.L. (1999) The tuna–dolphin controversy. In: *Conservation and Management of Marine Mammals* (eds J.R. Twiss & R.R. Reeves). Smithsonian Institution Press, Washington.

Goulet, A.M., Hammill, M.O. & Barrette, C. (2001) Movements and diving of grey seal females (*Halichoerus grypus*) in the Gulf of St. Lawrence, Canada. *Polar Biology*, **24**, 432–9.

Grant, J.W.A. (1993) Self thinning in stream-dwelling salmonids. In: *Production of Juvenile Atlantic salmon, Salmo salar, in Natural Waters* (eds R.J. Gibson & R.E. Cutting). Canadian Special Publication in Fisheries and Aquatic Sciences no. 118, pp. 99–102.

Grant, J.W.A. & Kramer, D.L. (1990) Territory size as a predictor of the upper limit to population density of juvenile salmonids in streams. *Canadian Journal of Fisheries and Aquatic Sciences*, **47**, 1724–37.

Greenstreet, S.P.R., Morgan, R.I.G., Barnett, S. & Redhead, P. (1993) Variation in the numbers of shags *Phalacrocorax aristotelis* and common seals *Phoca vitulina* near the mouth of an Atlantic salmon *Salmo salar* river at the time of the smolt run. *Journal of Animal Ecology*, **62**, 565–76

Hammill, M.O. & Stenson, G.B. (2000) Estimating prey consumption by harp seals (*Phoca groenlandica*), hooded seals (*Cystophora cristata*), grey seals (*Halichoerus grypus*) and harbour seal (*Phoca vitulina*) in Atlantic Canada. *Journal of Northwestern Atlantic Fisheries Science*, **26**, 1–23.

Hammond, P.S., Hall, A.J. & Rothery, P. (1994) Consumption of fish prey by grey seals. In: *Grey Seals in the North Sea and Their Interaction with Fisheries* (eds P.S. Hammond & M.A. Fedak). Final report to Ministry of Agriculture Food and Fisheries, London, pp. 35–69.

Hansen, L.P. (1993) Movements and migration of salmon at sea. In: *Salmon in the Sea and New Enhancement Strategies* (ed. D.H. Mills). Fishing News Books, Oxford.

Hansen, L.P. & Jacobsen, J.A. (2000) Distribution and migration of Atlantic salmon, *Salmo salar* L., in the sea. In: *The Ocean Life of Atlantic Salmon: Environmental and Biological Factors Influencing Survival* (ed. D.H. Mills). Fishing News Books, Oxford, pp. 75–87.

Harwood, J. & Greenwood, J.J.D. (1985) Competition between British grey seals and fisheries. In: *Interactions Between Marine Mammals and Fisheries* (eds J.R. Beddington, R.J.H. Beverton, & D.M. Lavigne). George Allen and Unwin, London, pp. 153–69.

Hastie, G.D. (2000) *Fine-scale aspects of habitat use and behaviour by bottlenose dolphins (*Tursiops truncatus*)*. PhD Thesis, University of Aberdeen, Scotland.

Hawkins, A.D. (2000) Problems facing salmon in the sea – summing up. In: *The Ocean Life of Atlantic Salmon: Environmental and Biological Factors Influencing Survival* (ed. D.H. Mills). Fishing News Books, Oxford, pp. 211–22.

Hilton, J., Welton, J.S., Clarke, R.T. & Ladle, M. (2001) An assessment of the potential for the application of two simple models to Atlantic salmon, *Salmo salar*, stock management in chalk rivers. *Fisheries Management and Ecology*, **8**, 189–295.

Hollowed, A.B., Bax, N., Beamish, R., et al. (2000) Are multispecies models an improvement on single-species models for measuring fishing impacts on marine ecosystems? *ICES Journal of Marine Science*, **57**, 707–19.

Hooker, S.K., Iverson, S.J., Ostrom, P. & Smith, S.C. (2001) Diet of northern bottlenose whales inferred from fatty-acid and stable-isotope analyses of biopsy samples. *Canadian Journal of Zoology*, **79**, 1442–54.

Jacobsen, J.A., Lund, R.A., Hansen, L.P. & O'Maoileidigh, N. (2001) Seasonal differences in the origin of Atlantic salmon (*Salmo salar* L.) in the Norwegian Sea based on estimates from age structures and tag recaptures. *Fisheries Research*, **52**, 169–77.

Janik, V.M. (2000) Food-related bray calls in wild bottlenose dolphins (*Tursiops truncatus*). *Proceedings of the Royal Society of London B*, **267**, 923–7.

Krebs, J.R. & Davies, N.B. (1993) *An Introduction to Behavioural Ecology*, 3rd edn. Blackwell Scientific Publications, Oxford.

Lawson, J.W., Stenson, G.B. & McKinnon, D.G. (1995) Diet of harp seals (*Phoca groenlandica*) in nearshore waters of the northwest Atlantic during 1990–1993. *Canadian Journal of Zoology*, **73**, 1805–18.

Lindroth, A. (1962) Baltic salmon fluctuations 2: porpoise and salmon. *Institute of Freshwater Research, Drottingholm*, **44**, 105–12.

Lindstrom, U., Harbitz, A., Haug, T. & Nilssen, K.T. (1998) Do harp seals *Phoca groenlandica* exhibit particular prey preferences? *ICES Journal of Marine Science*, **55**, 941–53.

Livingston, P.A. & Jurado-Molina, J. (2000) A multispecies virtual population analysis of the eastern Bering Sea. *ICES Journal of Marine Science*, **57**, 294–9.

Lydersen, C., Angantyr, L.A., Wiig, O. & Ortisalnd, T. (1991) Feeding-habits of northeast Atlantic harp seals (*Phoca groenlandica*) along the summer ice edge of the Barents Sea. *Canadian Journal of Fisheries and Aquatic Science*, **48**, 2180–83.

Macdonald, D. (2001) *The New Encyclopedia of Mammals*. Oxford University Press, Oxford.

MacLean, J.C. & Walker, A.F. (2002) *The status of salmon and sea trout stocks in the west of Scotland*. Fisheries Research Services Report no. 01/02.

Mather, M.E. (1998) The role of context-specific predation in understanding patterns exhibited by anadromous salmon. *Canadian Journal of Fisheries and Aquatic Sciences*, **55**, (Suppl. 1), 232–46.

Matson, P.A. & Hunter, M.D. (1992) The relative contributions of top-down and bottom-up forces in population and community ecology. *Ecology*, **73**, 723.

Mills, D. (1989) *Ecology and Management of Atlantic Salmon*. Chapman & Hall, London.

Mohn, R. & Bowen, W.D (1996) Grey seal predation on the eastern Scotian shelf: modelling the impact on Atlantic cod. *Canadian Journal of Fisheries and Aquatic Sciences*, **53**, 2722–38.

Naylor, R.L. Goldburg, R.J., Primavera, J.H., et al. (2000) Effect of aquaculture on world fish supplies. *Nature*, **405**, 1017–24.

Northridge, S.P. & Hofman, R.J. (1999) Marine mammal interactions with fisheries. In: *Conservation and Management of Marine Mammals* (eds J.R. Twiss & R.R. Reeves). Smithsonian Institution Press, Washington.

Pace, M.L., Cole, J.J., Carpenter, S.R. & Kitchell, J.F. (1999) Trophic cascades revealed in diverse ecosystems. *Trends in Ecology and Evolution*, **14**, 483–8.

Pierce, G.J., Diack, J.S.W. & Boyle, P.R. (1989) Digestive-tract contents of seals in the Moray Firth area of Scotland. *Journal of Fish Biology*, **35** (Suppl. A), 341–3

Pierce, G.J., Boyle, P.R. & Thompson, P.M. (1990) Diet selection by seals. In: *Trophic Relationships in the Marine Environment* (eds R.N. Barnes & M. Gibson). Aberdeen University Press, Aberdeen, pp. 222–38.

Pierce, G.J., Boyle, P.R. & Diack, J.S.W. (1991) Digestive tract contents of seals in Scottish waters: comparison of samples from salmon nets and elsewhere. *Journal of Zoology, London*, **255**, 670–76.

Pitcher, K.W. (1980) Stomach contents and faeces as indicators of harbor seal, *Phoca vitulina*, foods in the Gulf of Alaska. *Fisheries Bulletin*, **78**, 544–9

Power, G. & Gregoire, J. (1978) Predation by freshwater seals on the fish community of Lower Seal Lake, Quebec. *Journal of the Fisheries Research Board of Canada*, **35**, 844–50.

Pratten, D.J. & Shearer, W.M. (1983) The migrations of North Esk sea trout. *Fisheries Management*, **14**, 99–113.

Punt, A.E. & Butterworth, D.S. (1995) The effects of future consumption by the Cape fur seal on catches and catch rates of the Cape hakes. 4. Modelling the biological interaction between Cape fur seals *Arctocephalus pusillus pusillus* and the Cape hakes *Merluccius capensis* and *M. paradoxus*. *South African Journal of Marine Science*, **16**, 255–86.

Rae, B.B. (1960) Seals and Scottish fisheries. *Marine Research*, **2**, 1–39.

Rae, B.B. (1965) The food of the common porpoise (*Phocaena phocaena*). *Journal of Zoology, London*, **146**, 114–22.

Rae, B.B. (1968) The food of seals in Scottish waters. *Marine Research*, **2**, 1–23.

Rae, B.B. (1973) Further observations on the food of seals. *Journal of Zoology, London*, **169**, 287–97.

Readshaw, J.L. (1973). The numerical response of predators to prey density. *Journal of Applied Ecology*, **10**, 342–51.

Recchia, C.A. & Read, A.J. (1989) Stomach contents of harbor porpoises, *Phocoena phocoena* (L.), from the Bay of Fundy. *Canadian Journal of Zoology*, **67**, 2140–46.

Reddin, D.G. & Friedland, K.D. (1999) A history of identification to continent of origin of Atlantic salmon (*Salmo salar* L.) at west Greenland, 1969–1997. *Fisheries Research*, **43**, 221–35.

Reddin, D.G. & Short, P.B. (1991) Postsmolt Atlantic Salmon (*Salmo salar*) in the Labrador Sea. *Canadian Journal of Fisheries and Aquatic Sciences*, **48**, 2–6.

Reynolds, J.E., Wells, R.S. & Eide, S.D. (2000) *The Bottlenose Dolphin: Biology and Conservation*. University Press of Florida, Gainsville.

Rice, D.W. (1998) *Marine Mammals of the World: Systematics and Distribution*. The Society for Marine Mammalogy Special Publication Number 4. Allen Press, Lawrence, Kansas.

Roffe, T.J. & Mate, B.R. (1984) Abundances and feeding habits of pinnipeds in the Rouge River, Oregon. *Journal of Wildlife Management*, **48**, 1262–74.

Santos, M.B., Pierce, G.J., Reid, R.J., Patterson, I.A.P., Ross, H.M. & Mente, E. (2001) Stomach contents of bottlenose dolphins (*Tursiops truncatus*) in Scottish waters. *Journal of the Marine Biological Association of the United Kingdom*, **81**, 873–8.

Saultis, E., Matkin, C., Barrett-Lennard, L., Heise, K. & Ellis, G. (2000) Foraging strategies of sympatric killer whale (*Orcinus orca*) populations in Prince William Sound, Alaska. *Marine Mammal Science*, **16**, 94–109.

Selzer, L.A., Early, G., Fiorelli, P.M., Payne, P.M. & Prescott, R. (1986) Stranded animals as indicators of prey utilization by harbor seals, *Phoca vitulina concolor*, in southern New England. *Fishery Bulletin*, **84**, 217–20.

Shelton, R.G.J., Turrell, W.R., Macdonald, A., McLaren, I.S. & Nicoll, N.T. (1997) Records of post-smolt Atlantic salmon, *Salmo salar* L., in the Faroe–Shetland Channel in June 1996. *Fisheries Research*, **31**, 159–62.

Skelton, P. (ed.) (1993) *Evolution: A Biological and Palaeontological Approach*. Addison-Wesley, Wokingham.

Smith, R.J., Hobson, K.A., Koopman, H.N. & Lavigne, D.M. (1996) Distinguishing between populations of fresh- and salt-water harbour seals (*Phoca vitulina*) using stable-isotope ratios and fatty acid profiles. *Canadian Journal of Fisheries and Aquatic Sciences*, **53**, 272–9.

Solomon, M.E. (1949) The natural control of animal populations. *Journal of Animal Ecology*, **18**, 1–35.

Stenson, G.B., Hammill, M.O., Kingsley, M.C.S., Sjare, B., Warren, W.G. & Myers, R.A. (2002) Is there evidence of increased pup production in northwest Atlantic harp seals, *Pagophilus groenlandicus*? *ICES Journal of Marine Science*, **59**, 81–92.

Svärdson, G. (1955) Salmon stock fluctuations in the Baltic Sea. *Institute of Freshwater Research, Drottningholm*, **36**, 226–62.

Thompson, D., Hammond, P.S., Nicholas, K.S. & Fedak, M.A. (1991) Movements, diving and foraging behavior of gray seals (*Halichoerus grypus*). *Journal of Zoology, London*, **224**, 223–32.

Thompson, P.M. & Wilson, B. (1994) *Bottlenose Dolphins*. Colin Baxter Photography Ltd., Grantown-on-Spey.

Thompson, P.M., McConnell, B.J., Tollit, D.J., MacKay, A., Hunter, C. & Racey, P.A. (1996) Comparative distribution, movements and diet of harbour and grey seals from the Moray Firth, N.E. Scotland. *Journal of Applied Ecology*, **33**, 1572–84.

Tollit, D.J. (1996) *The diet and foraging ecology of harbour seals* (Phoca vitulina) *in the Moray Firth, Scotland*. PhD Thesis, University of Aberdeen, Scotland, 224pp.

Wathne, J.A., Haug, T. & Lydersen, C. (2000) Prey preference and niche overlap of ringed seals *Phoca hispida* and harp seals *P. groenlandica* in the Barents Sea. *Marine Ecology Progress Series*, **194**, 233–9.

Wilson, B., Thompson, P.M. & Hammond, P.S. (1997) Habitat use by bottlenose dolphins: seasonal distribution and stratified movement patterns in the Moray Firth, Scotland. *Journal of Applied Ecology*, **34**, 1365–74.

Woodward, A. (2002) Seals of disapproval. *Trout and Salmon*, **June 2002**, 50–52.

Yodzis, P. (1998) Local trophodynamics and the interaction of marine mammals and fisheries in the Benguela ecosystem. *Journal of Animal Ecology*, **67**, 635–58.

Yodzis, P. (2001) Must top predators be culled for the sake of fisheries? *Trends in Ecology and Evolution*, **16**, 78–84.

Yurk, H. & Trites, A.W. (2000) Experimental attempts to reduce predation by harbor seals on outmigrating juvenile salmonids. *Transactions of the American Fisheries Society*, **129**, 1360–66.

Chapter 6
Predation on Post-Smolt Atlantic Salmon by Gannets: Research Implications and Opportunities

W.A. Montevecchi[1] and D.K. Cairns[2]

[1] *Biopsychology Programme, Memorial University of Newfoundland, St. John's, Newfoundland, A1B 3X9, Canada*
[2] *Science Branch, Department of Fisheries and Oceans, Box 1236, Charlottetown, Prince Edward Island, C1A 7M8, Canada*

Abstract: Population declines of Atlantic salmon have been linked to juvenile mortality at sea. It has, however, proven difficult to document predation on marine-phase salmon. In the north-west Atlantic, gannets exhibited very low levels of predation on post-smolts during the late 1970s and 1980s. Following a regime shift in the pelagic food web in the north-west Atlantic during the 1990s, gannets markedly increased post-smolt consumption. This predation has the potential to negatively influence North American populations of Atlantic salmon. Migrating salmon pass through foraging ranges around gannet colonies, providing research opportunities for broad-scale quantification of avian predation on marine phase Atlantic salmon. Aspects of seabird research can also provide important information on the behaviour and ecology of post-smolts in the marine environment. Many conservation concerns are attributed to avian predators, such as cormorants, rather than to the circumstances that create these symptomatic interactions, e.g. temporarily restricted, mass releases of hatchery-reared fishes. Predatory influences that do exist likely contribute to cumulative effects with other sources of mortality, such as aquacultural practices, hydro-electric and other land-use activities that change flow regimes and river inputs, climate change and pesticide use.

Introduction

Declines of salmon populations in both the Atlantic and Pacific Oceans have been linked to the returns of 1 sea-winter fish (1 SW) and therefore to juvenile mortality at sea (Hansen & Quinn, 1998; Potter & Crozier, 2000; O'Maoiléidigh, 2002). Closures of commercial fisheries have not ameliorated the precipitous decline of Atlantic salmon (*Salmo salar* L.) populations (O'Maoiléidigh, 2002; Chase, this vol.). Hence, concern has arisen about predation on marine-phase Atlantic salmon. Most predation on Atlantic salmon has been recorded in rivers and estuaries (Shearer et al., 1987; Hvidsten & Lund, 1988; Kennedy & Greer, 1988; Cairns, 1998; Diepernik et al., 2002). It has proven difficult to document predation and natural mortality of salmon at sea (Hislop & Shelton, 1993; Friedland, 1998; Middlemas et al., this vol.).

The most significant predation on marine-phase Atlantic salmon documented to date has been attributed to northern gannets (*Morus bassanus* Viellot; Montevecchi & Cairns, 2002; Montevecchi et al., 2002). Owing to the scientific elusiveness of data on mortality at sea, we overview what is known about predation by gannets on post-smolt Atlantic salmon. Exploratory analyses are presented and the population implications associated with this predation are considered.

The range of northern gannets overlaps with that of Atlantic salmon throughout the north Atlantic (Montevecchi et al., 1988; Reddin, 1988; Holst et al., 2000; Nelson, 2002), and we propose broad-scale research opportunities to quantify avian predation over the entire range of post-smolt salmon by non-invasive sampling of gannet diets. These dietary collections can also be used to provide samples of post-smolts that can be used in size (growth), scale pattern and DNA analyses. Most importantly, light-weight data storage tags (DSTs) that record temperature, depth, salinity and geo-positioning data can be attached to free-ranging gannets in order to maximize information on the distribution, movements, species associations, behavioural ecology and marine habitat use of post-smolt Atlantic salmon as well as their avian predators (e.g. Garthe et al., 2000). This information has been extraordinarily difficult to obtain by conventional research means (Mills, 2000).

Scope of the problem

Most conservation concerns for salmon that are attributed to avian predators are directly associated with hatchery-reared and mass-released young fishes (Wood, 1985; Reitan et al., 1987; Roby et al., 2000; Montevecchi et al., 2002) which are more vulnerable to predators than wild fishes (Crozier & Kennedy, 1993; Mills, 2000; Diepernik et al., 2001). These 'enhancement' activities often produce circumstances, e.g. tight temporal concentrations of small fish abundance (i.e., 'bird feeder' situations) that attract avian and other predators to release rivers and estuaries (e.g., Woods, 1985). Such scenarios can create the problems that they seek to rectify (Levin et al., 2001). These approaches are then often followed with inappropriate actions directed at symptoms, e.g., culling predators, rather than at ecosystem aspects of the problem (Yodzis, 2001).

Strong pressures from angling interest groups and lobbies often focus exclusively on catches and direct inordinant attention at avian and other predators. At the same time, some often fail to grasp compelling environmental problems including destruction of wetlands, damming, deforestation and other land-use changes within drainage basins that influence runoffs, sedimentation and pollutant inputs (Sherwood et al., 1990; Bisbal & Connaha, 1998).

Much consideration has been given to a smolt's initial transition from fresh to salt water that is mediated in estuaries. Post-smolts are highly vulnerable to predation in these circumstances (Larsson, 1985; Järvi, 1990; Potter & Crozier, 2000; Diepernik et al., 2001, 2002) but also during the marine phase of their life cycle (Holst et al., 2000; MacLean et al., 2000). Radio-tagged Atlantic salmon smolts and other salmonids have been found to be subject to intense predation within hours of entering estuaries (Diepernik et al., 2001, 2002). Smolts that move into low-salinity surface waters likely increase their vulnerability to avian predators. Small fishes are also more vulnerable than large ones, reflecting the anti-predator aspects of increasing size (Hvidsten & Lund, 1988; Crozier & Kennedy, 1993; Potter & Crozier, 2000) but also raising the possibility that effects associated with radio tags could also have elevated predation rates in tracking studies (e.g., Diepernik et al., 2002).

Predation at sea

Knowledge of predation of marine Atlantic salmon is for the most part based on anecdotal information (Larsson, 1985). This is likely because salmon are rare dietary items of marine fishes, mammals and birds. It appears that the abundance and density of post-smolts at sea is so low that only generalist and opportunistic predators would take them as opportunities arise. It is possible that the cumulative effects of predation on post-smolts by different taxa and in combination with anthropogenic effects impacting salmon could be detrimental to salmon stocks without any one of the component influences appearing to be significant.

In this chapter, we present information on a long-term dietary study (1977–2001) of northern gannets at a large colony on Funk Island off the northeast coast of Newfoundland in the north-west Atlantic. This research has indicated that the birds have the potential to impose negative population effects on North American post-smolt Atlantic salmon (Montevecchi & Cairns, 2002; Montevecchi et al., 2002). We examine the implications of these findings and recommend research strategies to be carried out at gannet colonies throughout the north Atlantic to enhance understanding of avian predation on post-smolt Atlantic salmon. We further recommend research strategies focused on applying DSTs to free-ranging gannets to improve understanding of post-smolt behavioural ecology and habitat use at sea.

Predation on marine-phase Atlantic salmon by gannets

Gannets are the largest marine birds that breed in the north Atlantic and are highly opportunistic predators that prey on surface-schooling pelagic fishes and squids. They forage in inshore and offshore waters with ranges of ~180 km (Kirkham et al., 1985; Garthe, Benvenuti & Montevecchi, unpublished data) or longer (Hamer et al., 2000) around colonies. Their north Atlantic ranges overlap with those of post-smolt Atlantic salmon. Prey of gannets include mackerel (*Scombrus scomber*), herring (*Clupea harengus*), Atlantic saury (*Scomberesox saurus*), short- and long-finned squid (*Illex illecebrosus* and *I. loligo*), capelin (*Mallotus villosus*), sand lance (*Ammodytes* spp.), saithe (*Pollachius virens*; Montevecchi & Barrett, 1987; Montevecchi & Myers, 1996, 1997; Nelson, 2002) as well as discards that they scavenge near fishing vessels (Garthe et al., 1996).

Post-smolt Atlantic salmon have been a minor component of the diets of gannets during the chick-rearing period in August on Funk Island. Through the late 1970s and 1980s, salmon comprised on average < 1% (range = 0–2 %) of the mass of the gannets' diet (Montevecchi et al., 2002). During the 1990s, however, consumption levels of salmon increased by an order of magnitude to more than 2.5% (range = 0–6%) of dietary mass.

These consumption levels that appear to be of little consequence for the opportunistic-feeding gannets could hold population implications for Atlantic salmon in the north-west Atlantic (Montevecchi et al., 2002). To explore potential influences of predation by northern gannets on post-smolt Atlantic salmon, we examined exploratory correlations between the gannets' relative levels of consumption and the numbers of returning small 1 SW fish in the subsequent year for as many years as possible from 1977 through 2001. Positive correlations would be consistent with the possibility that inter-annual fluctuations in predation levels by gannets are indicative of fluctuations in post-smolt abundances in the Newfoundland region as has been demonstrated for mackerel and short-finned squid (Montevecchi and Myers, 1995). Conversely, negative correlations would be consistent with the possibility that predation on post-smolts by gannets was having a negative influence on 1 SW salmon returning to rivers during the following year (Potter & Crozier, 2000). Correlations between the relative levels of predation by gannets on Funk Island with estimates of total North American post-smolt abundance were also run.

Methods

Dietary sampling from gannets

Gannets regurgitate food to their chicks (Fig. 6.1), and as a fear response and in an effort to take flight at the approach of a human, gannets at breeding colonies regurgitate food loads. We exploited this behaviour to sample the diets of gannets on Funk Island by approaching birds at roosts and nests from which we collected regurgitations. We supplemented these samples with opportunistic collections and observations at nests, including some regurgitated by chicks. We identified prey species and, when possible with fresh samples, we measured and weighed specimens.

Regurgitations were collected when gannets were provisioning large, rapidly growing chicks primarily during August 1977 to 2001, except 1981 when it was not possible to land on Funk Island: 7081 regurgitations were recorded with an annual range of from 53 to 561 samples; 609 regurgitations were sampled at other gannet colonies: Baccalieu Island (81 in 1977–79), Cape St. Mary's (245, 1987–90) and Great Bird Rock (283, 1979, 1987–88). Most regurgitations consisted of multiple prey items of a single species, and regurgitations of capelin, the smallest prey, contained the most individual prey (Montevecchi et al., 2002; see also Montevecchi & Barrett, 1987). The regurgitations contained more than 30 000 prey items. Forty-one percent of all regurgitations contained more than one prey species. Colony locations are shown in Fig. 6.2.

Fig. 6.1 Northern gannet regurgitating a prey load to its chick. (Photo: W.A. Montevecchi.)

Newfoundland rivers at which we examined small salmon returns are (1) Exploits, (2) Salmon Brook, (3) Gander, (4) Middle Brook, (5) Lower Terra Nova, (6) Campbellton, all of these northeastern Newfoundland rivers combined except the Campbellton $(1+2+3+4+5)$ and all of these rivers combined except the Campbellton and Gander $(1+2+4+5)$ that had the shortest data series in the region (1993–99 and 1989–99, respectively), (6) Biscay Bay, (7) Trepassey, (8) Rocky, (9) Northeast Placentia, (10) Little, (11) Conne, all of these southern Newfoundland rivers combined $(6+7+8+9+10+11)$, (12) Humber, (13) Lomond, (14) Torrent, (15) Western Arm Brook, these four western Newfoundland rivers combined $(12+13+14+15)$ and excluding the Humber that had the shortest data series (1978–99; $13+14+15$), and all rivers combined except the Campbellton and Gander $(1+2+4+5+6+7+8+9+10+1+13+14+15$; Fig. 6.3).

Total North American 1 SW returns and two estimates of total North American wild smolts were obtained from Tables 6.2, 6.3 and 6.4 in Amiro (1998). These estimates were extrapolated from respective returns to and production in the Miramichi River, New Brunswick, using a maximum smolt production of 3.5 smolt/100/m^2 of river.

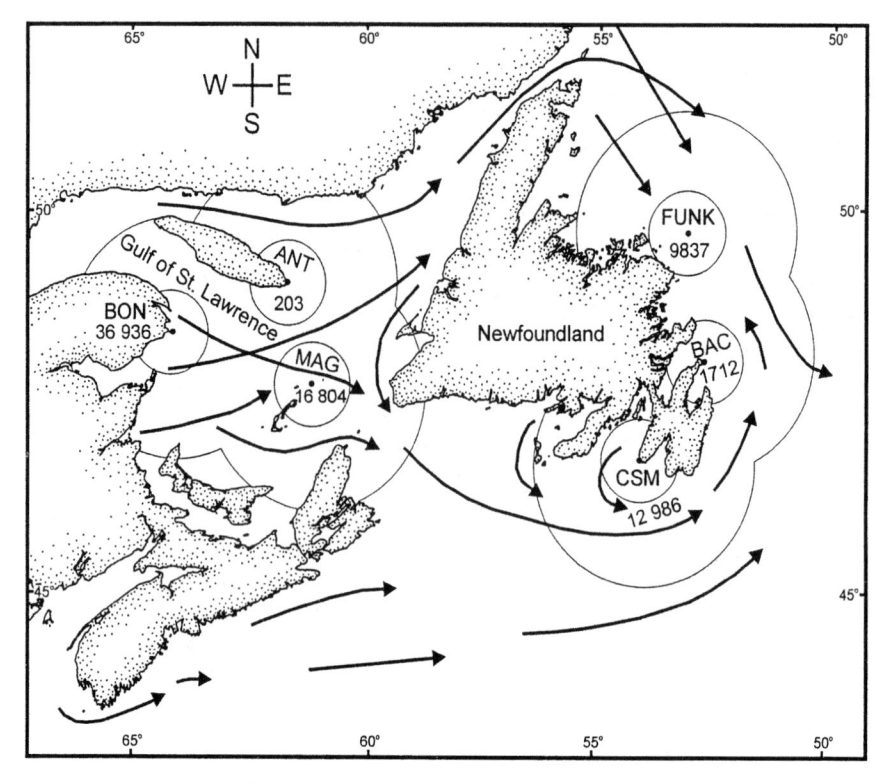

Fig. 6.2 Location of Funk Island and five other gannet colonies in North America: Baccalieu Island (BAC), Cape St. Mary's (CSM), Bird Rocks Magdalen Islands (MAG), Bonaventure Island (BON) and Anticosti Island (ANT). Numbers of breeding pairs at each colony from a 1999 census (Chardine, 2000; G. Chapdelaine, unpublished data) are given. Circles around colonies represent median (60 km) and maximum (180 km) foraging ranges of gannets. Post-smolt migration routes inferred from Montevecchi et al. (1988) and Reddin (1988) indicated by arrows. (From Montevecchi et al., 2002.)

Results

Dietary sampling from gannets

During 2001, the gannets' level of consumption of salmon increased by an order of magnitude from all previous sampling during the 1970s, 1980s and 1990s to more than 25% of the mass of the prey consumed (Fig. 6.4).

Of the correlations run between the relative levels of gannet predation on post-smolts and 1 SW returns, two of six rivers on the north-east coast of Newfoundland had significant positive correlations (Table 6.1). A significant positive correlation was also obtained for the north-east Placentia River on the south coast, and on the west coast, 1 SW returns on the Lomond and Torrent Rivers exhibited high significant positive correlations with levels of post-smolt predation. Biscay Bay and Conne Rivers on the south coast showed negative correlations (Table 6.1).

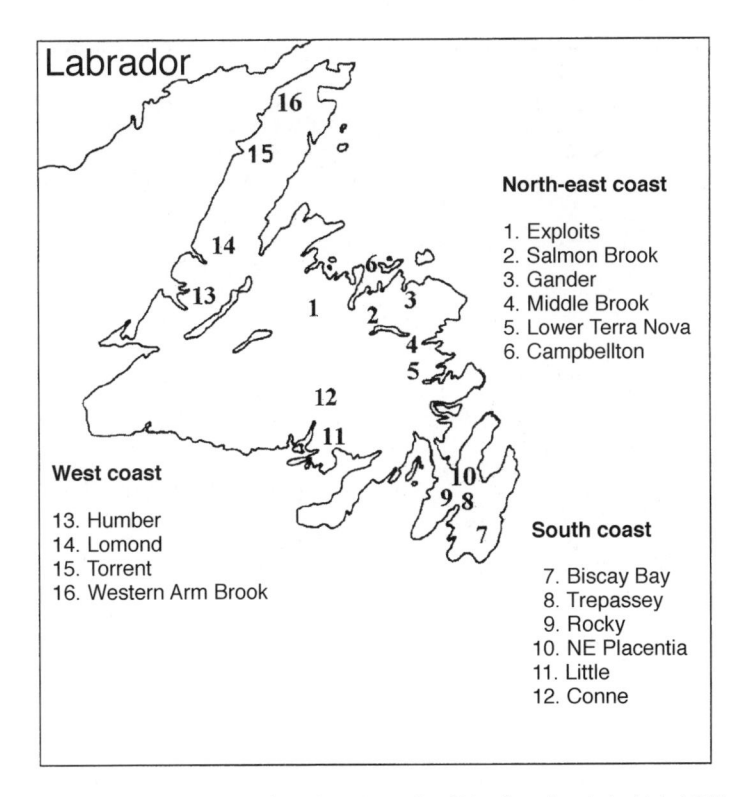

Fig. 6.3 Rivers on the north-east, south and west coasts of Newfoundland at which 1 SW returns were correlated with post-smolt predation by gannets at Funk Island during the previous summer. Names of rivers correspond to numbers in Table 6.1.

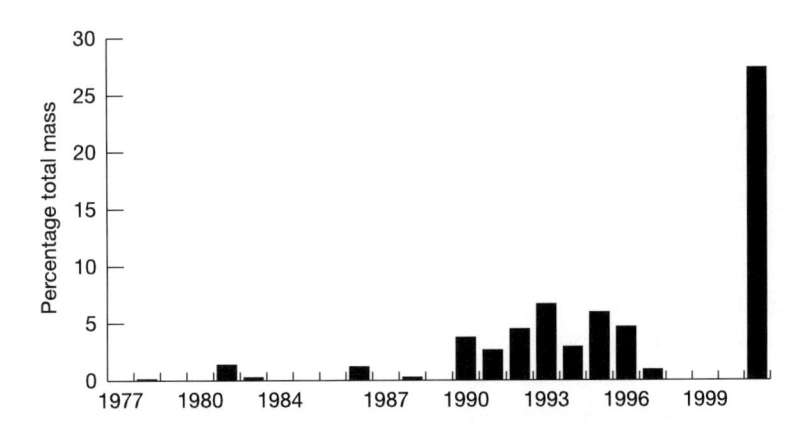

Fig. 6.4 Consumption of post-smolt Atlantic salmon by gannets on Funk Island as a percentage of the total mass of prey consumed, 1977–2001.

Table 6.1 Correlations between relative annual predation of post-smolt Atlantic salmon by gannets breeding on Funk Island and returns of small (1 sea-winter) fishes during the subsequent year in Newfoundland rivers and all monitored rivers in North America 1978–96 (Amiro, 1998) and 1978–2001 (ICES, 2002). Statistically significant relationships indicated with asterisks (* $P < 0.05$, ** $P < 0.01$, *** $P < 0.01$).

Rivers	Years	R	n
Northeast coast			
1 Exploits	1978–2001	0.384*	23
2 Salmon Brook	1978–2001	−0.072	23
3 Gander	1989–1999	0.269	10
4 Middle Brook	1978–2001	0.074	23
5 Lower Terra Nova	1978–2001	0.499***	23
6 Campbellton	1993–1999	−0.173	7
1+2+3+4+5	1989–2001	0.347	13
1+2+4+5	1978–2001	0.379*	23
South coast			
6 Biscay Bay	1983–1996	−0.501**	14
7 Trepassey	1984–2001	−0.226	18
8 Rocky	1987–2001	0.013	15
9 NE Placentia	1978–2001	0.476***	23
10 Little	1987–2001	0.180	15
11 Conne	1986–2001	−0.443**	16
6+7+8+9+10+11	1987–1996	−0.345	10
West coast			
12 Humber	1978–1999	0.266	21
13 Lomond	1978–2001	0.517****	23
14 Torrent	1978–2001	0.557****	23
15 Western Arm Brook	1978–2001	0.135	23
12+13+14+15	1978–1999	0.361*	21
13+14+15	1978–2001	0.529****	23
All rivers except Gander and Campbellton	1987–1996	0.466	10
All rivers except Gander, Campbellton and Biscay	1987–1996	0.473	10
All rivers except Gander, Campbellton, Biscay and Humber	1987–1996	0.583	10
Total North American returns	1978–1996	−0.148	18
Total North American returns	1978–2001	0.090	22

Of regional combinations of rivers, significant positive correlations between relative levels of gannet predation on post-smolts and 1 SW returns were obtained at rivers 1+2+4+5 on the northeast coast, at rivers 12+13+14+15 and at rivers 13+14+15 on the west, and the correlation for all Newfoundland rivers except the Gander and Campbellton combined was 0.466 ($P < 0.10$) with only 10 years of available concurrent data (1987–96). Correlation between relative annual levels of gannet predation on post-smolts and estimated North American 1 SW returns were non-significant ($n = 18$; 1978–96; $n = 18$; 1978–2001). Correlations between the gannet predation indices and two estimates of North American smolt numbers were highly significant at 0.587 and 0.588 ($n_s = 20$, $P_s < 0.005$; 1977–97).

Twenty-seven percent (49 of 181) regurgitations that contained post-smolts also contained other species. Most of these also contained capelin (20) followed by short-finned squid (9), herring (8), mackerel (8), Atlantic saury (6) and cod (1).

Discussion

Tag recoveries from post-smolts indicate that gannets breeding on Funk Island off the north-east coast of Newfoundland preyed on post-smolts that originated in the Penobscot River in Maine, the St. John River in New Brunswick in the inner Bay of Fundy, the Lower Clyde and LaHave Rivers in Nova Scotia (Montevecchi et al., 1988), the Miramichi River in New Brunswick in the Gulf of St. Lawrence and the Exploits River on the north-east coast of Newfoundland. They very likely prey on post-smolts from other rivers of origin also, but very few tags have been applied since the late 1980s, so there has been no means to assess this. All of the Newfoundland rivers included in the exploratory correlations between 1 SW returns and annual levels of post-smolt predation by gannets on Funk Island could be origins of post-smolts preyed on by gannets. The positive correlations associated with rivers of the north-east and west coasts of Newfoundland are consistent with the suggestion that gannet predation on post-smolts can provide indication of 1 SW returns the following year, as is the positive but non-significant correlation for all Newfoundland rivers ($P < 0.10$). The negative correlations obtained for the south coast rivers contradict this contention and are consistent with the suggestion that predation by gannets could have a negative influence on Atlantic salmon survival at sea. Overall, however, the positive correlations for individual rivers, regional river systems and for all Newfoundland rivers combined far outweighed the negative correlations from the south coast. Furthermore, the highly significant positive correlations between North American post-smolt abundance and gannet predation levels support the hypothesis that levels of gannet predation are a function of post-smolt abundance and hence could be used as a natural assay of post-smolt availability and abundance. The consumption indices of gannets on Funk Island do reflect abundance levels of mackerel and short-finned squid at multiple spatial and temporal scales throughout the Newfoundland region (Montevecchi & Myers, 1995).

Predation by gannets on Funk Island has the potential to influence population levels of Atlantic salmon (Montevecchi et al., 2002). The rarity of negative correlations between gannet predation levels and 1 SW returns in the present exercise suggests that gannets are not having significant population impacts on North American salmon. Further, the numerous positive correlations, including those with estimates of North American post-smolts, are consistent with the idea that gannets could provide a potentially informative index of cohort size the summer before 1 SW returns.

The level of gannet predation on Atlantic salmon post-smolts was an order of magnitude higher in 2001 than in any previous year in our 24-year data series. We recognize that this finding could simply be a consequence of a limited sampling regime trying to capture the dimensions of episodic predation. The result leads to two oppposite predictions about 1 SW returns in 2002. If gannets provide indication of post-smolt abundance, then returns of 1 SW to Newfoundland rivers would be expected to be very good during 2002. Conversely, if gannets on Funk Island are having a negative effect, then returns of 1 SW to Newfoundland rivers would be expected to be very poor during 2002. Avian indication of prey condition is most robust when prey are at low abundance/availability levels rather than when attempting to assess levels of moderate and high prey abundance (Montevecchi & Myers, 1995). It appears, however, on the basis of initial evidence that 1 SW returns in Newfoundland rivers were favourable during 2002, consistent with the suggestion that levels of gannet predation on post-smolts are indicative of 1 SW returns the following summer. It should also be pointed out that the gannets' level of predation on post-smolts during 2002 was similar to that of 2001, suggesting that predation is again increasing by another order of magnitude since the 1990s. This increase could be related to decreases in the availability of alternative prey, including mackerel and capelin.

Whether or not the correlations obtained are robust remains to be determined as more analyses are conducted and as longer time series of data are produced. Further, the north-west Atlantic has been subjected to major natural and human-induced biophysical changes during the 1990s (Montevecchi & Myers, 1996; Carscadden et al., 2002) that are consistent with a regime shift (Steele, 1996). So any outcomes regarding predation on marine-phase Atlantic salmon have to be considered within this context.

Post-smolt associations with other species

The percentage of post-smolts that occurred in regurgitations containing other prey species was about eight times higher than the percentage of all mixed prey regurgitations. A number of implications follow from this result. First, post-smolts could occur in association with other pelagic fishes and squid at sea. Second, post-smolt aggregations are likely diffuse compared to those of other schooling pelagic fishes and squids. Third, post-smolts might be vulnerable to surface trawl fisheries directed at these other species, as has been implicated for mackerel (Mills, 1993; Hansen et al., this vol.).

Regime shifts and Atlantic salmon in the north-west Atlantic

Declining population trajectories of salmon have been associated with and attributed to regime shifts in the north Pacific (e.g. Welch et al., 1998, 2000). Population declines of Atlantic salmon could also be related to a regime shift in the north Atlantic.

During the 1970s, 1980s and 1990s, substantive changes in oceanographic conditions (Drinkwater, 1996) and in fisheries activities occurred in the north-west Atlantic (Carscadden et al., 2002). A significant cold-water event during 1991 was followed by numerous striking and prolonged changes in biological oceanography (e.g. Regehr & Montevecchi, 1997; Frank et al., 1996; Carscadden et al., 2001). Capelin, the primary forage fish of the large vertebrate food web in the north-west Atlantic (Lavigne, 1996; Carscadden & Nakashima, 1997), exhibited southerly distributional shifts, significantly delayed inshore spawning migration, reduced spawning on beaches and decreased size of prey delivered to common murre (*Uria aalge*) chicks on Funk Island, the species' largest colony (Frank et al., 1996; Carscadden et al., 2001, 2002; Davoren & Montevecchi, 2003). Capelin are also an important food for Atlantic salmon (Hislop & Shelton, 1993; Jacobsen & Hansen, 2000).

Gannets exhibited a marked dietary shift from migratory, warm-water prey (mackerel, squid, saury) during the late 1970s and 1980s to cold-water prey (capelin, herring) through the 1990s. These dietary changes are indicative of larger-scale shifts in pelagic food webs (Montevecchi & Myers, 1996). Gannets at the Funk Island colony consumed an order of magnitude more salmon during August in the 1990s than during the 1980s and 1970s (Montevecchi et al., 2002). Owing to the relative scarcity of mackerel, squid and saury during the 1990s, and of capelin after 2000, Atlantic salmon could have been more vulnerable to opportunistic episodic predation by the gannets.

Predation on marine-phase Atlantic salmon by gannets

The migratory pathways of post-smolt Atlantic salmon pass through the foraging ranges of the six North American gannet colonies, three of which are in the Gulf of St. Lawrence and three off eastern Newfoundland (Montevecchi et al., 1988; Reddin, 1988; Fig. 6.2). North American populations of gannets have been increasing throughout the 20th century and most especially during recent decades (references in Montevecchi et al., 2002), and another colony that was extirpated during the 19th century might be re-establishing (Corrigan & Diamond, 2001).

Non-invasive, dietary sampling of gannets is an ecologically sound, cost-effective and efficient means of sampling and of obtaining information about the natural mortality, distribution and behavioural ecology of post-smolt Atlantic salmon. There has, however, been very little sampling of gannet diets at colonies other than Funk Island, and even on Funk Island the intra-annual temporal extent of sampling is limited. Systematic sampling of the diets of gannets over longer intra-annual time frames and at other colonies will greatly help quantify avian predatory mortality and will enhance efforts to assay the distributions, movements, abundance and habitat relationships of post-smolts (below). It would be most informative to expand the very limited dietary sampling programmes at gannet colonies in Norway (Montevecchi & Barrett, 1987), Iceland, the British Isles and France (Fig. 6.5) where post-smolts move through the gannets' foraging ranges (Hansen, 1993; Holst et al., 2000) and are occasionally preyed on (Wanless, 1984).

Recent findings indicate that at least some Atlantic salmon of North American origin occur in the north-east Atlantic (Tucker et al., 1999; Hansen & Jacobsen, 2000). Hence, dietary sampling from gannets at colonies in the north-east Atlantic could also be relevant for the survival of post-smolts from North American rivers and

72   *Salmon at the Edge*

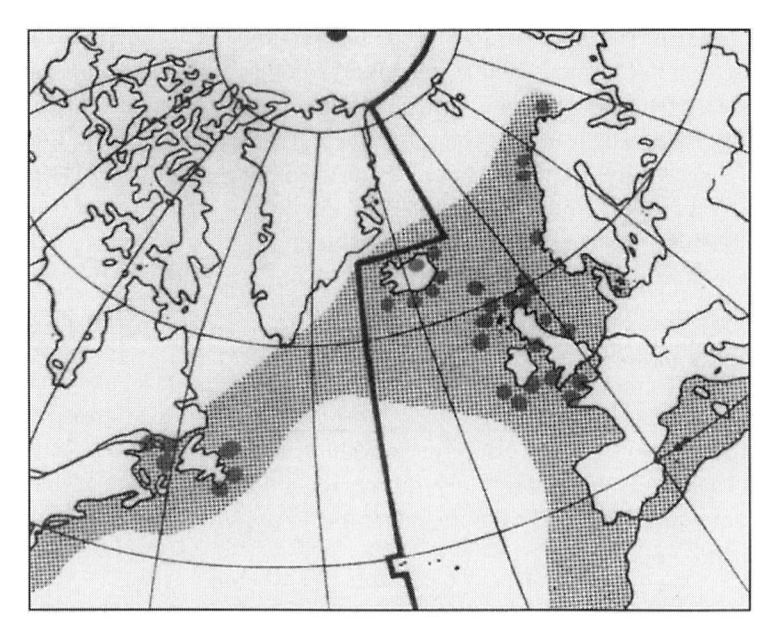

Fig. 6.5 Colonies of northern gannets in the north Atlantic. (From Nelson, 2002.)

vice versa. It would be extremely informative if post-smolts sampled from gannets could be linked to rivers of origin by some genetic or other biochemical or physiological means.

Variation in inter-annual and inter-colony predation by gannets on post-smolts could reflect changes in fish abundance at rivers of origin and subsequently within and beyond avian foraging ranges at sea. Such associations have been demonstrated for seabird consumption of other pelagic species in the north-west and north-east Atlantic (Montevecchi & Myers, 1995; Barrett, 2002).

Avian predation and cumulative effects on Atlantic salmon populations

Predation on Atlantic salmon by gannets and other predators is no doubt additive mortality that cumulates with other sources of mortality. Atlantic salmon is likely rare in the diets of all of its predators with the exception of local, periodic specialists. Salmon populations exhibit striking south to north latitudinal gradients that are inversely related to coastal human population density in the western and in the eastern Atlantic (O'Maoiléidigh, 2002), as well as in the north-east Pacific. These clines of human population density are strongly associated with general degradation of freshwater and marine ecosystems. Atlantic salmon from different natal rivers negotiate different environmental circumstances, and predatory influences on salmon populations have to be considered in the context of cumulative effects.

Predation at sea and in estuaries and rivers cumulates and interacts with other sources of mortality, such as aquacultural practices (Gross, 1998; Butler & Watt, this vol.; Goode & Whoriskey, this vol.; Holst et al., this vol.), hatchery activities (Levin et al., 2001), fishing (Chase, this vol.) including by-catch (O'Maoiléidigh, 2002; Hansen et al., this vol.), pesticide use and other sources of pollution (Fairchild et al., 1999; Mawle & Milner, this vol.), damming and agriculture and forestry that change flow regimes and river inputs (Sherwood et al., 1990; Bisbal & Connaha, 1998), and climate change (Dunbar, 1993; Welch et al., 1998, Hughes & Turrell, this vol.) and variability (Koslow et al., 2002).

Remote-sensing foraging gannets to assess the behavioural ecology and habitat of post-smolt Atlantic salmon

Research with marine birds is being used to collect data on the movements and habitat use of Atlantic salmon at sea. Data obtained from light-weight temperature storage tags attached to gannets are being used to enhance knowledge of the behavioural ecology and habitat use of post-smolt Atlantic salmon.

Previous attempts to delineate the marine thermal habitats of salmon have suggested that the fish may be more vulnerable to predation in some thermal regimes rather than others. In the Labrador Sea, salmon have been found to occur in sea surface temperatures (SST) between 3 and 13°C with peak occurrences between 7 and 9°C (Reddin & Friedland, 1993). The relationship between SST and salmon occurrences suggests that salmon might modify their movements in relation to water temperature. Reddin & Friedland (1993) presented evidence for the hypothesis that the distribution of sea ice that lowers SST is associated with southerly movements of salmon in the Labrador Sea, i.e. the salmon avoid ice and cold water. No relationship was found between SST and post-smolt occurrences, possibly due to the limited range of water temperatures over which sampling was conducted (Reddin & Friedland, 1993). Post-smolts presumably seek thermal environments that maximize opportunities for feeding and growth and minimize predation risks (Saunders, 1986; Lima, 1987). Friedland et al. (1998) defined a thermal habitat field for marine-phase salmon that was negatively related to the strength of the North Atlantic Oscillation (NAO). The area of the thermal habitat was related to 1 SW returns. Friedland and Reddin (1993) suggested that the marine habitat for Atlantic salmon during July through September in the north-west Atlantic falls within the 4–10°C range. Interestingly, of five post-smolts that were captured by Funk Island gannets equipped with temperature loggers during 2001, all were in 12–13°C water (S. Garthe and W.A. Montevecchi, unpublished data).

Gannets equipped with stomach thermal sensors, externally attached compass recorders and miniaturized temperature and pressure storage tags (e.g. Benvenuti et al., 1998; Garthe et al., 2000) are being used to obtain data on the locations of post-smolt Atlantic salmon in the Labrador Sea and the SSTs at these locations. For example, the spatial and temporal movement patterns and thermal habitats of the post-smolts can be derived from directional, water temperature and depth data obtained from gannets that catch salmon. Given the difficulty of tracking post-smolts at sea (Montevecchi et al., 1988; Ritter, 1989; Hutchinson & Mills, 2000; Moore et al., 2000) and the promising potential of these research opportunities (e.g. Wilson et al., 2002), these investigations warrant expansion.

Recommendations

(1) Expand long-term research on the dietary sampling of gannets to other colonies in North America, Iceland, Norway, the British Isles and France.
(2) Reinstate programmes that attach small plastic tags to the dorsal fins of smolts and younger salmon in rivers.
(3) Design seabird research involving data storage tags (DSTs) that record depth, external and stomach temperatures and geographic position to access information about the movements, species associations and thermal habitats of post-smolt Atlantic salmon.
(4) Establish systematic observer programmes on fisheries for surface-schooling pelagic fishes and squids to assess the by-catch of post-smolts and of other species.

Acknowledgements

We thank many people for help with the research. Iain Stenhouse, Stefan Garthe, Gail Davoren, April Hedd, Nicholas Montevecchi, Janet Russell and Laura Dominguez crewed our field expeditions since 1990. Brian Dempson provided data on salmon returns and advice. Chantelle Burke assisted with data management, analysis and figure preparations. Conrad Mullins and Peter Amiro offered constructive suggestions. Dr. John Anderson and Jeremy Read were extremely helpful and courteous with symposium arrangements. Research was supported by long-term research grants from the National Science and Engineering Research Council of Canada.

References

Amiro, P.G. (1998) Recruitment of the North American stock of Atlantic salmon (*Salmo salar*) relative to annual indices of smolt production and winter habitat in the northwest Atlantic. *Fisheries and Oceans Canada Research Document*, 98/45.

Barrett, R.T. (2002) Atlantic puffin *Fratercula arctica* and common guillemot *Uria aalge* chick diet and growth as indicators of fish stocks in the Barents Sea. *Marine Ecology Progress Series*, **230**, 275–87.

Benvenuti, S., Bonadonna, F., Dall'Antonia, L. & Gudmundsson, G.A. (1998) Foraging flights of breeding thick-billed murres in Iceland as revealed by bird-borne direction recorders. *Auk*, **115**, 57–66.

Bisbal, G.A. & Connaha, W.E. (1998) Consideration of ocean conditions in the management of salmon. *Canadian Journal of Fisheries and Aquatic Sciences*, **55**, 2178–86.

Cairns, D.K. (1998) Diet of cormorants, mergansers, and kingfishers in northeastern North America. *Canadian Technical Report of Fisheries and Aquatic Sciences*, **2225**.

Carscadden, J.E. & Nakashima, B.S. (1997) *Abundance and changes in distribution, biology, and behavior of capelin in response to cooler waters of the 1990s.* In: Proceedings Conference Forage Fishes in Marine Ecosystems, Anchorage, Alaska, 1996, AK-SG-97-01, pp. 457–68.

Carscadden, J.E., Frank, K.T. & Leggett, W.C. (2001) Ecosystem changes and the effects on capelin (*Mallotus villosus*), a major forage species. *Canadian Journal of Fisheries and Aquatic Sciences*, **58**, 73–85.

Carscadden, J.E., Montevecchi, W.A., Davoren, G.K. & Nakashima, B.S. (2002) Trophic relationships among capelin (*Mallotus villosus*) and marine birds in a changing ecosystem. *ICES Journal of Marine Science*, **59**, 1027–33.

Chardine, J.W. (2000) Census of northern gannet colonies in the Atlantic region. *Canadian Wildlife Service Atlantic Region Technical Report Series*, **361**.

Corrigan, S. & Diamond, A. (2001) Northern gannet, *Morus bassanus*, nesting on Whitehorse Island, New Brunswick. *Canadian Field-Naturalist*, **115**, 176.

Crozier, W.W. & Kennedy, G.J.A. (1993) Marine survival of wild and hatchery-reared Atlantic salmon (*Salmo salar*) from the River Bush, Northern Ireland. In: *Salmon in the Sea and New Enhancement Strategies* (ed. D. Mills). Fishing News Books, Oxford, pp. 139–62.

Davoren, G.K. & Montevecchi, W.A. (2003) Forage fish under stress: signals from seabirds. *Marine Ecology Progress Series* (in press).

Diepernik, C., Pedersen, S. & Pedersen, M.I. (2001) Estuarine predation on radiotagged sea trout (*Salmo trutta* L.). *Ecology of Freshwater Fishes*, **10**, 177–83.

Diepernik, C., Bak, B.D., Pedersen, L.-F., Pedersen, M.I. & Pedersen, S. (2002) Predation on Atlantic salmon and sea trout during their first days as post-smolts. *Journal of Fish Biology*, **61** (3), 848–52.

Drinkwater, K.F. (1996) Atmospheric and oceanic variability in the northwest Atlantic during the 1980s and early 1990s. *Journal of Northwest Atlantic Fisheries Science*, **18**, 77–97.

Dunbar, M.J. (1993) The Salmon at sea – oceanographic oscillations. In: *Salmon in the Sea and New Enhancement Strategies* (ed. D. Mills). Fishing News Books, Oxford, pp. 163–70.

Fairchild, W.L., Swanberg, E.O., Arsenault, J.T. & Brown, S.B. (1999) Does an association between pesticide use and subsequent declines in catch of Atlantic salmon (*Salmo salar*) represent a case of endocrine disruption? *Environment Health Perspective*, **107**, 349–57.

Frank, K.T., Carscadden, J.E. & Simon, J.E. (1996) Recent excursions of capelin (*Mallotus villosus*) to the Scotian Shelf and Flemish Cap during anomalous hydrographic conditions. *Canadian Journal of Fisheries and Aquatic Sciences*, **53**, 1473–86.

Friedland, K.D. (1998) Ocean climate influences on critical Atlantic salmon (*Salmo salar*) life history events. *Canadian Journal of Fisheries and Aquatic Sciences*, **55** (Suppl. 1), 119–30.

Friedland, K.D. & Reddin, D.G. (1993) Marine survival of Atlantic salmon from indices of post-smolt growth and sea temperature. In: *Salmon in the Sea and New Enhancement Strategies* (ed. D. Mills). Fishing News Books, Oxford, pp. 119–38.

Friedland, K.D., Hansen, L.P. & Dunkley, D.A. (1998) Marine temperatures experienced by survival post-smolts and the survival of Atlantic salmon (*Salmo salar* L.) in the North Sea area. *Fisheries Oceanography*, **7**, 22–34.

Garthe, S., Camphuysen, C.J. & Furness, R.W. (1996) Amounts of discards by commercial fishes and their significance as food for seabirds in the North Sea. *Marine Ecology Progress Series*, **136**, 1–11.

Garthe, S., Benvenuti, S. & Montevecchi, W.A. (2000) Pursuit-plunging by northern gannets (*Sula bassana*) feeding on capelin (*Mallotus villosus*). *Proceedings of the Royal Society of London*, **267**, 1717–22.

Gross, M.R. (1998) One species with two biologies: Atlantic salmon (*Salmo* salar) in the wild and in aquaculture. *Canadian Journal of Fisheries and Aquatic Science*, **55** (Suppl. 1), 131–44.

Hamer, K.C., Phillips, R.A., Wanless, S., Harris, M.P. & Wood, A.G. (2000) Foraging ranges, diets and feeding locations of gannets in the North Sea: evidence from satellite telemetry. *Marine Ecology Progressive Series*, **200**, 257–64.

Hansen, L.P. (1993) Movement and migration of salmon at sea. In: *Salmon in the Sea and New Enhancement Strategies* (ed. D. Mills). Fishing News Books, Oxford, pp. 26–39.

Hansen, L.P. & Jacobsen, J.A. (2000) Distribution and migration of Atlantic salmon, *Salmo salar* L., in the sea. In: *The Ocean Life of Atlantic Salmon* (ed. D. Mills). Fishing News Books, Oxford, pp. 75–87.

Hansen, L.P. & Quinn, T.P. (1998) The marine phase of the Atlantic salmon (*Salmo salar*) life cycle, with comparisons to Pacific salmon. *Canadian Journal of Fisheries and Aquatic Sciences*, **55** (Suppl. 1), 104–18.

Hislop, J.R.G. & Shelton, R.G.J. (1993) Marine predators and prey of Atlantic salmon (*Salmo salar* L.). In: *Salmon in the Sea and New Enhancement Strategies* (ed. D. Mills). Fishing News Books, Oxford, pp. 139–62.

Holst, J.C., Shelton, R., Holm, M. & Hansen, L.P. (2000) Distribution and possible migration routes of post-smolt Atlantic salmon in the north-east Atlantic. In: *The Ocean Life of Atlantic Salmon* (ed. D. Mills). Fishing News Books, Oxford, pp. 65–74.

Hutchinson, P. & Mills, D.H. (2000) Executive summary. In: *The Ocean Life of Atlantic Salmon* (ed. D. Mills). Fishing News Books, Oxford, pp. 7–18.

Hvidsten, N.A. & Lund, R.A. (1988) Predation on hatchery-reared and wild smolts of Atlantic salmon, *Salmo salar* L., in the estuary of River Orkla, Norway. *Journal of Fish Biology*, **33**, 121–6.

ICES (2002) Report of working group on North Atlantic salmon. ICES CM 2002/ACFM:14.

Jacobsen, J.A. & Hansen, L.P. (2000) Feeding habits of Atlantic salmon at different life stages at sea. In: *The Ocean Life of Atlantic Salmon* (ed. D. Mills). Fishing News Books, Oxford, pp. 170–92.

Järvi, T. (1990) Cumulative acute physiological stress in Atlantic salmon smolts: the effects of osmotic imbalance and the presence of predators. *Aquaculture*, **89**, 337–50.

Kennedy, G.J.A. & Greer, J.E. (1988) Predation by cormorants, *Phalacrocorax carbo* (L.), on the salmonid population of an Irish river. *Aquaculture and Fisheries Management*, **19**, 159–70.

Kirkham, I.R., Mclaren, P. & Montevecchi, W.A. (1985) The food habits and distribution of northern gannets off eastern Newfoundland and Labrador. *Canadian Journal of Zoology*, **63**, 181–8.

Koslow, J.A., Hobday, A.J. & Boehlert, G.W. (2002) Climate variability and marine survival of coho salmon (*Oncorhynchus kisutch*) in the Oregon production area. *Fisheries and Oceanography*, **11**, 65–77.

Larsson, P.O. (1985) Predation on migrating smolt as a regulating factor in Baltic salmon, *Salmo salar* L., populations. *Fisheries Society of Biologists*, **26**, 391–7.

Lavigne, D.M. (1996) Untangling marine food webs. In: *Studies of High-Latitude Seabirds. 4. Trophic Interactions* (ed. W.A. Montevecchi). Canadian Wildlife Service Occasional Paper, Ottawa.

Levin, P.S., Zabel, R.W. & Williams, J.G. (2001) The road to extinction is paved with good intentions: negative association of fish hatcheries with threatened salmon. *Proceedings of the Royal Society of London*, **268**, 1153–8.

Lima, S.L. (1987) Vigilance while feeding and its relation to the risk of predation. *Journal of Theoretical Biology*, **124**, 303–16.

MacLean, J.C., Smith, G.W. & Whyte, B.D.M. (2000) Description of marine growth checks observed on the scales of salmon returning to Scottish home waters in 1997. In: *The Ocean Life of Atlantic Salmon* (ed. D. Mills). Fishing News Books, Oxford, pp. 37–48.

Mills, D.H. (1993) Control of marine exploitation. In: *Salmon in the Sea and New Enhancement Strategies* (ed. D. Mills). Fishing News Books, Oxford, pp. 119–38.

Mills, D.H. (ed.) (2000) *The Ocean Life of Atlantic Salmon*. Fishing News Books, Oxford.

Montevecchi, W.A. & Barrett, R.T. (1987) Prey selection by gannets at breeding colonies in Norway. *Ornis Scandinavica*, **18**, 233–48.

Montevecchi, W.A. & Cairns, D.K. (2002) Episodic predation on post-smolt Atlantic salmon (*Salmo salar*) by northern gannets (*Morus bassanus*) in the northwest Atlantic. *North Pacific Anadromous Fish Commission Technical Report*, **4**, 48–50.

Montevecchi, W.A. & Myers, R.A. (1995) Prey harvests of seabirds reflect pelagic fish and squid abundance on multiple spatial and temporal scales. *Marine Ecology Progressive Series*, **117**, 1–9.

Montevecchi, W.A. & Myers, R.A. (1996) Dietary changes of seabirds indicate shifts in pelagic food webs. *Sarsia*, **80**, 313–22.

Montevecchi, W.A. & Myers, R.A. (1997) Centurial and decadal oceanographic influences on changes in northern gannet populations and diets in the north-west Atlantic: implications for climate change. *ICES Journal of Marine Science*, **54**, 608–14.

Montevecchi, W.A., Cairns, D.K. & Birt, V.L. (1988) Migration of post-smolt Atlantic salmon, *Salmo salar*, off northeastern Newfoundland, as inferred by tag recoveries in a seabird colony. *Canadian Journal of Fisheries and Aquatic Sciences*, **45**, 568–71.

Montevecchi, W.A., Cairns, D.K. & Myers, R.A. (2002) Predation on marine-phase Atlantic salmon (*Salmo salar*) by gannets (*Morus bassanus*) in the northwest Atlantic. *Canadian Journal of Fisheries and Aquatic Sciences*, **59**, 602–12.

Moore, A., Lacroix, G.L. & Sturlaugsson, J. (2000) Tracking Atlantic salmon post-smolts in the sea. In: *The Ocean Life of Atlantic Salmon* (ed. D. Mills). Fishing News Books, Oxford, pp. 49–64.

Nelson, B. (2002) *The Atlantic Gannet*. Fenix Books, Norfolk.

O'Maoiléidigh, N. (2002) What's happening to Atlantic salmon? ICES Newsletter, **39**, 8–11.

Potter, E.C.E. & Crozier, W.W. (2000) A perspective on the marine survival of salmon. In: *The Ocean Life of Atlantic Salmon* (ed. D. Mills). Fishing News Books, Oxford, pp. 19–36.

Reddin, D.G. (1988) Ocean life of Atlantic salmon (*Salmo salar* L.) in the northwest Atlantic. In: *Atlantic Salmon: Planning for the Future* (eds D. Mills & D. Piggins). Croom Helm, London, pp. 488–511.

Reddin, D.G. & Friedland, K.D. (1993) Marine environmental factors influencing the movement and survival of Atlantic salmon. In: *Salmon at Sea and New Enhancement Strategies* (ed. D. Mills). Fishing News Books, Oxford, pp. 79–103.

Regehr, H.M. & Montevecchi, W.A. (1997) Interactive effects of food shortage and predation on breeding failure of black-legged kittiwakes: indirect effects of fisheries activities and implications for indicator species. *Marine Ecology Progressive Series*, **155**, 249–60.

Reitan, O., Hvidsten, N.A., & Hansen, L.P. (1987) Bird predation on hatchery-reared Atlantic salmon smolts, *Salmo salar* L., released in the River Eira, Norway. *Fauna Norvegica Series A*, **8**, 5–38.

Ritter, J.A. (1989) Marine migration and natural mortality of North American Atlantic salmon (*Salmo salar* L.). *Canadian Manuscript Report of Fisheries and Aquatic Sciences*, **2041**.

Roby, D.D., Craig, D.P., Lyons, D. & Collis, K. (2000) *Caspian tern conservation and management in the Pacific northwest: who's tern is next?* Proceedings of the American Ornithololologists' Union Meeting, St. John's, Newfoundland, Canada.

Saunders, R.L. (1986) The thermal biology of Atlantic salmon: influence of temperature on salmon culture with particular reference to constraints imposed by low temperature. *Report of Institute of Freshwater Research, Drottningholm*, **63**, 68–81.

Shearer, W.M., Cook, R.M., Dunkley, D.A., MacLean, J.A. & Shelton, R.G.J. (1987) A model to assess the effect of predation by sawbill ducks on the salmon stocks of the River North Esk. *Scottish Fisheries Research Report*, **37**, 1–12.

Sherwood, C.R., Jay, D.A., Harvey, R.B. & Hamilton, P. (1990) Historical changes in the Columbia River Estuary. *Progressive Oceanographer*, **25**, 299–352.

Steele, J.H. (1996) Regime shifts in marine ecosystems. *Ecological Applications*, **8**, S33–S36.

Tucker, S., Pazzia, I., Rowan, D. & Rasmussen, J.B. (1999) Detecting pan-Atlantic migration in salmon (*Salmo salar*) using ^{137}Cs. *Canadian Journal of Fisheries and Aquatic Sciences*, **56**, 2235–9.

Wanless, S. (1984) The growth and food of young gannets on Ailsa Craig. *Seabird*, **7**, 62–70.

Welch, D.W., Ishida, Y. & Nagasawa, K. (1998) Thermal limits and ocean migrations of sockeye salmon (*Oncorhynchus nerka*): long-term consequences of global warming. *Canadian Journal of Fisheries and Aquatic Sciences*, **55**, 937–48.

Welch, D.W., Ward, B.R., Smith, B.D. & Eveson, J.P. (2000) Temporal and spatial responses of British Columbia steelhead (*Oncorhynchus mykiss*) populations to ocean climate shifts. *Fisheries and Oceanography*, **9**, 17–32.

Wilson, R.P., Grémillet, D., Syder, J., et al. (2002) Remote-sensing systems and seabirds: their use, abuse and potential for measuring marine environmental variable. *Marine Ecology Progressive Series*, **228**, 241–61.

Wood, C.C. (1985) Aggregative response of common mergansers (*Mergus merganser*): predicting flock size and abundance on Vancouver Island salmon streams. *Canadian Journal of Fisheries and Aquatic Science*, **42**, 1259–71.

Yodzis, P. (2001) Must top predators be culled for the sake of fisheries? *Trends in Evolution and Ecology*, **16**, 78–84.

Chapter 7
Progress in Ending Mixed-Stock Interceptory Fisheries: United Kingdom

A. Whitehead

North Atlantic Salmon Fund (UK), Cleish, Kinrosshire, KY13 0LN, UK

Abstract: An account is given of the origins of proposals to close all mixed-stock salmon net fisheries around the coast of the UK and of the lobbying by the North Atlantic Salmon Fund (NASF)(UK) and the government response and funding. The north-east coast drift-net fishery and its effects on wild fish stocks are described. A summary of negotiations to date is given and the cost of funding and the choices faced by NASF(UK).

Introduction

In April 1998, the UK Minister for Agriculture, Fisheries and Food and the Secretary of State for Wales formed a small independent group to review existing policies and legislation in England and Wales concerning the management and conservation of salmon, trout, eels and freshwater fish. The Review Group submitted its report in February 2000, and the Salmon and Freshwater Fisheries Review was published in March.

A very wide range of individuals and organizations contributed to the Review Group's deliberations, and over 700 responses were made to the Review. Among the evidence presented to the Review Group, and submitted directly to Government, was a paper prepared jointly by the Atlantic Salmon Trust (AST) and NASF(UK), in 1999. This paper had received a broad measure of support from organizations to whom it was sent, ahead of its submission to Government and the Review Group.

The AST/NASF(UK) paper was entitled '*Interceptory Exploitation of Salmon*'. In it, we argued that:

- Interceptory mixed stock salmon net fisheries around the coast of the UK should close down, with suitable compensation being paid to the licensees.
- While the Government should pay 50% of this compensation, the balance should be collected from the river catchments that most stand to benefit from the enhanced stocks that should eventually result, plus a public appeal.

The Review Group made almost 200 recommendations, the majority of which were accepted by Government. Among them were two that specifically reflected those made in our joint paper. They were:

(1) The phase out of mixed stock salmon net fisheries in England and Wales should be accelerated, and to achieve this compensation should be offered to netsmen to encourage them to leave these fisheries as soon as possible.
(2) The Government should provide substantial pump priming funds to launch compensation arrangements designed to accelerate the phase out of mixed stock salmon net fisheries on a voluntary basis, and should take the lead in setting up these arrangements.

Government response

The Chairmen of AST and NASF(UK) met Mr Elliot Morley MP, the UK Fisheries Minister, in autumn 2000 to press our case further. We were naturally pleased when just before Christmas 2000, he announced that up to £750 000 of public money would be set aside in two slices, in 2002/2003 and 2003/2004, subject to matching funding from the private sector, for the buy out, on a voluntary basis, of mixed stock nets, starting with the north-east drift-nets. We immediately began discussions with the netsmen's representatives, who indicated that they were ready to discuss a voluntary buy-out.

The north-east drift-nets

There are currently 70 netsmen licensed to fish for salmon and sea trout off the Northumberland and Yorkshire coast. The fishery is the subject of a Net Limitation Order, which defines the conditions that a netsman must meet in order to obtain or renew an annual licence. The effect of this is that if any fisherman leaves the fishery for whatever reason, he may not subsequently apply for a renewal of licence; nor indeed may anyone else apply in his place. Thus in theory the number of netsmen in the fishery gradually reduces. Over the 10 years from 1992 during which the Net Limitation Order has been in existence, the number of nets has reduced from 140 to 70, though in fact the overall catch has remained broadly steady or has shown a small increase. The fishery takes a reported catch of between 18 000 and 53 000 salmon per year, and between 22 000 and 54 000 sea trout. The average, over the years 1993–2000, is 33 815 salmon and 35 927 sea trout.

The fish caught by the north-east drift-nets are mainly heading for rivers in the north-east of England and in the east of Scotland, possibly as far north as the Ythan, and almost certainly the Aberdeenshire Dee. It is believed that up to 80% of the catch is from Scottish rivers. In spite of evidence that suggests that salmon stocks along the east coast of England and Scotland are either declining or improving only very slightly, and from a low starting position, the Environment Agency has some difficulty with the notion that there is a strong conservation case for closing this mixed stock net fishery immediately. It is widely accepted that mixed stock fisheries represent bad salmon management, but the fact that the north-east drift-net catch itself is holding up, or even increasing, in spite of a reduced numbers of nets, is held to be adequate justification for doing no more than allowing the Net Limitation Order to be reintroduced for a further 10 years and for the natural wastage from the fishery to be sufficient to meet Government expectations. In 10 years from now, only 17 of the existing 70 licence holders will be over 70. Many netsmen fish very successfully well beyond that age. Moreover, the small rise in rod catches in five rivers in England and two in Scotland is adduced as evidence that the net fishery is *not* having a damaging effect on stocks of wild fish in those rivers. Not surprisingly, that view is not shared by NASF(UK), nor indeed many other organizations. As everyone in this audience will know, rod catches on their own are not a reliable measurement of the health of stocks.

Other nets

Other smaller mixed stock operations around the UK have in some cases successfully been bought out. They include nets on the Rivers Taw and Torridge in the south west, and on the Rivers Usk and Wye. The situation in Ireland is confused, with the Republic's Government announcing the introduction of quotas for the very large drift net fishery, which was widely welcomed, until the size of the quotas was announced, when they were quite definitely not. There is even the suspicion in some quarters that the size of the quota varies according to the size of the potential vote for one particular political party.

Negotiations

The first meetings to discuss the buy out of the north-east coast drift-nets (NEDN) took place in London and Newcastle in the spring of 2001. Acting as facilitator was the Environment Agency, and attending were representatives from the National Federation of Fishermen's Organizations, the Department of the Environment, Food and Rural Affairs (DEFRA), together with Orri Vigfusson (the President of NASF) and myself.

Once it became clear that the majority of the 70 netsmen were willing to consider a buy-out, various methods of arriving at an agreed price were examined. The first method adopted, and favoured by DEFRA, was by means of a confidential letter from the Environment Agency to each net licensee. This simply invited them to value their own business. In other words, each was asked to state the sum of money for which he would surrender his licence. Perhaps not surprisingly, this produced a result that was so far in excess of what was affordable and realistic as to be not worth further consideration. A different method was therefore tried.

For some time, NASF(UK) had been making clear its belief that a figure of around £2 000 000 was the maximum likely to be available for the buy out. This sum included the Government pump priming money and was influenced by what we believed we could reasonably expect to raise from the private sector, indeed what was reasonable to expect the private sector to contribute. A hypothetical calculation was carried out by NASF(UK), whereby £2 000 000 was divided among 70 netsmen, first on the basis of their reported average catches over 5 years, and then on their fishing effort expressed in days per season. These two figures for each netsman were then averaged, to produce a pay out figure for use in negotiation. DEFRA then carried out a similar exercise, but using the fisherman's age and his fishing effort as parameters. We had earlier dismissed age as a reliable factor, since the figures show that while some 65 year olds catch a large number of fish, some 40 year olds catch very few. However, five payment bands were derived, and each licensee was placed in one.

A new letter was dispatched to each netsman in February 2002, inviting him on this occasion to declare whether or not he would surrender his licence on the basis of the sum declared in his payment band. The results were disappointing. Twenty-eight netsmen agreed to the hypothetical deal, while 42 either said no, or did not reply. NASF(UK) has made it clear that this does not constitute an arrangement that we could recommend to our sponsors, since apart from closing only 40% of the fishery, it offers no guarantees about the activities of the remaining 42 netsmen. Historical figures give no grounds for optimism. In the year 2000, 70 nets were catching more salmon and sea trout than were 124 nets 8 years earlier.

A meeting was held on 19 April 2002, in London, between DEFRA and Environment Agency officials and NASF(UK), to attempt to move the issue forward. The outcome was inconclusive. It appeared to us that Government officials were keen for NASF(UK) to accept the 28 out of 70 arrangement, even though it only closed 40% of the fishery and offered no reassurance as to the activities of the 42 netsmen who would continue fishing. It was made clear to us that not only was there no more Government money, but such as there was stood at risk of being withdrawn if a deal was not concluded by the end of the financial year 2002/2003. This approach is disappointing, and calls into question in our minds the extent of Government's real commitment to the early closure of the fishery. It has not been possible so far to convince officials that what was being done at present, namely an extension of the Net Limitation Order for a further 10 years, was quite inadequate in terms both of the management of stocks and of conservation of the species. The Environment Agency advice remained that there was no conservation argument for closing the fishery at a rate faster than that which would be achieved by the Net Limitation Order. This view does of course take little account of the impact of the fishery on Scottish stocks, beyond some observations about the level of rod catches on the Tweed and the Tay, since the Environment Agency's remit does not extend beyond the border.

As a result of this impasse, it was decided to make one last attempt to reach a deal, by visiting the netsmen in small groups, face-to-face. Consequently, a series of six separate meetings were held with groups of netsmen in their ports of operation, resulting in the Secretary of NASF(UK) meeting 67 out of the 70 licensees. The purpose of the meetings was simple; to ensure that the netsmen understood NASF(UK)'s point of view, and that we understood theirs. The results were revealing and interesting. It emerged that there was widespread misunderstanding about the sums of money that might be involved, equally widespread irritation over the anomalies generated by including a licensee's age in the banding calculations and a general lack of information about the progress of our negotiations.

What has also emerged is a better understanding on the part of NASF(UK) of attitudes among the netsmen, what they might or might not accept, and thus the prospect of some kind of deal. There now appears to be an encouraging degree of trust between the netsmen and NASF(UK). At the same time we were able to explain to them what would represent an acceptable result to us, namely a target of at least 50 out of 70 nets, at a price we could afford or could reasonably expect to raise. There is little doubt that most if not all of the licensees would leave the fishery if the sum offered to them was acceptable to them, and that such a sum is not an absurdly excessive one. If you were to think in terms of an average pay out of between £50 000 and £75 000, you would not be too wide of the mark.

Calculations have now been completed and have been laid before the full board of NASF(UK). Clearly it would be inappropriate to declare those now, but the board are being asked to respond to the following questions:

- Is it worth it?
- Firstly, are we getting value for money?
- And secondly, is the sum reasonable that the private sector is being asked to contribute?
- Finally, can we raise the money?

Without being aware of the sums of money that might be involved it is of course impossible for you to answer these questions now, and I am not asking you to do so. Suffice it to say that in my view the opportunity to buy out the nets, or a substantial number of them, is one that may not come our way again for some time, if at all, and we should grasp it with both hands, expensive though it may appear to be. The cost needs to be measured against the long-term benefit to the species in general, and to individual rivers in particular. It also needs to be considered in the political context; we surely do not wish to be isolated from initiatives in other Atlantic countries to bring to an end mixed stock fisheries.

The netsmen recognize our objectives and now understand our methodology, which is to arrive at a price that they can accept and we can afford. I have been continually impressed at the agreeable and co-operative attitudes I have experienced in talks with them.

But I cannot conceal my disappointment and dismay at the response of Government, both in their reluctance to acknowledge the importance of the problem, and in their refusal to provide meaningful sums of money to solve it. It is hard to ignore the simple conclusion that a Government that was fully committed to the removal of a mixed stock fishery for wild salmon and sea trout would not in the first place have handed the problem to the voluntary sector to solve, together with an extremely modest financial contribution towards the solution.

I very much hope that it is not too late for good sense to prevail and for real progress to be made. The international and national political and environmental benefits will be enormous.

Post-symposium update – May 2003

As a result of the meetings with netsmen and of a re-examination of the factors involved in assessing the level of compensation payments for the surrender of licences, a revised offer was made by NASF(UK) to individual licensees in January 2003. This offer was based on a re-estimation of the value and allocation of the payment bands on a basis that was considered to be more equitable and realistic.

By the end of February, 52 of the remaining 69 netsmen had indicated their willingness to accept the revised offer. This represented 75% of the licensees who have caught salmon and sea trout between Whitby and Holy Island, and, taking into account previous performances, was estimated to save 80% of the catch – the average of total annual catches reported between 1993 and 2001 was 33 655 salmon and grilse and 33 907 sea trout.

This catch reduction represents a level that was assessed as effective. An agreement on these terms was calculated to cost a total of £3.34 million. In the light of an increase to £1.25 million in the contribution offered by the Fisheries Minister on behalf of the English Department for the Environment, Food and Rural Affairs, and of the funding of over £1 million already given or pledged by District Salmon Fishery Boards, river owners and private donors, NASF(UK) has decided to commit formally to the buy-out process. This requires signing an agreement with each licensee who has accepted the revised terms and, if possible, with some of those who previously refused but have since shown interest. These agreements were being prepared at the time this update was written. The compensation payments will be paid in two equal tranches, separated by twelve months. The first payment is to be made by the end of May 2003, so that all those who have accepted will not fish in the 2003 netting season (which opens on 1 June) and thereafter.

A public appeal to raise the balance of £1 million has been launched and the initial response has been most encouraging. In the event that insufficient funds have been raised when the second payment is due, NASF(UK) has obtained confirmation of the availability of a bank loan to make up the shortfall. Although the agreement will leave a part of the fishery still in operation, NASF(UK) believes that its implementation will have achieved a major step forward in salmon conservation, since the north-east drift-net fishery is the last significant mixed stock netting operation in English or Welsh waters (drift-netting was banned in Scotland in 1962). Moreover, this buy out will act as an example to others. Finally, when the deal is completed and paid for, it will leave NASF(UK) free to continue to support the wider international objectives of the North Atlantic Salmon Fund as it has done for the last ten years, including efforts to consolidate the agreement with Greenland netsmen, to secure a lasting and sustainable settlement with the Faroese, and to support a buy out of the very large Irish drift-net fishery.

Chapter 8

Closing the North American Mixed-Stock Commercial Fishery for Wild Atlantic Salmon

S. Chase

Atlantic Salmon Federation, PO Box 5200, St. Andrews, New Brunswick, E5B 3S8, Canada

Abstract: In 1998, after several years of effort on the part of the Government of Canada, the provinces and the Atlantic Salmon Federation (ASF), the commercial fishery for wild Atlantic salmon effectively ended in Canada. Steps to end the fishery began in 1966 and proceeded through closures, retirement of licenses, and compensation programmes involving hundreds of commercial salmon fishermen in Quebec and the Atlantic provinces. This occurred at immense political and financial cost [$72 million (Canadian)]. In New England, the last elements of the commercial salmon fishery were terminated in 1948 following a catch of 47 salmon.

In 1993–94 ASF and the North Atlantic Salmon Fund (NASF) negotiated a two-year compensated conservation agreement with commercial salmon fishermen in Greenland. Following this, the salmon fishery re-emerged, although in a reduced state due to limitations resulting from the North Atlantic Salmon Conservation Organization (NASCO) treaty. A subsistence fishery was permitted to allow for local needs.

More recently, however, the commercial salmon fishery in Greenland has resumed harvest of salmon from North American and European waters. In Labrador, a marine harvest of salmon has also re-emerged through the resident food fishery that intercepts salmon bound for rivers in the Atlantic provinces, Quebec and the USA. Another interceptory fishery in St. Pierre presents risks for salmon in Newfoundland rivers. All of these fisheries threaten the success of restoration of salmon populations in North America to their former abundance.

This paper reviews the several measures that were adopted in North American waters to terminate the commercial salmon fishery and the results of these measures. It outlines the many challenges faced by government regulators and non-government organizations in achieving termination of the commercial salmon fishery.

Introduction

The Atlantic salmon represents many things to many people. It is certain, over and above its food and commercial value, that the wild salmon enhances our quality of life in many ways. It has been described as 'the supreme symbol of a healthy ecosystem' (Lawrence Felt, correspondence to ASF, 1994). Importantly, in Canada, it represents the underpinning of an important recreational fishery that continues to provide environmentally sustainable jobs and income for thousands of individuals, businesses and rural communities. There is no question that the wild Atlantic salmon has helped define the culture and identity of the people that share its river valleys.

For hundreds of years the wild Atlantic salmon sustained a significant commercial fishery that was second only in value to the cod. In the nineteenth and twentieth centuries as salmon populations began to decline precipitously, however, a *perception* emerged within some circles of the salmon conservation movement that the commercial salmon fishery was the most significant factor affecting the survival of the species. This emerged even though it was well known that dams, pollution, habitat alteration or destruction and poaching stood alongside the commercial fishery as causes of the decline. This perception, and the efforts to deal with the commercial fishery issue, came to take precedence over much needed work on the several other adverse impacts also threatening the future viability of the resource, in both the marine and freshwater environments.

From early times in colonial North America the various Atlantic salmon populations experienced losses from curtailed access to spawning and nursery habitat due to dams and impoundments or habitat damage from industrial practices. Many eastern North American rivers, especially in the New England states, were dammed at various mill sites, or impoundments were created for agricultural purposes. Forestry operations and the clearing of land resulted in sedimentation, elevated water temperature and other damage to habitat. In the twentieth century, pollution, expanded road building, urbanization and other poor land-use practices created additional profound impacts on juvenile production, which is critical to the viability of the resource.

In the marine environment some causes of the decline were, and remain, less certain, although they definitely result from a combination of natural and anthropogenic factors. Some of these factors include: predation by marine mammals, birds and fish; capture as by-catch in fisheries targeted at other species; lack of food, or loss of forage species to harvest; exposure to diseases or parasites from aquaculture; poor oceanographic conditions from global warming; changes in marine migration routes; and other causes yet to be identified. Clearly, with reduced overall numbers of salmon, individual salmon stocks and the species as a whole become more susceptible to extirpation.

It is certainly true that the commercial fishery is not the only problem affecting survival of the wild Atlantic salmon. It is, however, an obvious and significant one that can be addressed by concrete and measurable performance by governments and individuals. Now that we have largely eliminated the commercial salmon fishery in North America the question becomes: *Is there sufficient time left to address the other issues and prevent the extinction of the species*?

This paper will outline a brief history of the commercial Atlantic salmon fishery in North America, steps taken to close it, and the results of those actions. It will also outline the additional action necessary if we are to succeed in conserving the wild Atlantic salmon.

Historical context

From the beginning of human habitation in North America, the wild Atlantic salmon has been a significant element of the social and cultural identity of Aboriginal people and the non-Aboriginal people that followed. Over 10 000 years ago, people began to capture wild salmon through several means, primarily through traps and spears, in rivers and estuaries as a means of supplementing other wild animals harvested for food, social and ceremonial purposes.

The coming of European settlers brought with it new ways to harvest the salmon, mainly through nets, weirs and new forms of traps set in rivers and estuaries. Eventually, as fishing techniques became more advanced, nets in larger vessels moved further offshore into the bays and eventually to the high seas. By the mid-20th century salmon fishing had evolved into a major commercial enterprise, with fish caught in rural and remote parts of Atlantic Canada and north-eastern USA, dried or salted and sold in towns and cities across North America. The salmon fishery stood second only to cod in terms of its commercial value.

At various times over several generations, however, alarms began to sound over the impact of the commercial salmon harvest and other problems on the overall populations of salmon in North American rivers as commercial fishermen and recreational anglers alike noted the continuing decline in numbers of fish returning to the rivers each year. In his several books in which he outlines the early period in North America, Anthony Netboy cited the problems represented by industrialization, dams, pollution and loss of habitat, as well as commercial fishing in the loss of salmon in both Canada and the USA. In light of his and many of his predecessors' clearly stated concerns, we should consider it unfortunate that action was not taken to address the impacts of forestry, agriculture, dams, pollution, sedimentation, poaching and predation with the same degree of effort given to the commercial salmon fishery.

History of North American closures

It was not until the late nineteenth century, however, that regulation of the commercial fishing of salmon began to be implemented by governments in Canada and the USA. By that time, the catch and the corresponding effort of the commercial salmon fishery had already begun to decline. In Canada, as early as the 1840s, records indicate that shipments of salmon from each of the present-day provinces declined sharply (Dunfield, 1985). Moreover, not only was the catch declining, but also the average weight of commercially caught salmon was decreasing (Dunfield, 1985). In Maine, by the mid-1840s, the catch had been reduced by nearly 50% (Baum, 1997). The previously 'unlimited resource' was beginning to come under some very clear limitations.

Although the trends were evident, the picture in the USA contrasted somewhat with that in Canada. According to Baum, commercial Atlantic salmon landings steadily declined in concert with the diminishing resource, although throughout history most of the catch continued to occur on the Penobscot River. There were few 'good' years reported during the early 1900s; however, the fishery was finally closed in 1948 after a reported catch of only 40 salmon in the Penobscot River (Baum, 1997).

In Canada, by the late 1860s it became evident to government authorities that overexploitation of the salmon was a major cause in the decline in Atlantic salmon populations. Several reports commissioned by governments shortly after Confederation in 1867 pointed to the need to adopt effective regulation of the salmon fishery. Canada's first fishery act therefore introduced prohibitions on the catch of kelt, smolt and small salmon, the taking of salmon roe and fishing during the spawning season. The regulations also specified salmon net size, net location and use of fishways as commercial fishing devices (Dunfield, 1985).

It was not until 1966, however, that a series of new measures involving closure of various commercial fisheries, license retirements and transfer restrictions, bans on driftnets, reduced seasons and other measures progressively became implemented. This resulted largely from the concerns of many people such as Wilfred Carter (1985), Anthony Netboy (1968, 1973, 1980) and Richard Buck (1993), with others, who urged that the commercial salmon fishery be closed for good.

In March 1992, the governments of Canada and Newfoundland took the historic and courageous step finally imposing a 5-year moratorium and license buy-out offer for the commercial salmon fishery on the island of Newfoundland. This came after decades of lobbying by the Atlantic Salmon Federation and its two founding organizations, the Canadian-based Atlantic Salmon Association and the USA-based International Atlantic Salmon Foundation.

By the end of 1992, 96% of commercial salmon fishing license holders on Newfoundland had accepted the buy-out offer. Shortly thereafter, in 1993, the offer was extended to Labrador fishermen, with 60% uptake. These efforts continued to 1998 when the closure of the commercial salmon fishery in Canada was finally achieved. The total cost to Canadian taxpayers was approximately $72 million, mostly in the latter part of the period and due to buy-out of commercial salmon licenses.

This victory did not occur easily because of the enormous political and social clout of the commercial salmon fishery. Commercial salmon fishing represented a big part of the culture, lifestyle and income for fishermen in many parts of Atlantic Canada. Closure of this part of the fishery was pitted against the recreational fishery, which was frequently portrayed as a pursuit of wealthy foreigners.

The results of these closures provided temporary improvement in salmon runs in most rivers. It did not, however, result in the sustained increases in salmon abundance that most observers had expected. Closure of the commercial exploitation of salmon was necessary, but did it occur in time to allow action on the other issues contributing to the possible collapse of the species? Was it a clear case of too late ... too little?

New challenges

The closures of the domestic American commercial fishery in 1948 and the Canadian commercial fishery by the end of the 1990s are major achievements. But by no means did they eliminate the threat of commercial fishing for North American Atlantic salmon. Today, some significant challenges in eliminating the commercial harvest of wild salmon in Canadian and in international waters remain. These fisheries need to be addressed to assist in the restoration of salmon populations back to abundance in North American rivers.

Labrador resident food fishery

The resident food fishery for salmon in Labrador is recognized today as a legal fishery largely because of the difficulty faced by Canada's Department of Fisheries and Oceans (DFO) in effectively regulating closure of the Labrador commercial salmon fishery after 1998. With the Labrador moratorium several individuals, including many Métis residents, and others, began to fish salmon off the headlands of Labrador. This effectively reintroduced, on a somewhat smaller scale, the commercial salmon fishery that had intercepted salmon bound for other North American rivers that had just been closed.

ASF and its affiliate, the Salmonid Council of Newfoundland and Labrador (SCNL), protested this fishery, leading DFO to launch an effort to stop the illegal fishery. The enforcement actions, however, were not successful largely due to the numbers of individuals involved and the difficulty in catching them with limited enforcement resources along a very long and geographically difficult coastline.

In 2001, DFO, therefore, entered into a self-regulatory agreement for Labrador that permitted the establishment of a resident food fishery. The reported aggregate results for the resident food fishery and communal native fisheries of the Innu and Inuit of Labrador for 2001 stand at 4730 small salmon and 2334 large salmon weighing a total of 18.7 tonnes.

ASF is concerned at the magnitude of this fishery, although it does not dispute the rights or needs of individual Aboriginal people to a subsistence food fishery. ASF has proposed tighter regulation of this fishery by DFO to ensure that only Aboriginal people recognized under Canada's *Indian Act* are able to participate and that the fishery be restricted to rivers and estuaries to avoid capture of salmon headed for other North American rivers. This issue remains unresolved and it is expected to remain so indefinitely.

St. Pierre and Miquelon salmon fisheries

In 2001, the commercial mixed-stock salmon fishery off the French islands of St. Pierre and Miquelon reported a harvest of 2.155 tonnes. This was a minor decrease from the reported catch of the preceding 3-year period but higher than the 15-year mean catch at 2 tonnes. Up to 52 individuals were licensed for this fishery for 2001, an increase of over 20% over the previous year.

The continuance of a commercial salmon fishery in St. Pierre and Miquelon is a major problem for salmon headed for the rivers of southern Newfoundland and for other rivers in Atlantic Canada, Quebec and Maine. St. Pierre holds a 200-mile limit that cuts across salmon migration routes to these areas and most certainly intercepts fish native to those areas. It should be noted that no salmon are native to St. Pierre or to Miquelon.

To date, France, in respect of St. Pierre and Miquelon, has been un-cooperative in finding solutions to limit this fishery in accordance with NASCO policy or out of respect for Canada or the USA. This issue is gaining prominence in concern expressed by ASF and SCNL and the governments in Canada and the USA. It is anticipated that additional pressure and diplomacy will have to be applied to effect a solution to this fishery. The issue has been discussed several times during the proceedings of the North American Commission of NASCO without results. Canada and the USA and NASCO representatives have attempted to gain concessions from France, including agreement to a cooperative biological sampling programme, without success.

West Greenland fishery

The fishery for wild Atlantic salmon off the west coast of Greenland is a relatively recent phenomenon compared to the commercial fisheries in the North American and European waters. It was not until the late 1950s that the potential of a salmon fishery was identified when it became understood that the smolt migrating to sea from rivers on both sides of the Atlantic mixed in the feeding grounds off Greenland. The proportion of fish from North American and from European rivers is typically in the percent range 60/40 respectively.

This discovery brought with it a large multi-national fleet and the harvest quickly rose from the 40-tonne level of 1958 to a peak of 2689 tonnes by 1971. The Greenland fishery quickly became the largest commercial salmon fishery of all, far outstripping the commercial fisheries in European countries and the Faeroe Islands, threefold or more. The sheer scale of this fishery led many salmon conservationists to lobby governments for limitations on the Greenland fishery but it was not until the early 1980s that reduced catches began to manifest themselves.

The *International Atlantic Salmon Foundation* (IASF) *Newsletter* (1985) explained that (North American salmon) 'are first exploited by drift netting near Greenland, followed by heavy inshore netting as they move through waters near Newfoundland (approximately 600,000 annually), then in the area of the Atlantic Provinces and Québec, where they face a maze of cod traps, mackerel nets, groundfish gill nets, herring nets, and multi-species traps that take an additional 250,000 annually'. Anthony Netboy (1980) noted that the decline of Canadian salmon runs was perceptibly accelerated, as in the British Isles, by the upsurge in the Greenland fishery.

An important outcome of the lobby efforts was the eventual formation of NASCO in 1984. A most influential person in this effort was Richard Buck and his organization, Committee on the Atlantic Salmon Emergency (CASE), which promoted the control on the Greenland fishery. The tremendous efforts and years of lobbying the USA and Canadian governments by Buck, Carter and many others finally resulted in NASCO's formation and, along with it, the possibility of internationally sanctioned controls on the commercial salmon fishery and other problems affecting the salmon.

Clearly, however, the prevailing notion was that the west Greenland fishery represented a major problem that had to be solved following the multiple measures invoked by Canada up to 1992 to address many of the above-noted problems. In the 11-year period between 1982 and 1992, the reported landings of European and North American salmon averaged over 200 000 salmon annually, with several years in the 300 000 range or higher.

Accordingly, ASF and its partners the North Atlantic Salmon Fund (NASF) and the Atlantic Salmon Trust, with funding provided by the USA National Fish and Wildlife Foundation, developed a 2-year conservation agreement with the KNAPK (Organization of Fishermen and Hunters in Greenland). In return for compensation amounting to $400 000 USD each year, KNAPK members agreed to suspend fishing for salmon except for subsistence purposes. This agreement was not renewed for 1995 largely due to its expense while efforts to address the other issues were struggling with limited resources.

From 1995 to 2000, inclusive, the west Greenland fishery was severely restricted as a result of a quota agreed to by NASCO based on pre-fishery abundance advice from the International Council for the Exploration of the Sea (ICES). In that 6-year period, the reported catch ranged from a high of 21 400 North American salmon in 1995 to a low of 3100 North American salmon in 1998.

In 2001, however, the information available to NASCO from ICES, using the same formula, indicated a much greater abundance of salmon available for harvest by Greenland fishermen. The prospect of a return to a harvest exceeding 60 000 salmon resulted in concerns for the impact on conservation and restoration efforts, particularly in the USA. As a result, a management agreement was developed and ratified by NASCO parties that resulted in an adaptive management system with in-season quota adjustments and based on catch per unit effort (CPUE) measured during the season. The harvest level was to be determined by a progressive quota adjusted during three separate harvest periods according to CPUE. The maximum allowable catch under this agreement was 200 tonnes.

The results of the 2001 fishery were a reported catch of 42.5 tonnes consisting of 34.5 tonnes commercial catch and 8 tonnes subsistence fishery. An additional 10 tonnes was estimated as the unreported catch. These figures were surprisingly low to some but appeared to corroborate the low abundance anticipated by ICES and being experienced in North American and European rivers.

For 2002 a somewhat different scenario emerged, with ICES recommending a zero quota for the Greenland commercial fishery based on pre-fishery abundance estimates and a 75% confidence level of attaining conservation requirements over the 167-tonne quota that would result based on the previous 50% confidence level. This resulted in a new round of negotiations at NASCO that created a new adaptive management system, similar to the 2001 agreement, but with two distinct fisheries periods and allowing a maximum commercial harvest of 55 tonnes in addition to the subsistence fishery.

As noted earlier, in 1993 and 1994, ASF and its partner conservation organizations entered into a conservation agreement with KNAPK to end the commercial salmon fishery in Greenland. NASF and ASF believe the Greenland commercial salmon fishery to be a resolvable problem facing conservation of North American wild Atlantic salmon populations. This closure will be of benefit to southern European stocks, although they will remain threatened until the major commercial salmon fisheries in the Faeroe Islands and in Irish waters are also closed. These closures are a necessary step and an immediate 'do-able' that salmon conservationists hope will buy time to find a resolution of the environmental issues also affecting the well-being of wild Atlantic salmon.

NASF's and ASF's intention is that a long-term conservation agreement with KNAPK must be durable and bring clear benefits to the fishermen who give up their rights to fish, today and for the future. One very effective way of achieving this goal would be to provide support to development of alternative fisheries that may be more lucrative and provide greater long-term benefits to those involved.

In August 2002, NASF, in partnership with ASF and NFWF (National Fish and Wildlife Foundation), successfully concluded a conservation agreement with KNAPK to end the commercial salmon fishery in Greenland waters. The agreement is renewable for at least 5 years to include commercial fishing seasons from 2002–2006, inclusive. Excellent potential exists to renew thereafter, but funds need to be secured for this purpose.

Broadly, the key terms and conditions of the conservation agreement are:

(1) All commercial fishing within Greenland territorial waters is suspended. In addition, commercial sale of salmon is suspended. The agreement does not, however, prohibit a limited subsistence fishery.
(2) A trust fund is established to enable development of projects to benefit KNAPK and the Greenland inshore fishing industry, especially development of new fisheries (lumpfish and snow crab) and other economic activities. An economic adjustment fund is also established to help fishermen purchase gear and equipment or find alternative income.
(3) The parties agree to seek additional sources of funding to strengthen the objects of the trust fund.
(4) Provision is made for accounting of the funds expended under the agreement.

Conclusions

The wild Atlantic salmon is tightly interwoven into the culture, outlook and economy of those who live within its North American range. For over 300 years the people of New England, the maritime provinces, Quebec, Newfoundland and Labrador pursued a commercial fishery for Atlantic salmon that was important to the development of coastal communities.

As a result of a combination of factors, salmon populations began to decline precipitously after the mid-nineteenth century. At that time, one of the greatest single factors at the root of the decline was overexploitation of the resource by the commercial salmon fishery. It was not the only factor, but it certainly compounded reduced reproductive capacity from habitat loss, pollution and many other factors that have contributed to the regression in salmon populations since that period.

Unfortunately, it was only in the relatively recent mid-twentieth century that action to limit and eventually close the commercial fishery was taken. This became a single-focused activity that experience indicates has presented its results too late to provide the much needed respite to enable a genuine recovery. Closure of commercial salmon fisheries does provide increased returns to some rivers in the short term, but the North American experience also indicates that any rebound has not been sustained.

This is not to say that a return to abundant populations of wild Atlantic salmon to our rivers is an impossible task. Rather, it underscores the necessity of maintaining determined action on many fronts addressing several issues simultaneously to promote the desired, and *sustained*, returns.

This underscores the importance in current times of a partnership among those interested in conservation of wild Atlantic salmon: non-government organizations, governments, universities and individuals. Time is of the essence. The resource continues to decline. We must 'get on the same page', get the priorities identified and tackle them with our respective strengths. No one sector has the monopoly on the ideas and the solutions. But through partnership and concerted, hard work we can return the wild Atlantic salmon to abundance.

If we do not follow this course of action the goal will continue to elude us.

References

Baum, E. (1997) *Atlantic Salmon: A National Treasure*. Atlantic Salmon Unlimited, Hermon, Maine.
Buck, R. (1993) *The Silver Swimmer*, Lyons & Burford, New York.
Carter, W. (1985) Editorial. In: *International Atlantic Salmon Foundation Newsletter*. IASF, Montreal, Quebec.
Dunfield, R.W. (1985) *The Atlantic Salmon in the History of North America*. Minister of Supply and Services, Canada.
Netboy, A. (1968) *The Atlantic Salmon: A Vanishing Species*. Faber & Faber, London.
Netboy, A. (1973) *The Salmon: Their Fight for Survival*. Houghton & Miflin, Boston.
Netboy, A. (1980) *The World's Most Harassed Fish*. André Deutsch, London.

Chapter 9
Assessing and Managing the Impacts of Marine Salmon Farms on Wild Atlantic Salmon in Western Scotland: Identifying Priority Rivers for Conservation

J.R.A. Butler[1] and J. Watt[2]

[1] *Spey Fishery Board Research Office, 1 Nether Borlum Cottage, Knockando, Morayshire, AB38 7SD, Scotland, UK*
[2] *Lochaber and District Fisheries Trust, Arieniskill Cottage, Lochailort, Invernesshire, PH38 4LZ, Scotland, UK*

Abstract: The Scottish salmon farming industry became established in the early 1980s. However, impacts of farm salmon on wild Atlantic salmon (*Salmo salar*) were not predicted and many farms were located near the mouths of rivers. This paper assesses 38 salmon rivers in the centre of the west coast salmon farming zone, and compares the status of stocks in those with and without salmon farms in their sea lochs. Juvenile surveys carried out at 230 sites in 35 rivers in 1997, 1999 and 2001 revealed that rivers with farms had 62–82% and 44–62% lower mean abundances of salmon fry and parr, respectively, and the differences were statistically significant. Calibration of juvenile abundance indices against smolt counts at two rivers with fish traps demonstrated that 86% of predicted smolt runs were depleted in rivers with farms, versus 26% in rivers without farms. Severe stock collapses were evident in 14 (50%) of rivers with farms, where only remnant populations remain. Applying the NASCO Rivers Database classifications, nine of these are considered 'threatened with loss', and five may be 'lost'. In 1990–2001 escaped farm salmon were significantly more prevalent for rod catches in rivers with farms (mean 9%) than in those without (mean 2%), resulting in a greater threat of genetic introgression. Epidemics of sea lice (*Lepeophtheirus salmonis*) found on sea trout (*Salmo trutta*) suggest that salmon smolts emigrating from sea lochs with farms are also suffering lethal infestations. It is proposed that in sea lochs with farms the combined effects of genetic introgression and lice infestation have suppressed smolt survival rates to < 8% during the 1990s, resulting in stock collapse. Invoking the requirements of the Habitats and Birds Directives, and NASCO's 'precautionary principle', the 15 largest and potentially most genetically diverse salmon rivers on the west coast are identified. Measures are proposed to avoid further deleterious effects of marine salmon farms on these rivers, including the establishment of exclusion zones in their sea lochs.

Introduction

The Scottish marine salmon farming industry is centred on the west coast, where fjordic sea lochs provide sheltered moorings for cage installations (Fig. 9.1). Since the

94 *Salmon at the Edge*

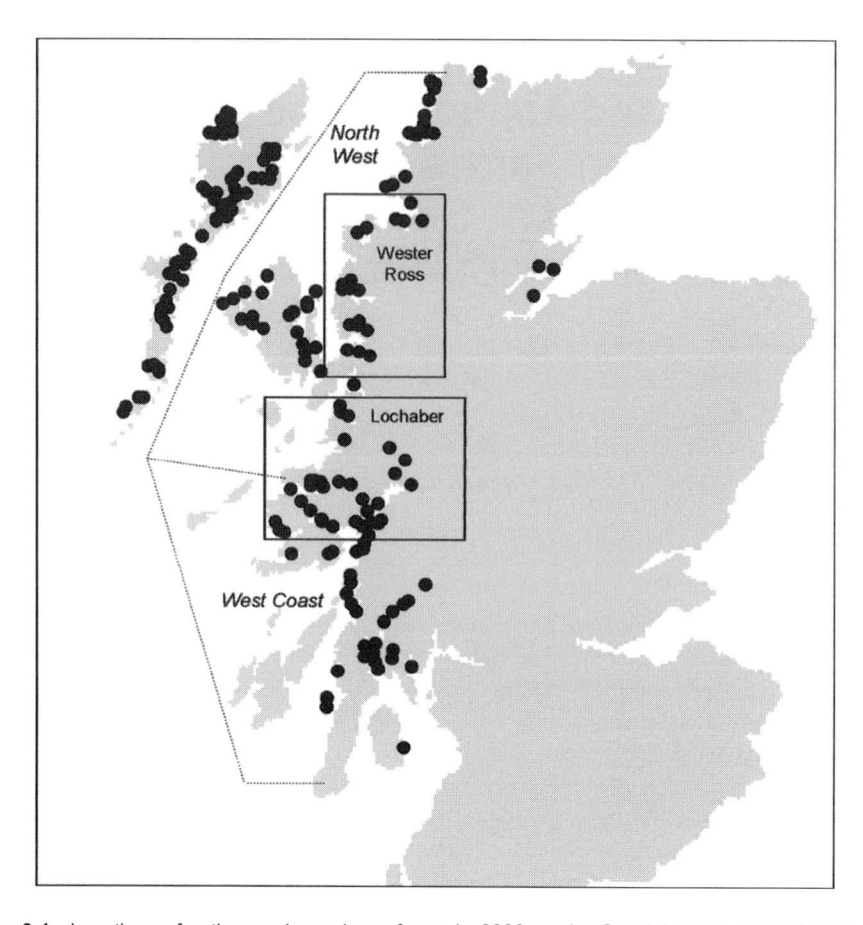

Fig. 9.1 Locations of active marine salmon farms in 2000 on the Scottish west coast (from Anon., 2001a), the North West and West Coast Statistical Regions, and the Wester Ross and Lochaber study areas. Farm sites are not to scale.

inception of the industry in the early 1980s production has expanded rapidly to 120 000 tonnes in 2001 (Anon., 2001a) (Fig. 9.2a). During the 1980s regulatory authorities took no account of the potential impacts of farm salmon on wild Atlantic salmon (*Salmo salar*), and consequently many sites were established near the mouths of rivers.

The expansion of the industry has coincided with marked declines in rod catches of wild salmon in the North West and West Coast Statistical Regions (Fig. 9.2a), which cover the bulk of the salmon farming zone (Fig. 9.1). Average catches in the 1990s were 41% lower than those in the 1970s, prior to the establishment of the industry (Fig. 9.2a). Over the same period, rod catches outside the salmon farming zone in eastern and northern Scotland have increased by 30% (Fig. 9.2b). This trend has raised concerns that salmon farming is impacting upon wild salmonid populations (Walker, 1994; Anon., 1995a; Northcott & Walker, 1996; Hawkins, 2000; Butler,

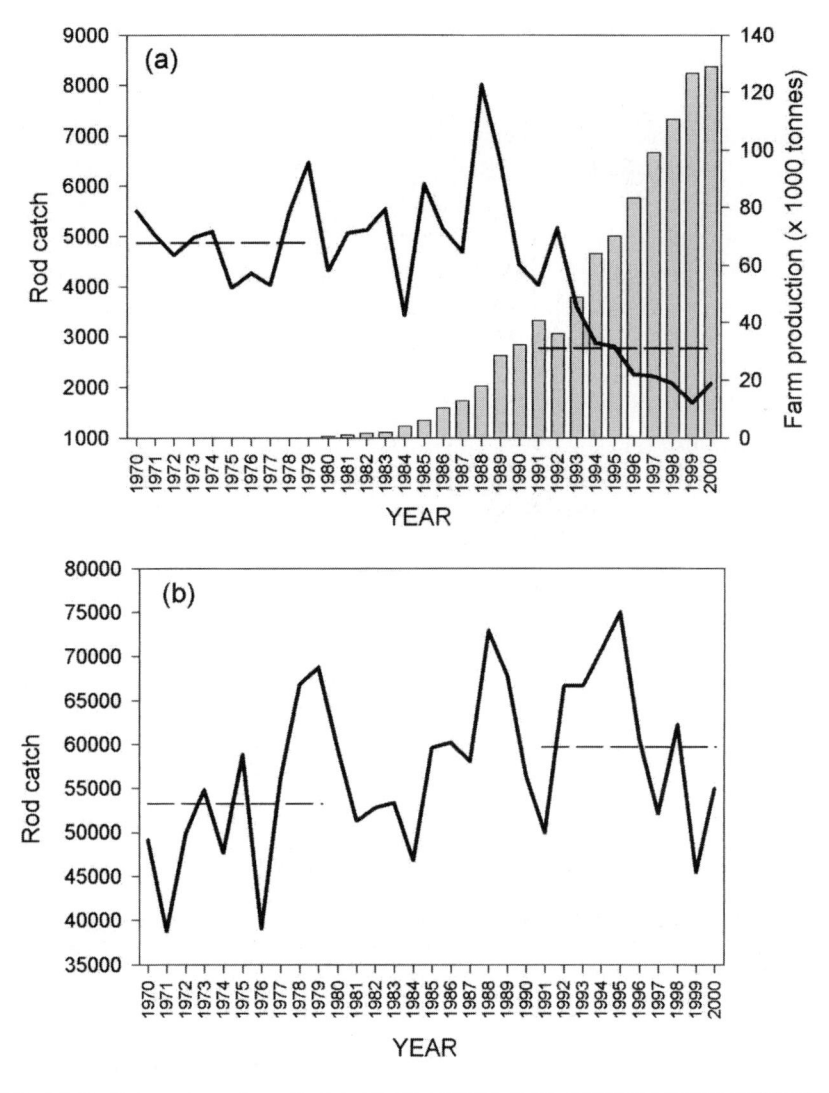

Fig. 9.2 (a) Scottish marine salmon farm production relative to the combined declared wild salmon rod catch for the North West and West Coast Statistical Regions, 1970–2000, and (b) declared wild salmon rod catch from the East, Moray, North, North East and North Statistical Regions, 1970–2000. Dashed lines represent average catches for 1970–79 and 1991–2000. (Data are kindly provided by the Scottish Executive Fisheries Research Services.)

2002a). Similar problems have been identified in the salmon farming zones of western Ireland (Anon., 1995b; Gargan, 2000; Gargan et al., this vol.) and Norway (Anon., 1999a).

Marine salmon farms potentially have three deleterious effects on wild salmonids. First, they may induce an elevated incidence of pathogens (McVicar, 1997). The best documented impact is increased infection levels of the copepodid sea louse

Lepeophtheirus salmonis, a marine parasite specific to salmonids. Lice epidemics have been recorded on wild sea trout (*Salmo trutta*) following the establishment of salmon farms in Scotland, Ireland and Norway (Anon., 1997) because the large numbers of farmed hosts boost production of larval lice above natural background levels (Tully & Whelan, 1993; Heuch & Mo, 2001; Butler, 2002a). Amongst sea trout such epidemics result in impaired osmoregulation and many return to fresh water where they cease feeding, lose growth and condition, and up to 20% may die (Tully et al., 1993a, b; Birkeland, 1996; Birkeland & Jakobsen, 1997; Tully et al., 1999; Gargan et al., this vol.). In western Scotland at least 14–40% of sea trout carried potentially lethal lice infections in June 1998–2000 (Butler et al., 2001; Butler, 2002a), and in Norway lice epidemics potentially cause the mortality of up to 47% of sea trout smolts (Bjorn et al., 2001). Where epidemics occur on sea trout, they are also likely to affect emigrating wild salmon smolts (Whelan, 1993; Butler, 2002a). This is evident in salmon farming areas of Norway (Holst & Jakobsen, 1998; Finstad et al., 2000), where up to 95% of wild smolts die due to lice infestations (Holst et al., this vol.). In western Scotland, returning adult salmon have also been observed with unusually high lice burdens (Butler, 2002b).

Second, escaped farm salmon successfully interbreed with wild conspecifics, diluting genetically defined local adaptions and reducing population fitness [reviewed by Youngson et al., (1998, 2001)]. In Scotland escapes are caused by storm or predator damage to cages, operational accidents or vandalism (Anon., 2000a). In 2000 an estimated 411 433 salmon escaped from marine cages (Anon., 2001a), versus 255 000 in 1999 and 95 000 in 1998 (Butler, 2000). The annual presence of farm salmon in west-coast netting station catches since 1989 suggests that unreported escapes have occurred regularly (Youngson et al., 1997; Butler, 1999a). After escaping, most farm fish migrate to oceanic feeding grounds, and when mature return to the region of the coast where they were liberated (Hansen & Jonsson, 1991). In Scotland, farm salmon have been observed successfully breeding in a river near the site of an escape (Webb et al., 1991, 1993a), and juveniles of farmed parentage were found in 11 of 12 rivers surveyed within the fish farming zone in 1991 (Webb et al., 1993b). In all of these cases the numbers of escapees and the extent of hybridization are likely to have been underestimated (Webb et al., 1993b; Youngson et al., 2001).

Third, the location of farms near river mouths may attract concentrations of predators, potentially increasing predation of emigrating smolts and returning adults. In western Scotland sea fish congregate around cages, along with escaped salmonids, resulting in high densities of shags (*Phalacrocorax aristotelis*) and cormorants (*Phalacrocorax carbo*) (Carss, 1990, 1993a, b). Both grey seals (*Halichoerus grypus*) and harbour seals (*Phoca vitulina*) are abundant in west coast waters, and are attracted to marine salmon farms where they are the primary cause of stock losses (Quick et al., 2002).

The impacts of salmon farming on wild populations are of wider concern because the species has been listed in Annex II of the 1992 European Union Habitats and

Birds Directives (Anon., 2000b). Under the Directives, the wild Atlantic salmon is considered threatened in a European context and governments are required to improve the species' conservation status. This includes the designation of candidate Special Areas of Conservation (cSACs), two of which are being established on the west coast for the Little Gruinard and Grimersta river systems. The UK government is also a contracting party to the North Atlantic Salmon Conservation Organization (NASCO). In 2000 NASCO passed a resolution on the adoption of the 'precautionary approach', whereby imperfect scientific knowledge should not preclude the implementation of conservation measures. This was preceded in 1994 by the Oslo Resolution, which aims to minimize the impacts of aquaculture on wild salmon stocks (NASCO, 2001).

This study assesses 38 wild salmon populations during 1997–2001 in areas of western Scotland with and without marine salmon farming. Although a recent survey of juvenile salmon stocks was undertaken by the Scottish Executive's Fisheries Research Services (FRS) in five north and west coast rivers, the results were inconclusive because so few rivers were surveyed, and no comparative data exist for years prior to the establishment of salmon farming (MacLean & Walker, 2002). Based on our results, and by applying NASCO's principles of the precautionary approach and the Oslo Resolution, recommendations are made for the prioritization of salmon rivers for conservation. Sea lochs are identified where the location and management of salmon cages should be reviewed in order to minimize the risk of deleterious effects on important wild salmon populations.

Study areas and methods

Study areas

The study was undertaken on 38 rivers in the districts of Wester Ross and Lochaber, located in the centre of both the salmon farming zone and the North West and West Coast Statistical Regions (Figs. 9.1, 9.3). The historical presence of wild salmon prior to 1995 was confirmed for all rivers by catch records, for 32 rivers by the NASCO Rivers Database (NASCO, pers. comm.), and for 12 by juvenile surveys (Table 9.1).

Including still waters and tributaries, the range of river lengths accessible to salmon was 1–75 km (Table 9.1). The underlying geology in both Wester Ross and Lochaber is mildly acidic granites and schists (Doughty, 1990). Water quality assessments have been made by the Scottish Environment Protection Agency (SEPA) for 24 of the rivers. Under the River Classifications Scheme all were Biological Class A or B in 1995, and A1, A2 or B in 1999 and 2000 (Table 9.1), confirming an absence of pollution (I. Milne, SEPA, pers. comm.). Freshwater salmon cages or hatcheries occurred in eight of the rivers, and hydro-electric dams within accessible areas existed on three (Fig. 9.3).

98 Salmon at the Edge

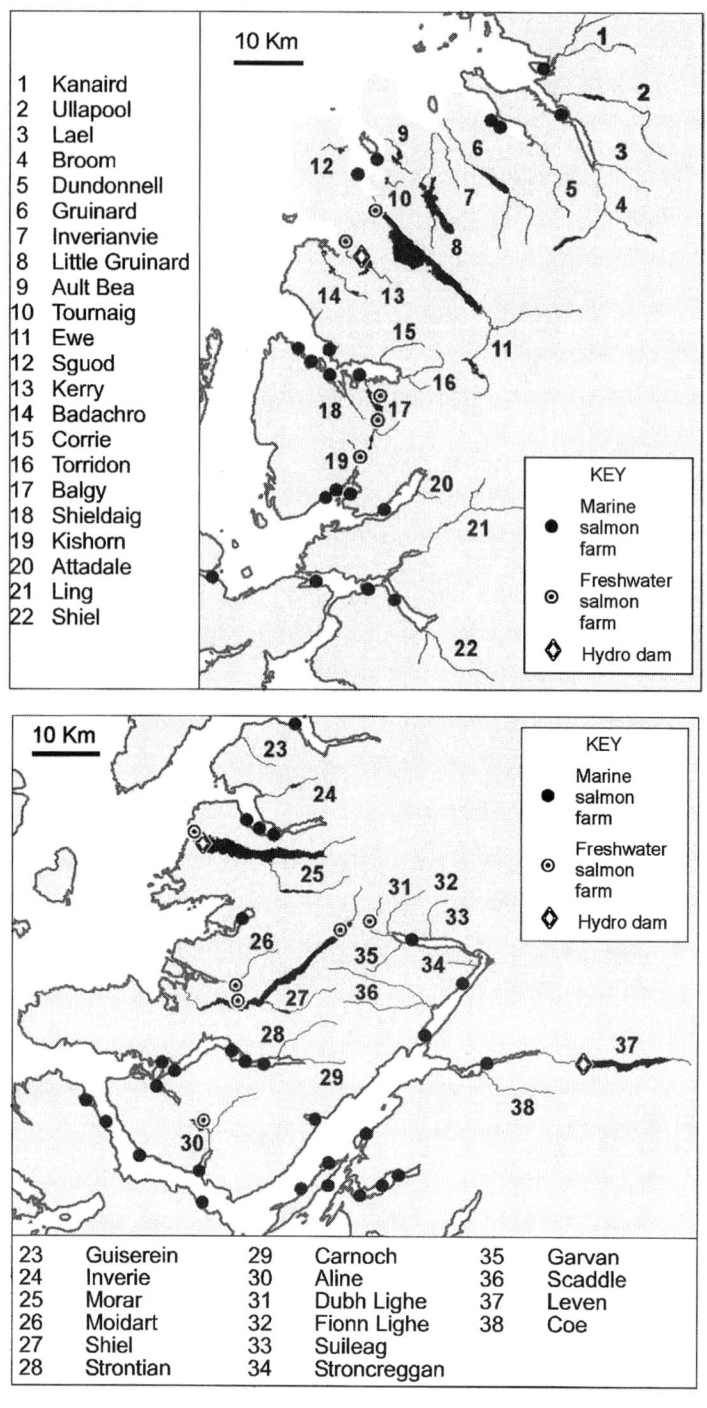

Fig. 9.3 Locations of the 38 rivers in Wester Ross (above) and Lochaber (below), relative to active marine salmon farms, freshwater salmon farms and hydro-electricity dams in 2001.

Table 9.1 Details of the 38 salmon rivers, subdivided into those (a) with and (b) without marine salmon farms in their sea lochs. 'NASCO' refers to the NASCO Rivers Database (data reproduced by kind permission of NASCO), 'CR' denotes 'catch records' held by owners of the rivers concerned, and 'JS' denotes 'juvenile survey'. River Classification Scheme Biological Class data are provided by SEPA; 'ns' denotes 'not surveyed'.

River	Accessible length (km)	Evidence of pre-1995 salmon population	River Classification Scheme Biological Class		
			1995	1999	2000
(a) With farms					
1. Kanaird	19	NASCO, CR	A	ns	A2
2. Ullapool	17	NASCO, CR	A	ns	ns
3. Lael	3	CR	ns	ns	ns
4. Broom	10	NASCO, CR, JS[1]	A	ns	ns
5. Dundonnell	9	NASCO, CR, JS[2]	A	ns	ns
9. Ault Bea	2	CR	ns	ns	ns
10. Tournaig	5	CR, JS[3]	ns	ns	ns
11. Ewe	75	NASCO, CR, JS[4]	A	ns	ns
12. Sguod	5	CR, JS[5]	ns	ns	ns
15. Corrie	1	CR	ns	ns	ns
16. Torridon	11	NASCO, CR, JS[6]	A	ns	A1
17. Balgy	17	NASCO, CR	A	ns	A2
18. Shieldaig	5	NASCO, CR	ns	ns	ns
19. Kishorn	5	NASCO, CR	A	ns	A2
20. Attadale	4	NASCO	A	ns	A2
21. Ling	13	NASCO, CR	A	ns	A1
22. Shiel	15	NASCO, CR	A	ns	A2
24. Inverie	11	NASCO, CR	ns	ns	ns
28. Strontian	7	NASCO, CR	A	ns	ns
29. Carnoch	8	NASCO, CR	A	ns	ns
30. Aline	26	NASCO, CR, JS[7]	ns	A2	ns
31. Dubh Lighe	5	NASCO, CR	ns	ns	ns
32. Fionn Lighe	2	NASCO, CR	ns	ns	ns
33. Suileag	6	CR	ns	ns	ns
34. Stroncreggan	4	NASCO, CR	ns	ns	ns
35. Garvan	4	NASCO, CR	ns	ns	ns
36. Scaddle	23	NASCO, CR	B	B	ns
37. Leven	3	NASCO, CR	A	A2	ns
38. Coe	11	NASCO, CR	A	A2	ns
(b) Without farms					
6. Gruinard	38	NASCO, CR, JS[2,6,8]	A	ns	ns
7. Inverianvie	3	NASCO, CR, JS[2]	ns	ns	ns
8. Little Gruinard	25	NASCO, CR	A	ns	ns
13. Kerry	7	NASCO, CR, JS[6]	A	ns	A2
14. Badachro	7	NASCO, CR	A	ns	A2
23. Guiserein	9	NASCO, CR	ns	ns	ns
25. Morar	34	NASCO, CR, JS[9]	A	A2	ns
26. Moidart	13	NASCO, CR	A	ns	ns
27. Shiel	58	NASCO, CR, JS[10]	A	A2	ns

[1] 1990, 1991, 1993, 1994 (R. Gardiner, FRS, in Butler, 1998)
[2] 1984 (R. Gardiner, FRS, unpublished data)
[3] 1991 (C. West, FRS, in Butler & Starr, 2001)
[4] 1992 (Walker et al., 1993a)
[5] 1991 (A.F. Walker, FRS, in Butler, 1998)
[6] 1991 (Webb et al., 1993b)
[7] 1995 (Laughton & Bray, 1995)
[8] 1991 (Walker & Stewart, 1993)
[9] 1992 (Walker et al., 1993b, 1993 (Walker et al., 1993c)
[10] 1994 (Walker et al., 1994)

Presence of marine salmon farms

Rivers were classified according to whether or not active salmon farms were present in their sea lochs (Table 9.1). The outer limit of a sea loch was defined by a straight line taken between the most seaward points of each fjord or bay. In all sea lochs with farms, sites had been present since 1988 (SEPA, pers. comm.).

Surveys of juvenile salmon

Survey design

In July–October 1997, 1999 and 2001 juvenile surveys were undertaken on 35 rivers using electric-fishing equipment, and each river was surveyed at least once (Table 9.2). In total, 230 sites were used, some of which were established by previous surveys (Table 9.1). Where repeat surveys were undertaken the same sites were revisited. However, inclement weather sometimes restricted the number of sites revisited, resulting in variable numbers used between years. Sites were selected according to the Scottish Fisheries Coordination Centre's (SFCC) standard protocol (SFCC, 1998, 2001), whereby each survey provided a representative coverage of juvenile riverine habitat in the accessible area of a river. Including still waters, the median nearest-neighbour distance of sites within each catchment was 700 m (range 100–8000 m) (Fig. 9.4), and median altitude above sea level was 35 m (range 1–175 m). Limited salmon stock enhancement had been carried out historically on some rivers. However, in 1996–2001 only sites on the Ewe (one), Morar (four) and Shiel (one) were directly influenced by stocking of hatchery-reared juveniles, and these were excluded from our analyses.

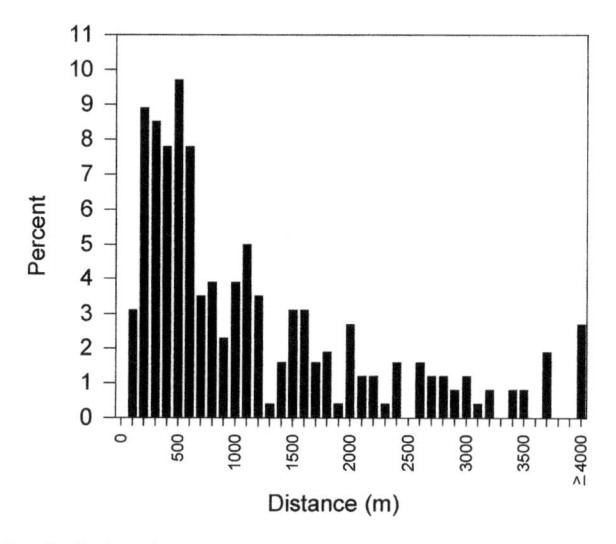

Fig. 9.4 Relative distribution of nearest-neighbour distances of the 230 juvenile survey sites. Distances were measured along watercourses.

Table 9.2 Juvenile survey sites and salmon fry and parr indices (fish/100 m^2) for rivers surveyed in 1997, 1999 and 2001; 'ns' denotes 'not surveyed'.

River	1997			1999			2001		
	Sites	Fry	Parr	Sites	Fry	Parr	Sites	Fry	Parr
(a) With farms									
1. Kanaird	8	2.6	15.3	8	7.3	6.1	ns	—	—
2. Ullapool	6	0.8	4.4	6	7.5	4.2	ns	—	—
3. Lael	ns	—	—	1	1.2	0	ns	—	—
4. Broom	6[1]	6.1	10.9	6[1]	10.2	10.5	6[1]	12.7	7.2
5. Dundonnell	5	0.2	3.4	7	6.2	6.2	ns	—	—
9. Ault Bea	ns	—	—	3	40.3	2.0	ns	—	—
10. Tournaig	5	20.9	3.2	12	14.0	3.7	12	0	2.4
11. Ewe	27	4.1	4.8	26	7.2	4.0	26	10.2	5.7
12. Sguod	3	0	0.7	3	0	0	ns	—	—
15. Corrie	1	0	1.0	ns	—	—	ns	—	—
16. Torridon	5	4.2	5.4	ns	—	—	ns	—	—
17. Balgy	4	0.2	5.0	4	2.2	2.5	ns	—	—
18. Shieldaig	10[2]	0	0	10[2]	0	0.3	10[2]	0	0
19. Kishorn	1	0	6.3	ns	—	—	ns	—	—
20. Attadale	2	0	0	ns	—	—	ns	—	—
24. Inverie	ns	ns	ns	12	14.4	8.3	10	16.9	12.1
28. Strontian	6	0	1.2	ns	—	—	4	0	0.6
29. Carnoch	7	2.5	1.1	ns	—	—	3	10.3	5.5
30. Aline	12	2.9	8.2	10	2.6	5.3	6	22.5	6.5
31. Dubh Lighe	ns	—	—	1	11.4	7.3	ns	—	—
32. Fionn Lighe	ns	—	—	1	0	3.6	1	6.6	1.8
33. Suileag	ns	—	—	4	6.7	2.7	2	0	2.8
34. Stroncreggan	ns	—	—	ns	—	—	4	0	0
35. Garvan	ns	—	—	1	0	0	ns	—	—
36. Scaddle	ns	—	—	5	0	0.8	9	10.1	1.0
37. Leven	4	3.8	12.5	ns	—	—	ns	—	—
38. Coe	3	7.3	4.8	3	0	2.4	4	21.5	1.0
(b) Without farms									
6. Gruinard	10	29.6	4.3	10	18.0	10.6	ns	—	—
7. Inverianvie	ns	—	—	1	8.7	3.5	ns	—	—
8. Little Gruinard	8	23.0	8.3	10	17.9	13.3	10	20.2	10.4
13. Kerry	5	3.6	15.3	5	12.1	7.5	5	17.3	9.7
23. Guiserein	ns	—	—	ns	—	—	6	13.5	1.5
25. Morar	8	10.7	8.7	9	53.0	19.8	9	27.5	16.9
26. Moidart	10	25.2	8.8	3	28.5	5.0	14	22.5	6.8
27. Shiel (LA)	14	14.5	7.3	14	15.7	7.4	11	24.7	7.9

[1] Data for four sites provided by R. Gardiner, Scottish Executive Fisheries Research Services
[2] Data provided by A.F. Walker, D. Hay & M. McKibben, Scottish Executive Fisheries Research Services

Indices of juvenile abundance

Sampling methodology followed the SFCC's standard protocol (SFCC, 1998, 2001). Only data collected from the first fishing at each site were used. Juvenile salmon were counted, measured for fork length, and scales taken for ageing as fry (0+ years) or parr ($\geq 1+$ years). Sites were fished at medium or low flows, and the wetted area of each site was calculated. An index of fry and parr abundance was derived as follows for each river:

$$\text{Index} = \frac{n_1 + n_2 + n_3 + n_4 \ldots n_n}{a_1 + a_2 + a_3 + a_4 + \ldots a_n}$$

where n_1 = the number of fry or parr caught at site 1 and a_1 = the wetted area of site 1. Indices were expressed as fish/100 m².

Data analysis

Within each year, surveyed rivers were grouped as those with and without marine salmon farms, and fry and parr indices of each group were then compared.

Fish traps

To calibrate fry and parr indices with stock status, upstream–downstream fish traps located on the Tournaig and Shieldaig river systems in Wester Ross (Fig. 9.3) were used in conjunction with annual juvenile surveys. Both rivers are of similar size and character, and have been monitored with similar techniques:

- *Tournaig*: The catchment has an area of 9 km² and an accessible length of 5 km (Table 9.1), including a 20-ha loch 700 m from the sea. A temporary trap was installed at the seaward extremity of the system in 1999, and is operated from April–December (inclusive) to cover the smolt and adult run. Juvenile surveys have been carried out annually in the accessible area above the trap at five sites in 1997, and 12 sites in 1998–2002. Based on the accessible area of habitat, optimum salmon smolt production is estimated to be 1100 (Butler & Starr, 2001).
- *Shieldaig*: The catchment has an area of 11 km², and an accessible length of 5 km (Table 9.1), including a 13-ha loch 700 m from the sea. FRS began temporary trapping of smolts and retuning adults in 1995 using a fyke net at the outflow of the loch. This was superceded in 1999 by a permanent trap located at the seaward extremity of the system. Juvenile surveys have been undertaken annually in the accessible area above the trap at 10 sites in 1995–2001 (Walker et al., 1998, 1999; McKibben & Hay, 2002). Potential optimum salmon smolt production is not known.

Salmon stocks in both systems have collapsed, whereby numbers of spawning adults have decreased to the extent that juvenile production is related directly to egg deposition, and hence smolt production has fallen below the potential optimum (Walker et al., 1998, 1999; Butler & Starr, 2001). Because both systems are of similar accessible length, data were combined to calibrate fry and parr indices against counts of anadromous spawners and subsequent smolt production. These calibrations were then used to assess the stock status of other surveyed rivers in 1997, 1999 and 2001.

Prevalence of escapees

To assess the relative prevalence of escaped farm salmon in surveyed rivers, rod catches were made available by river owners and angling clubs and analysed. Complete records were only available for 14 rivers, including the Badachro, Ling and Shiel (Wester Ross), where juvenile surveys were not undertaken (Table 9.1). The proportion of each river's total rod catch in 1990–2001 consisting of escapees (identified by physical appearance) was calculated, and comparisons were made between rivers with and without salmon farms.

Results

Juvenile abundance

Fry indices for rivers with farms were significantly lower than for those without farms in 1997 (Mann Whitney; $W = 179.0$, $P < 0.002$), 1999 ($W = 202.0$, $P < 0.002$) and 2001 ($W = 98.5$, $P < 0.01$) (Fig. 9.5a). Parr indices for rivers with farms were also significantly lower than for those without in 1997 ($W = 194.5$, $P < 0.05$), 1999 ($W = 210.0$, $P < 0.01$) and 2001 ($W = 104.0$, $P < 0.05$) (Fig. 9.5b). Overall, mean fry and parr indices for rivers with farms were 60–82% and 44–62% less than for rivers without farms, respectively.

Status of stocks

Combined Tournaig and Shieldaig data showed a positive and significant linear correlation between anadromous adults trapped and fry indices the subsequent year (Fig. 9.6a), between fry indices and parr indices the following year (Fig. 9.6b), and also between parr indices and smolts the following year (Fig. 9.6c).

The largest sub-optimal run of 701 smolts was recorded at Tournaig, and was equivalent to a parr index of $7.3/100$ m^2 (Fig. 9.6c). Parr indices of ≤ 7.3 were recorded by 43 (86%) of 50 surveys on rivers with farms, compared to 5 (26%) of 19 surveys on rivers without farms, a significant difference ($\chi^2 = 5.1$, df $= 1$, $P < 0.05$).

Fry or parr indices of zero were recorded at least once in 14 (50%) of rivers with farms, but never for rivers without farms (Table 9.2). Tournaig and Shieldaig data demonstrate that such low indices are indicative of severe collapse, with only remnant numbers of anadromous adults and smolts present (Fig. 9.6a, c). Within the same survey, fry and parr indices of zero were recorded for Sguod, Shieldaig, Attadale, Stroncreggan and Garvan (Table 9.2), suggesting that these populations may have been lost (Table 9.3). This is confirmed for Shieldaig by trap data which show that the population is 'virtually extinct' (McKibben & Hay, 2002).

104 Salmon at the Edge

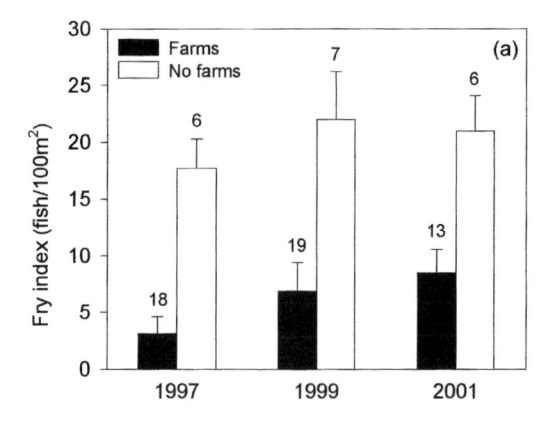

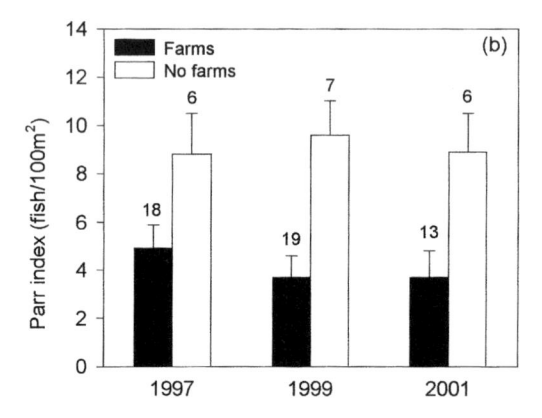

Fig. 9.5 Comparisons of (a) mean fry indices and (b) mean parr indices (+ SE) between rivers surveyed with and without marine salmon farms in 1997, 1999 and 2001. Figures above SE bars are sample sizes of rivers.

Prevalence of escapees

The mean proportion of escaped farm salmon recorded in the total rod catch of rivers with farms was 9% (4.2 ± SE, range 0.6–41.0), versus 2% for rivers without farms (0.5 ± SE, range 0–2.7) (Fig. 9.7). The proportion of escapees was significantly higher for rivers with farms compared to those without (two sample T-test on arcsine transformed data; $T = 2.456$, $df = 10.4$, $P < 0.05$). Excluding the result for the Shiel (Wester Ross) as an outlier, the difference remained significant ($T = 2.84$, $df = 12.9$, $P < 0.02$).

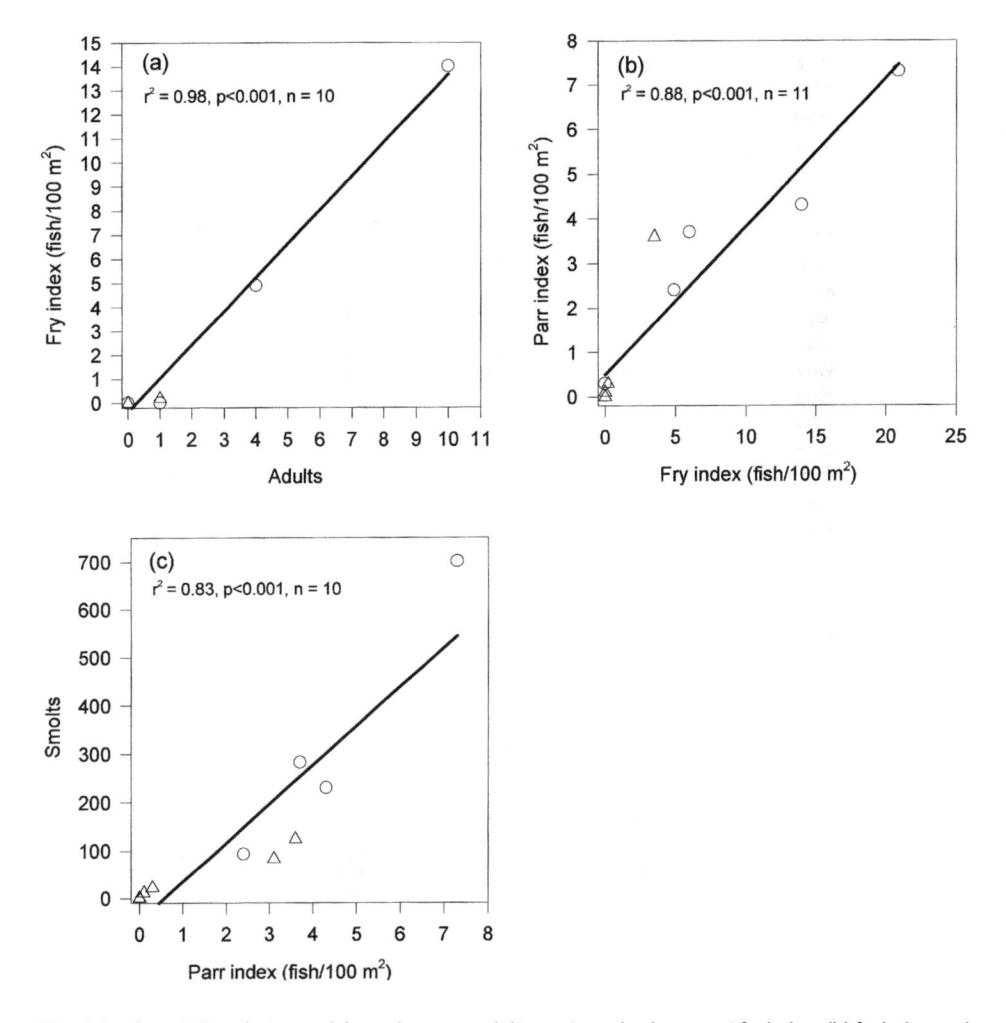

Fig. 9.6 Correlations between (a) anadromous adult counts and subsequent fry index, (b) fry index and subsequent parr index, and (c) parr index and subsequent smolt runs recorded by trap and juvenile survey data from Tournaig (circles) and Shieldaig (triangles), 1995–2002. (Data for Shieldaig are kindly provided by A.F. Walker, D. Hay & M. McKibben, The Scottish Executive, Fisheries Research Services.)

Discussion

Assessing the impacts of marine salmon farms

The abundance of wild Atlantic salmon has declined throughout the species' range since the early 1970s, largely due to sub-optimal oceanic feeding related to climatic change (Friedland & Reddin, 1993; Friedland et al., 1998; Hutchinson & Mills, 2000; Reddin et al., 2000). This has been reflected in declines in smolt marine survival, resulting in reduced adult returns to home waters (Hutchinson & Mills, 2000; Potter

Table 9.3 Reviewed status of the 14 rivers where severe stock collapses were identified in 1997–2001, relative to their 1995 classifications in the NASCO Rivers Database (data kindly provided by NASCO). Classifications are 'not threatened with loss' (category 5), 'threatened with loss' (category 4) and 'lost' (category 1); 'na' denotes 'not available'.

River	Accessible length (km)	1995 NASCO classification	Reviewed 1997–2001 classification
3. Lael	3	na	Threatened with loss
10. Tournaig	5	na	Threatened with loss
12. Sguod	5	na	Lost
15. Corrie	1	na	Threatened with loss
18. Shieldaig	5	Not threatened with loss	Lost
19. Kishorn	5	Not threatened with loss	Threatened with loss
20. Attadale	4	Not threatened with loss	Lost
28. Strontian	7	Not threatened with loss	Threatened with loss
32. Fionn Lighe	2	Not threatened with loss	Threatened with loss
33. Suileag	6	na	Threatened with loss
34. Stroncreggan	4	Not threatened with loss	Lost
35. Garvan	4	Not threatened with loss	Lost
36. Scaddle	23	Not threatened with loss	Threatened with loss
38. Coe	11	Not threatened with loss	Threatened with loss

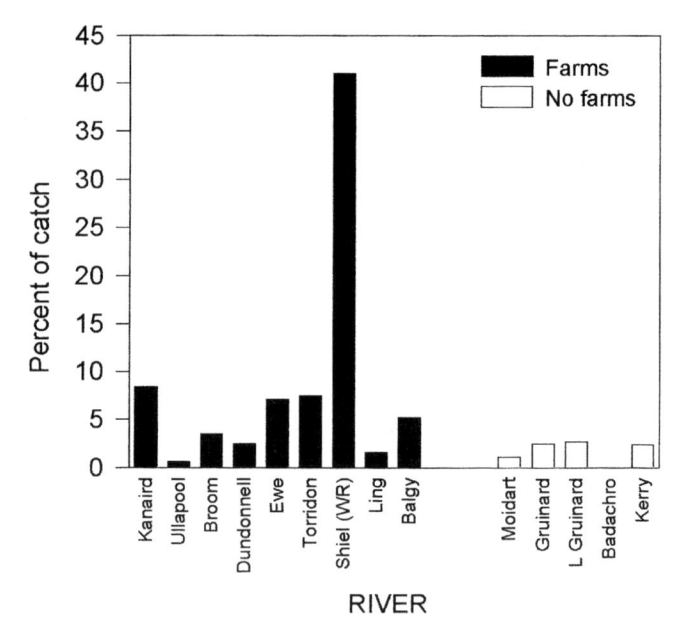

Fig. 9.7 Comparison of the proportion of escaped farm salmon in the total declared rod catch of rivers with ($n=9$) and without ($n=5$) salmon farms in their sea lochs, 1990–2001.

& Crozier, 2000; Anon., 2001b). In Scotland long-term monitoring of smolt survival has been undertaken in the River North Esk, on the east coast. Records show that although estimated marine survival has fallen by a factor of two, sufficient numbers of adults are returning to maintain optimal smolt production (Shearer, 1992; MacLean et al., 2000).

This study indicates that where marine salmon farms have been located in sea lochs with salmon rivers, they have exacerbated declines in local wild stocks. The degree of impact is quantified by mean fry and parr abundances which were 60–82% and 44–62% lower in rivers with farms than in those without, respectively. Calibration of juvenile indices at Tournaig and Shieldaig showed that 86% of smolt runs were suboptimal in rivers with farms, suggesting stock collapse in the majority of years. In rivers without farms only 26% of smolt runs were depleted, indicating adequate egg deposition by anadromous adults in most years.

Because the majority of sea lochs in the North West and West Coast Statistical Regions are occupied by salmon farms (Fig. 9.1), our data suggest that most wild salmon stocks will have been adversely affected. This may explain the trend in rod catches from these regions, which indicates that the abundance of returning adults has declined in the 1990s as farm production has increased (Fig. 9.2a). In 1999 adult abundance appeared to reach its lowest level since 1970, and this may have been reflected in the severe collapse of 14 (50%) of rivers surveyed with farms in 1997–2001. The rate of decline has been such that in 1995 nine of these rivers were considered 'not threatened with loss' by the NASCO Rivers Database, but by 1997–2001 we suggest that the status of five has become 'threatened with loss', and four may be 'lost' (Table 9.3). Overall, stocks in five small rivers of 5 km or less accessible length (Sguod, Shieldaig, Attadale, Stroncreggan and Garvan) may have been lost (Table 9.3).

The effects of marine salmon farms must be substantial if their influence can be detected amongst juvenile populations, since a range of factors could also have caused variation in abundance indices. SEPA monitoring in 1995, 1999 and 2000 suggests that pollution is not a limiting factor for salmon populations in Wester Ross and Lochaber. However, some upland tributaries of six surveyed rivers suffer from natural winter acid episodes which may limit productivity (Butler, 1999b; Watt & Bartels, 1998). Winter spates may also cause redd washout in watercourses with unstable beds, elevating mortality of eggs (Butler, 1999b, 2000). Otter (*Lutra lutra*) predation of spawning adults may also limit egg deposition in some headwater streams (Carss et al., 1990; Cunningham, 2001). Freshwater salmon cages and hatcheries were present on eight surveyed rivers, and dams for hydro power were located on three (Fig. 9.3). These may also have impacted upon freshwater salmon production via disease, escapes or restricted adult and juvenile migration caused by water abstraction or impoundment. Furthermore, many of the populations studied were small and consequently inherently vulnerable to stochastic events and acute fluctuations in abundance (Routledge & Irvine, 1999). Finally, enhancement with hatchery-reared stock had been carried out historically in some rivers, and if successful such exercises may have influenced juvenile populations in 1997–2001.

Sampling error during juvenile surveys may also have caused variation. Water temperature, depth, velocity and electrical conductivity are known to affect the catchability of juvenile salmon [reviewed by Cowx & Lamarque (1990)]. Some sites may have been located in areas of habitat best suited to trout, underestimating juvenile salmon abundance (MacLean & Walker, 2002). Despite these inherent limitations, electric-fishing surveys can still provide useful assessments of relative adult spawning success and potential smolt output (Cowx & Lamarque, 1990; Crozier & Kennedy, 1994, 1995), as verified at Tournaig and Shieldaig. Despite often severe logistical constraints the surveys in this study were all carried out in July–October, and provided good coverage of accessible areas with a median nearest-neighbour distance of 700 m between sites. Therefore in most cases the surveys are likely to have produced representative results. Furthermore, any sampling error probably affected results from all 35 surveyed rivers, and is not likely to have accounted for the consistently significant differences found between those with and without farms in 1997, 1999 and 2001.

Stock collapses in rivers with farms, and the regional declines in returning adults indicate that marine survival of wild salmon is being compromised by salmon farms. Smolt survival monitored at the North Esk has averaged 8% in 1994–98 (Anon., 2001b), and has been adequate to produce sufficient adults to maintain optimum egg deposition and smolt production (MacLean et al., 2000). The relatively healthy state of smolt runs in surveyed rivers without farms indicates that a similar scenario is probably occurring in areas of the west coast where salmon farms are absent. However, the collapse of 86% of smolt runs in rivers with farms indicates that insufficient adults are returning to maintain optimum smolt output. Therefore marine survival in many of these rivers may have fallen to less than 8% during the 1990s. This is corroborated by fish trap data from Tournaig and Shieldaig, where return rates are estimated to be less than 3% (Butler & Starr, 2001).

One probable cause is sea lice epidemics triggered by salmon farms. Due to their large numbers relative to wild salmonid hosts, farm salmon produce up to 98% of sea lice larvae in western Scotland (Butler, 2002a). Localized epidemics have been recorded on west coast sea trout in 1991–92 (Sharp et al., 1994), 1994 (Mackenzie et al., 1998) and 1997–2001 (Butler et al., 2001; Butler, 2002a). A review of these data indicates that the highest infection levels occur at rivers nearest salmon farms (Fig. 9.8). Unlike sea trout smolts which are coastal [reviewed by Butler (2002a)], salmon smolts probably depart sea lochs within a few hours of leaving their natal rivers, as in Norway (Hvidsten et al., 1992; Finstad et al., 1994; Hvidsten et al., 1995), before travelling north to the Norwegian Sea (Holst et al., 2000). Therefore the locus for infection by lice larvae is likely to be confined to the salmon smolt's natal sea loch. This conclusion is endorsed by Norwegian studies, which found heavy lice infestations on wild salmon smolts leaving fjords with salmon farms (Holst & Jakobsen, 1998; Finstad et al., 2000), capable of causing the mortality of 95% of fish (Holst et al., this vol.).

A second probable cause is the influence of escaped farm salmon on the fitness of wild stocks. During the 1990s escapees contributed a mean of 9% of rod-caught

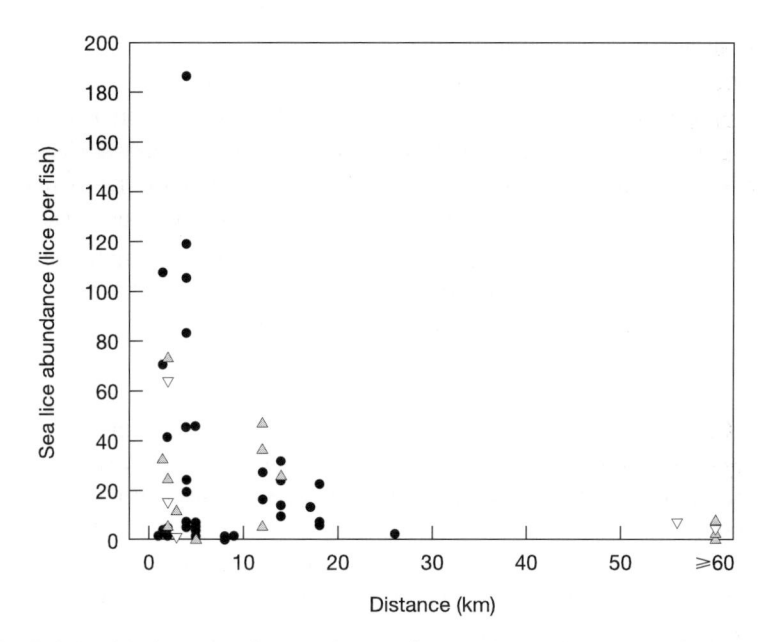

Fig. 9.8 Relationship between distance by sea from active marine salmon farms and sea lice (*Lepeophtheirus salmonis*) abundance on sea trout sampled at river mouths in 1991–2001. Symbols refer to data published by Sharp et al. (1994) (inverted triangles), Mackenzie et al. (1998) (triangles) and Butler et al. (2001) (circles).

salmon in rivers with salmon farms, but only 2% in rivers without. Such localized prevalence of escapees is supported by studies in northern Scotland, where after an escape at the mouth of the River Polla, farm salmon ascended and successfully spawned in both the first and second year after the escape (Webb et al., 1991, 1993a). While 54% of the Polla's rod catch consisted of farm fish, only a few escapees were reported in the neighbouring river systems of the Hope and Dionard, 7.5 and 29 km distant, respectively (Webb et al., 1991).

The impact of genetic introgression on wild salmon populations is complex. Juveniles of farm or hybrid parentage outcompete and displace native fry and parr, but having smolted, their marine survival and subsequent breeding success are poorer than for wild fish (McGinnity et al., 1997; Youngson et al., 1998, 2001; McGinnity et al., this vol.). Consequently, the fitness of wild populations may be eroded in a cumulative fashion, and the rate of genetic dilution will be greater for smaller, depleted stocks (Hutchings, 1991; Youngson et al., 2001). Since most west coast stocks are small, they may be inherently vulnerable to this problem. Furthermore, our data show that populations in rivers with farms are more depleted than those without, probably due to elevated marine mortality of smolts, and are thus more susceptible to genetic dilution.

The potential impact of escaped farm salmon is probably underestimated by our results, however. First, farm salmon recorded in catch records were identified by physical appearance, but anglers may have mistaken some for wild fish. For example, of 24 'wild' rod-caught fish radio-tagged in the Ewe in 2001, 8% were subsequently verified from growth patterns on scales as farm salmon which had escaped as smolts (Butler, 2002b). Second, mature escapees enter rivers later in the year when fisheries have closed and therefore their presence is underestimated by catches (Lund et al., 1991). Third, the regular presence of escapees in sea lochs with farms may attract predators such as seals, shags and cormorants, resulting in greater predation pressure on emigrating smolts and returning adults.

Managing the impact of marine salmon farms – prioritizing rivers for conservation

Due to increasing worldwide pressure on salmon stocks, initiatives have been taken to identify important populations, and to prioritize these for conservation action. In Norway, National Salmon Fjords have been proposed at the mouths of the largest salmon rivers (Anon., 1999a). In the Baltic Sea states, conservation areas have been established based on genetically defined *evolutionary significant units* (Koljonen, 2001). On the west coast of North America, stocks of Pacific salmonids have been prioritized according to their evolutionary and ecological importance (Allendorf et al., 1997; Wainwright & Waples, 1998). However, this approach has not yet been adopted by regulatory authorities in Scotland. Considering the potential impacts of marine salmon farms on wild stocks on the west coast such an exercise is timely, particularly since the UK government is obligated to conserve salmon under the Habitats and Birds Directive, and is a signatory to NASCO's precautionary principle and the Oslo Resolution.

Currently there are no statutory regulations on the location of farms relative to salmon rivers. In 1995 the West Highland Sea Trout and Salmon Group recommended that new sites should not be located within 5 km of river mouths (Anon., 1995a), but this was not enforced. Government guidelines for fish farm locations in 1999 also considered this issue, and recommended that no further finfish farm development should take place on the east and north coasts (Anon., 1999b). However, no prioritization of rivers in the salmon farming zone was undertaken, despite the establishment of cSACs on the Little Gruinard and Grimersta river systems. A new opportunity to address this issue now exists with the establishment of the Tripartite Working Group (TWG), chaired by the Scottish Executive, which seeks to develop and promote the implementation of measures for the restoration and maintenance of healthy stocks of wild and farmed fish (Anon., 2000c). The TWG focuses on improving fish health through local Area Management Agreements (AMAs) formed between farming and wild fishery interests. The major aims of AMAs are to improve sea lice control, minimize escapes and review the location of farms.

We assessed existing data on salmon rivers within the west coast salmon farming zone and subjectively ranked them to identify those populations of highest conservation value. Unlike similar exercises carried out on the Baltic and Pacific coasts, little detailed genetic information exists for west coast populations on which to formulate Evolutionary Significant Units or minimum viable populations (E. Verspoor, FRS, pers. comm.). In the absence of these data, the following criteria were used:

- *Population size*: The principle was followed that populations of 500 or more spawners are at least risk of extinction (Frankel & Soulé, 1981; Allendorf et al., 1997). Larger stocks are also likely to contain greater genetic and adaptive variation (Frankel & Soulé, 1981; Waples et al., 2001), and are more robust in absorbing the effects of genetic introgression from escaped farm salmon (Hutchings, 1991; Youngson et al., 1998, 2001). Populations of wild anadromous spawners in each river were extrapolated from the 2001 total rod catch, using the average exploitation rate of 15% found on the River Awe (Kettlewhite, 2000) (Fig. 9.9). No attempt was made to estimate numbers of mature male parr, which can contribute significantly to spawning populations (Taggart et al., 2001; Jones & Hutchings, 2002).
- *Sub-stocks*: There is considerable evidence to show that genetically distinct stock structuring occurs within rivers based on local adaption (Jordan et al., 1992, 1997; Verspoor, 1997). One example of the presence of sub-stocks is variable run-timing of anadromous adults [reviewed by Youngson et al., (2001)]. The presence of spring (March–May) and autumn (September–October) runs was taken as evidence of stock structuring, and hence conservation value.
- *Geographical coverage*: The definition of genetically distinct populations at a regional scale is problematic. However, it is likely that the greater distance between rivers, the greater the probability of adaptive differences between stocks (Verspoor, 1997). To take account of this, rivers were selected at regular intervals over a wide geographical area.

In total, 15 rivers flowing into 12 sea loch systems were identified to be of highest conservation value (Table 9.4, Fig. 9.9). Salmon farms occur in eight of the sea lochs, but not in four, namely the Kyle of Durness, Gruinard Bay, Loch Moidart and Lochs Long, Gare and Firth of Clyde. A presumption should be made against the installation of farms in these sea lochs, and in particular in Gruinard Bay where the Little Gruinard cSAC is located. In sea lochs with farms, improved management should be promoted through AMAs, including synchronized or alternative production strategies that have been shown to minimize the risk of lice infection to emigrating wild salmon smolts (Butler, 2002a). Initiatives to limit escapes of farm salmon have been formulated (Anon., 2000a) and should also be adopted.

However, in the long term the risks to wild salmon populations would be best negated by the exclusion of salmon farms in the 12 priority sea lochs. For those currently containing farms this would involve the relocation of sites out of the sea lochs, or the diversification of production into non-salmonid species. For lochs without farms, exclusion zones should be established immediately.

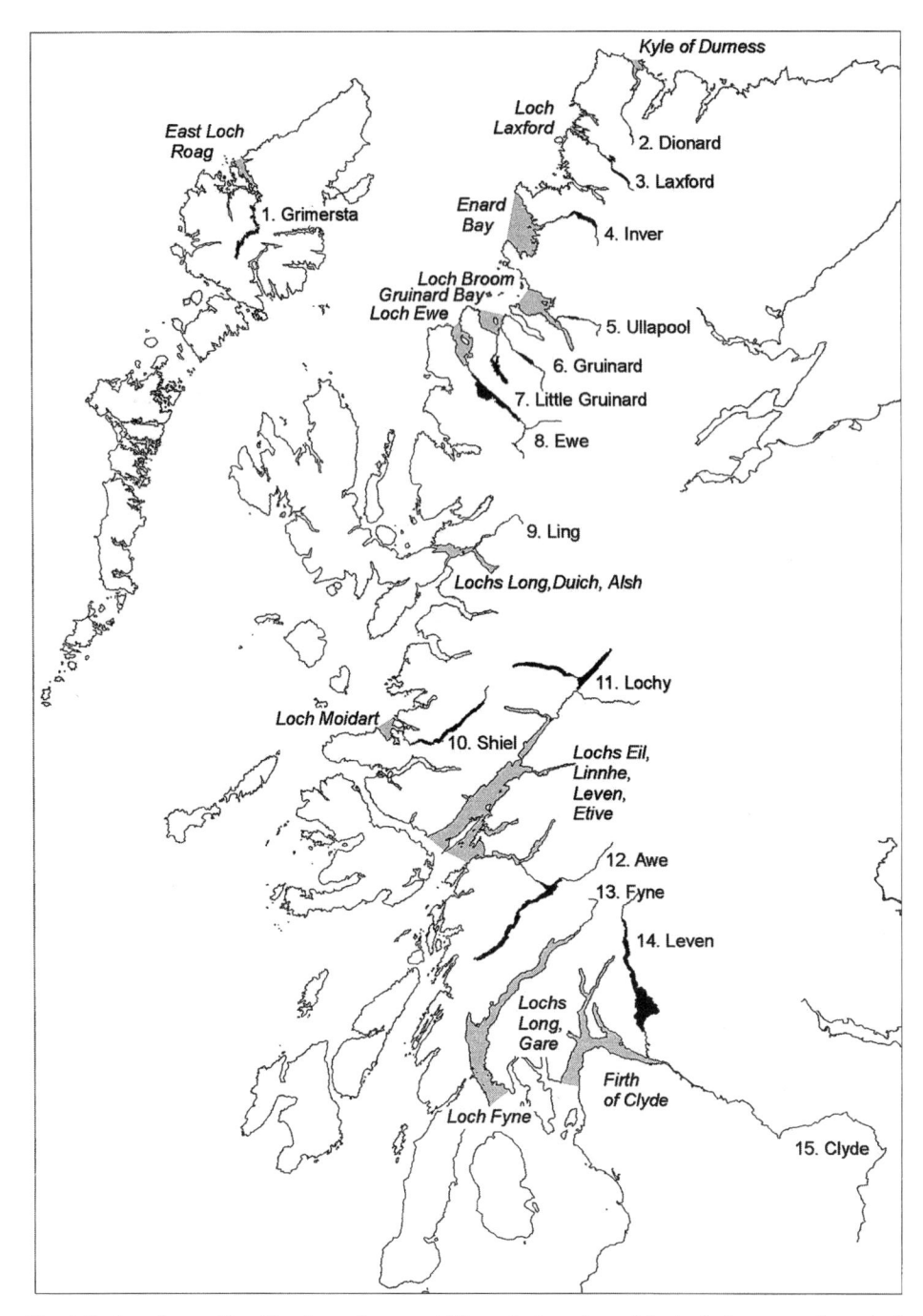

Fig. 9.9 Locations of the 15 salmon rivers and 12 sea loch systems (shaded) of conservation priority within the west coast salmon farming zone. See Table 9.4 for details.

Table 9.4 Details of the 15 rivers and 12 sea loch systems of conservation priority, listed north to south. Locations are shown in Fig. 9.9.

River	Accessible length (km)	Estimated adult population 2001	Sub-stocks	Sea loch	Salmon farms
1. Grimersta[1]	45	> 1000	Spring, autumn	East Loch Roag	Yes
2. Dionard	65	> 1000	Spring	Kyle of Durness	No
3. Laxford	22	600	Spring, autumn	Loch Laxford	Yes
4. Inver	62	500	Spring, autumn	Enard Bay	Yes
5. Ullapool	17	150	Spring	Loch Broom	Yes
6. Gruinard	38	950	Spring	Gruinard Bay	No
7. Little Gruinard[1]	25	500	Spring	Gruinard Bay	No
8. Ewe	75	750	Spring, autumn	Loch Ewe	Yes
9. Ling	13	100	Spring	Lochs Long, Alsh, Duich	Yes
10. Shiel	58	300	Spring	Loch Moidart	No
11. Lochy	130	4000	Spring, autumn	Lochs Eil, Linnhe, Leven, Etive	Yes
12. Awe	235	3000[2]	Spring	Loch Eil, Linnhe, Leven, Etive	Yes
13. Fyne	26	50	Spring	Loch Fyne	Yes
14. Leven	176	> 3000	Spring	Firth of Clyde, Lochs Long, Gare	No
15. Clyde	> 125	> 2000	Spring, autumn	Firth of Clyde, Lochs Long, Gare	No

[1] cSAC
[2] Fish count at Awe barrage (A. Kettlewhite, pers. comm.)

Acknowledgements

The work was undertaken while J.R.A.B. was Biologist for the Wester Ross Fisheries Trust. We are indebted to information and advice provided by Andy Walker, Ross Gardiner, David Hay, Maggie McKibben, Chris West, Ron Harriman, Eric Verspoor, Julian MacLean, Gordon Smith and Jackie Anderson (FRS), Ian Milne and Ross Doughty (SEPA), Peter Hutchinson (NASCO) and Dick Shelton (University of St. Andrews). Assessments would not have been possible without the assistance of the following fisheries trust biologists: Peter Cunningham and Karen Starr (Wester Ross), Mark and Helen Bilsby (Western Isles), Shona Marshall (West Sutherland) and Alan Kettlewhite (Argyll). David Nall-Cain, Michael Brady, Jeremy Inglis, Robin Bradford, Douglas Brown and the owners and angling clubs on all rivers concerned are also gratefully acknowledged. Andy Walker, Ross

Gardiner, Gordon Smith and Dick Shelton provided valuable comments on previous drafts.

References

Allendorf, F.W., Bayles, D., Bottom, D.L., et al. (1997) Prioritizing Pacific salmon stocks for conservation. *Conservation Biology*, **11**, 140–52.

Anon. (1995a) *West Highland Sea Trout and Salmon Group Report and Action Plan*. HMSO Scotland 9501114. HMSO, Edinburgh.

Anon. (1995b) *Report of the Sea Trout Working Group, 1994*. Department of the Marine, Dublin.

Anon. (1997) *Report of the Workshop on the Interactions Between Salmon Lice and Salmonids*, Edinburgh, Scotland, UK, 11–15 November 1996. ICES CM 1997/M:4, Ref.: F.

Anon. (1999a) *Report of the Norwegian Wild Salmon Committee*. PDC Tangan, Norgus Offentlige Utredninger 1999:9.

Anon. (1999b) *Locational guidelines for the authorisation of marine fish farms in Scottish waters*. Scottish Executive Rural Affairs Department, Edinburgh.

Anon. (2000a) *Report of the Working Group on Escapes of Farmed Fish*. Scottish Executive Rural Affairs Department, Edinburgh.

Anon. (2000b) *Nature conservation: implementation in Scotland of EC Directives on the conservation of natural habitats and of wild flora and fauna and the conservation of wild birds ('The Habitats and Birds Directives')*. Update of Circular no. 6/1995. Scottish Executive Rural Affairs Department, Edinburgh.

Anon. (2000c) *Tripartite Working Group, concordat and report: wild and farmed salmonids – ensuring a better future*. Scottish Executive Rural Affairs Department, Edinburgh.

Anon. (2001a) *Scottish fish farms annual production survey 2000*. Fisheries Research Services, Scottish Executive Environment and Rural Affairs Department, Edinburgh.

Anon. (2001b) *Report of the Working Group on North Atlantic Salmon*. ICES CM 2001/ACFM:15.

Birkeland, K. (1996) Consequences of premature return by sea trout (*Salmo trutta*) infested with the salmon louse (*Lepeophtheirus salmonis* Krøyer): migration, growth and mortality. *Canadian Journal of Fisheries and Aquatic Sciences*, **53**, 2808–13.

Birkeland, K. & Jakobsen, P.J. (1997) Salmon lice, *Lepeophtheirus salmonis*, infestations as a causal agent of premature return to rivers and estuaries by sea trout, *Salmo trutta*, juveniles. *Environmental Biology of Fishes*, **49**, 129–37.

Bjorn, P.A., Finstad, B. & Kristoffersen, K. (2001) Salmon lice infection of wild sea trout and Arctic char in marine and freshwaters: the effect of salmon farms. *Aquaculture Research*, **32**, 1–17.

Butler, J.R.A. (1998) *Wester Ross Fisheries Trust annual review, 1997*. Wester Ross Fisheries Trust, Gairloch, Rosshire.

Butler, J.R.A. (1999a) Escaped farm salmon on the west coast of Scotland: impacts and solutions. *The Salmon Net*, **30**, 23–6.

Butler, J.R.A. (1999b) *Wester Ross Fisheries Trust annual review, 1998–1999*. Wester Ross Fisheries Trust, Gairloch, Rosshire.

Butler, J.R.A. (2000) *Wester Ross Fisheries Trust annual review, 1999–2000*. Wester Ross Fisheries Trust, Gairloch, Rosshire.

Butler, J.R.A. (2002a) Wild salmonids and sea lice infestations on the west coast of Scotland: sources of infection and implications for the management of marine salmon farms. *Pest Management Science*, **58**, 595–608.

Butler, J.R.A. (2002b) *River Ewe fishery management plan, 2002–2006*. Wester Ross Fisheries Trust, Gairloch, Rosshire.

Butler, J.R.A. & Starr, K. (2001) *Tournaig trap report, 1999–2001*. Wester Ross Fisheries Trust, Gairloch, Rosshire.

Butler, J.R.A., Marshall, S., Watt, J., et al. (2001) *Patterns of sea lice infestations on Scottish west coast sea trout: survey results, 1997–2000*. Association of West Coast Fisheries Trusts, Gairloch, Rosshire.

Carss, D.N. (1990) Concentrations of wild and escaped fishes immediately adjacent to fish farm cages. *Aquaculture*, **90**, 29–40.

Carss, D.N. (1993a) Cormorants *Phalocrocorax carbo* at cage fish farms in Argyll, western Scotland. *Seabird*, **15**, 38–44.

Carss, D.N. (1993b) Shags *Phalocrocorax aristotelis* at cage fish farms in Argyll, western Scotland. *Bird Study*, **40**, 203–11.

Carss, D.N., Kruuk, H. & Conroy, J.W.H. (1990) Predation on adult Atlantic salmon, *Salmo salar* L., by otters, *Lutra lutra* (L.), within the River Dee system, Aberdeenshire, Scotland. *Journal of Fish Biology*, **37**, 935–44.

Cowx, I.G. & Lamarque, P. (1990) *Fishing with Electricity – Applications to Freshwater Fisheries Management*. Blackwell Scientific Publications, Oxford.

Crozier, W.W. & Kennedy, G.J.A. (1994) Application of semi-quantitative electro-fishing to juvenile salmonid stock surveys. *Journal of Fish Biology*, **45**, 159–64.

Crozier, W.W. & Kennedy, G.J.A. (1995) The relationship between a summer fry (0+) abundance index, derived from semi-quantitative electric fishing, and egg deposition of Atlantic salmon, in the River Bush, Northern Ireland. *Journal of Fish Biology*, **47**, 1055–62.

Cunningham, P.D. (2001) *The occurrence of Atlantic salmon (*Salmo salar*) carcasses along spawning streams in Scotland. 2: Shee Water (Tay system), Clunie Water (Dee system) and Kinlochewe River (Ewe system): October–December 2001.* Progress Report for the Atlantic Salmon Trust, Pitlochry, Perthshire.

Doughty, C.R. (1990) *Acidity in Scottish Rivers: a chemical and biological baseline survey*. Report prepared for the Department of the Environment (Research Contract no. PECD7/10/104). Scottish River Purification Boards/Department of the Environment, Edinburgh.

Finstad, B., Johnsen, B.O. & Hvidsten, N.A. (1994) Prevalence and mean intensity of salmon lice *Lepeophtheirus salmonis* Krøyer, infection on wild Atlantic salmon, *Salmo salar* L., postsmolts. *Aquaculture and Fisheries Management*, **25**, 761–4.

Finstad, B., Bjorn, P.A., Grimnes, A. & Hvidsten, N.A. (2000) Laboratory and field investigations of salmon lice (*Lepeophtheirus salmonis* Krøyer) infestation on Atlantic salmon (*Salmo salar* L.) post-smolts. *Aquaculture Research*, **31**, 795–803.

Frankel, O.H. & Soulé, M.E. (1981) *Conservation and Evolution*. Cambridge University Press, Cambridge.

Friedland, K.D. & Reddin, D.G. (1993) Marine survival of Atlantic salmon from indices of post-smolt growth and sea temperature. In: *Salmon in the Sea and New Enhancement Strategies* (ed. D. Mills). Fishing News Books, Oxford, pp. 119–38.

Friedland, K.D., Hansen, L.P. & Dunkley, D.A. (1998) Marine temperatures experienced by postsmolts and the survival of Atlantic salmon *Salmo salar* L. in the North Sea area. *Fisheries Oceanography*, **7**, 22–34.

Gargan, P. (2000) *The impact of the salmon louse (*Lepeophtheirus salmonis*) on wild salmonid stocks in Europe, and recommendations for effective management of sea lice on salmon farms*. Paper presented at Aquaculture and the Protection of Wild Salmon Workshop, Simon Fraser University, Vancouver, British Columbia, 1–3 March 2000.

Hansen, L.P. & Jonsson, B. (1991) The effect of timing of Atlantic salmon smolt and post-smolt release on the distribution of adult return. *Aquaculture*, **98**, 61–7.

Hawkins, A.D. (2000) Problems facing salmon in the sea – summing up. In: *The Ocean Life of Atlantic Salmon – Environmental and Biological Factors Influencing Survival*. (ed. D. Mills). Fishing News Books, Oxford, pp. 211–21.

Heuch, P.A. & Mo, T.A. (2001) A model of salmon louse production in Norway: effects of increasing salmon production and public management measures. *Diseases of Aquatic Organisms*, **45**, 145–52.

Holst, J.C. & Jakobsen, P.J. (1998) Dødelighet hos utvandrende postsmolt av laks som følge av lakselusinfeksjon. *Fiskets Gang*, **8**, 13–15.

Holst, J.C., Shelton, R., Holm, M. & Hansen, L.P. (2000) Distribution and possible migration routes of post-smolt Atlantic salmon in the north-east Atlantic. In: *The Ocean Life of Atlantic Salmon – Environmental and Biological Factors Influencing Survival* (ed. D. Mills). Fishing News Books, Oxford, pp. 65–75.

Hutchings, J.A. (1991) The threat of extinction to native populations experiencing spawning intrusions by cultured Atlantic salmon. *Aquaculture*, **98**, 119–32.

Hutchinson, P. & Mills, D. (2000) Executive summary. In: *The Oceanic Life of Atlantic Salmon – Environmental and Biological Factors Influencing Survival* (ed. D. Mills). Fishing News Books, Oxford, pp. 7–18.

Hvidsten, N.A., Johnsen, B.O. & Levings, C.D. (1992) At furd og ernaering hos utvandrende laksesmolt I Trondheimsfjorden. *NINA Oppdragsmelding*, **164**, 14.

Hvidsten, N.A., Johnsen, B.O. & Levings, C.D. (1995) Vandering og ernaering hos laksesmolt i Trondheimsfjorden og pa Frohavet. *NINA Oppdragsmelding*, **332**, 17.

Jones, M.W. & Hutchings, J.A. (2002) Individual variation in Atlantic salmon fertilization success: implications for effective population size. *Ecological Applications*, **12**, 184–93.

Jordan, W.C., Youngson, A.F., Hay, D.W. & Ferguson, A. (1992) Genetic protein variation in natural populations of Atlantic salmon (*Salmo salar*) in Scotland: temporal and spatial variation. *Canadian Journal of Fisheries and Aquatic Sciences*, **49**, 1863–72.

Jordan, W.C., Verspoor, E. & Youngson, A.F. (1997) The effect of natural selection on estimates of genetic divergence among populations of the Atlantic salmon. *Journal of Fish Biology*, **51**, 546–60.

Kettlewhite, A. (2000) *Awe Fisheries Trust annual report 1999*. Awe Fisheries Trust, Dalmally, Argyll.

Koljonen, M.-L. (2001) Conservation goals and fisheries management units for Atlantic salmon in the Baltic Sea area. *Journal of Fish Biology*, **59**, 269–88.

Laughton, R. & Bray, J. (1995) *River Aline juvenile survey 1996*. Spey Research Trust Report no. 3/95. Spey Fishery Board, Knockando, Morayshire.

Lund, R.A., Økland, F. & Hansen, L.P. (1991) Farmed Atlantic salmon in fisheries and rivers in Norway. *Aquaculture*, **98**, 143–50.

Mackenzie, K., Longshaw, M., Begg, G.S. & McVicar, A.H. (1998) Sea lice (Copepoda: Caligidae) on wild sea trout (*Salmo trutta* L.) in Scotland. *ICES Journal of Marine Science*, **55**, 151–62.

MacLean, J.C. & Walker, A.F. (2002) *The status of salmon and sea trout stocks in the west of Scotland*. Freshwater Fisheries Laboratory Report no. 01/02. Fisheries Research Services, Pitlochry, Perthshire.

MacLean, J.C., Smith, G.W. & Whyte, B.D.M. (2000) Description of growth checks observed on the scales of salmon returning to Scottish home waters in 1997. In: *The Ocean Life of Atlantic Salmon – Environmental and Biological Factors Influencing Survival* (ed. D. Mills). Fishing News Books, Oxford, pp. 37–47.

McGinnity, P., Stone, C., Taggart, J.B., et al. (1997) Genetic impact of escaped farmed Atlantic salmon (*Salmo salar* L.) on native populations: use of DNA profiling to assess freshwater performance of wild, farmed, and hybrid progeny in a natural river environment. *ICES Journal of Marine Science*, **54**, 998–1008.

McKibben, M.A. & Hay, D.W. (2002) *Shieldaig project review: January 2000–June 2001*. Fisheries Research Services Freshwater Fisheries Laboratory, Pitlochry, Perthshire.

McVicar, A.H. (1997) Interaction of pathogens in aquaculture with wild fish populations. *Bulletin of the European Association of Fish Pathology*, **17**, 197–200.

NASCO (2001) *Report on the activities of the North Atlantic Salmon Conservation Organization 2000–2001*. NASCO, Edinburgh.

Northcott, S.J. & Walker, A.F. (1996) *Farming salmon, saving sea trout; a cool look at a hot issue*. In: Proceedings of a Joint Meeting of the Scottish Association for Marine Science and the Challenger Society at Dunstaffnage Marine Laboratory, Aquaculture and Sea Lochs, June 1996, pp. 72–81.

Potter, E.C.E. & Crozier, W.W. (2000) A perspective on the marine survival of Atlantic salmon. In: *The Ocean Life of Atlantic Salmon – Environmental and Biological Factors Influencing Survival* (ed. D. Mills). Fishing News Books, Oxford, pp. 19–36.

Quick, N.J., Middlemas, S.J. & Armstrong, J.D. (2002) *The use of anti-predator controls at Scottish marine salmon farms*. Fisheries Research Services Report no. 03/02. Fisheries Research Services, Pitlochry, Perthshire.

Reddin, D.G., Helbig, J., Thomas, A., Whitehouse, B.G. & Friedland, K.D. (2000) Survival of Atlantic salmon (*Salmo salar* L.) related to marine climate change. In: *The Ocean Life of Atlantic Salmon – Environmental and Biological Factors Influencing Survival* (ed. D. Mills). Fishing News Books, Oxford, pp. 88–91.

Routledge, R.D. & Irvine, J.R. (1999) Chance fluctuations and the survival of small salmon stocks. *Canadian Journal of Fisheries and Aquatic Sciences*, **56**, 1512–19.

SFCC (1998) *A guide to the SFCC electrofishing protocol*. Scottish Fisheries Co-ordination Centre, Freshwater Fisheries Laboratory, Pitlochry, Perthshire.

SFCC (2001) *Electrofishing team leader training course manual*. Scottish Fisheries Co-ordination Centre, Freshwater Fisheries Laboratory, Pitlochry, Perthshire.

Sharp, L., Pike, A.W. & McVicar, A.H. (1994) Parameters of infection and morphometric analysis of sea lice from sea trout (*Salmo trutta* L.) in Scottish waters. In: *Parasitic Diseases of Fish* (ed A.W. Pike & J.W. Lewis). Tresaith: Samara Publishing, Dyfed, pp. 151–70.

Shearer, W.M. (1992) *The Atlantic Salmon, Natural History, Exploitation and Future Management.* Halsted Press, New York.

Taggart, J.B., McLaren, I.S., Hay, D.W., Webb, J.H. & Youngson, A.F. (2001) Spawning success in Atlantic salmon (*Salmo salar* L.): a long-term DNA profiling-based study conducted in a natural stream. *Molecular Ecology*, **10**, 1047–60.

Tully, O. & Whelan, K.F. (1993) Production of nauplii of *Lepeophtheirus salmonis* (Krøyer) (Copepoda: Caligidae) from farmed and wild salmon and its relation to the infestation of wild sea trout (*Salmo trutta* L.) off the west coast of Ireland in 1991. *Fisheries Research*, **17**, 187–200.

Tully, O., Poole, W.R. & Whelan, K.F. (1993a) Infestation parameters for *Lepeophtheirus salmonis* (Krøyer) (Copepoda: Caligidae) parasitic on sea trout, *Salmo trutta* L., off the west coast of Ireland during 1990 and 1991. *Aquaculture and Fisheries Management*, **24**, 545–55.

Tully, O., Poole, W.R., Whelan, K.F. & Merigoux, S. (1993b) Parameters and possible causes of epizootics of *Lepeophtheirus salmonis* (Krøyer) infesting sea trout (*Salmo trutta* L.) off the west coast of Ireland. In: *Pathogens of Wild and Farmed Fish: Sea Lice* (eds G.A. Boxshall & D. Defaye). Ellis Horwood, New York, pp. 202–13.

Tully, O., Gargan, P., Poole, W.R. & Whelan, K.F. (1999) Spatial and temporal variations in the infestation of sea trout (*Salmo trutta* L.) by the caligid copepod *Lepeophtheirus salmonis* (Krøyer) in relation to sources of infection in Ireland. *Parasitology*, **119**, 41–51.

Verspoor, E. (1997) Genetic diversity among Atlantic salmon (*Salmo salar* L.) populations. *ICES Journal of Marine Science*, **54**, 965–73.

Wainwright, T.C. & Waples, R.S. (1998) Prioritizing Pacific salmon stocks for conservation: response to Allendorf et al. *Conservation Biology*, **12**, 1144–47.

Walker, A.F. (1994) Sea trout and salmon stocks in the Western Highlands. In: *Problems with Sea Trout and Salmon in the Western Highlands.* Atlantic Salmon Trust, Pitlochry, Perthshire, pp. 6–18.

Walker, A.F. & Stewart, I. (1993) *An electrofishing survey of juvenile salmon stocks in the River Gruinard system, Wester Ross (5 August–13 September 1991).* Fisheries Research Services Report no. 3/93. Freshwater Fisheries Laboratory, Pitlochry, Perthshire.

Walker, A.F., Dora, S.J., Walker, A.M. & Thorne, A.E. (1993a) *An electrofishing survey of juvenile salmonid stocks in the River Ewe system, Wester Ross (24–30 July 1992).* Fisheries Research Services Report no. 6/93. Freshwater Fisheries Laboratory, Pitlochry, Perthshire.

Walker, A.F., Dora, S.J., Walker, A.M. & Thorne, A.E. (1993b) *A survey of juvenile salmonid stocks in the Loch Shiel system, Inverness-shire (11–20 August 1992).* Fisheries Research Services Report no. 7/93. Freshwater Fisheries Laboratory, Pitlochry, Perthshire.

Walker, A.F., Henderson, P. & Hoult, A. (1993c) *Juvenile salmonid stocks in the Loch Shiel system, Inverness-shire (4–11 August 1993).* Fisheries Research Services Report no. 18/93. Freshwater Fisheries Laboratory, Pitlochry, Perthshire.

Walker, A.F., Black, A., Heggarty, D., et al. (1994) *An electrofishing, netting and habitat survey in relation to juvenile salmonid fish densities in the Loch Morar system, Inverness-shire.* Freshwater Fisheries Laboratory, Pitlochry, Perthshire.

Walker, A.F., Northcott, S.J., Macdonald, A.I.M. & Thorne, A.E. (1998) *Shieldaig sea trout project progress and plans.* Fisheries Research Services Report no. 11/98. Freshwater Fisheries Laboratory, Pitlochry, Perthshire.

Walker, A.F., Northcott, S.J., Macdonald, A.I.M. & Thorne, A.E. (1999) *Shieldaig sea trout project annual report, 1998–99.* Freshwater Fisheries Laboratory, Pitlochry, Perthshire.

Waples, R.S., Gustafson, R.G., Weitkamp, L.A., et al. (2001). Characterizing diversity in salmon from the Pacific northwest. *Journal of Fish Biology*, **59**, 1–41.

Watt, J. & Bartels, B. (1998) *Lochaber and District Fisheries Trust 2nd annual report.* Lochaber and District Fisheries Trust, Lochailort, Invernesshire.

Webb, J.H., Hay, D.W., Cunningham, P.D. & Youngson, A.F. (1991) The spawning behaviour of escaped farmed and wild adult Atlantic salmon (*Salmo salar* L.) in a northern Scottish river. *Aquaculture*, **98**, 97–110.

Webb, J.H., McLaren, I.S., Donaghy, M.J. & Youngson, A.F. (1993a) Spawning of farmed Atlantic salmon, *Salmo salar* L., in the second year after their escape. *Aquaculture and Fisheries Management*, **24**, 557–61.

Webb, J.H., Youngson, A.F., Thompson, C.E., Hay, D.W., Donaghy, M.J. & McLaren, I.S. (1993b) Spawning of escaped farmed Atlantic salmon, *Salmo salar* L., in western and northern Scottish rivers: egg deposition by females. *Aquaculture and Fisheries Management*, **24**, 663–70.

Whelan, K.F. (1993) Decline of sea trout in the west of Ireland: an indication of forthcoming marine problems for salmon? In: *Salmon in the Sea and New Enhancement Strategies* (ed. D. Mills). Fishing News Books, Oxford, pp. 171–83.

Youngson, A.F., Webb, J.H., MacLean, J.C. & Whyte, B.M. (1997) Frequency of occurrence of reared salmon in Scottish salmon fisheries. *ICES Journal of Marine Science*, **54**, 1216–20.

Youngson, A.F., Hansen, L.P. & Windsor, M.L. (1998) *Interactions between salmon culture and wild stocks of Atlantic salmon: the scientific and management issues*. In: Proceedings Symposium organized by the International Council for the Exploration of the Sea (ICES) and North Atlantic Salmon Conservation Organization (NASCO), Bath, 18–22 April 1997. NINA, Trondheim, Norway.

Youngson, A.F., Dosdat, A., Saroglia, M. & Jordan, W.C. (2001) Genetic interactions between marine finfish species in European aquaculture and wild conspecifics. *Journal of Applied Ichthyology*, **17**, 153–62.

Chapter 10
Relationship Between Sea Lice Infestation, Sea Lice Production and Sea Trout Survival in Ireland, 1992–2001

P.G. Gargan[1], O. Tully[2] and W.R. Poole[3]

[1] Central Fisheries Board, Mobhi Road, Glasnevin, Dublin 9, Ireland
[2] Zoology Department, Trinity College, Dublin 2, Ireland
[3] Marine Institute, Newport, Co. Mayo, Ireland

Abstract: The relationship between sea lice infestation on sea trout with distance to salmon aquaculture sites for a broad geographic range of Irish rivers was examined over a 10-year period. Highest mean levels of total lice and juvenile (chalimus stages) lice were recorded at sites less than 20 km from farms. The mean total lice infestation was lower at sites less than 30 km from farms, and beyond 30 km, very low mean total lice levels were recorded. Chalimus lice stages dominated the sea lice population structure at distances of < 20 and < 30 km. At distances < 60 and < 100 km, chalimus and post-chalimus stages are equally represented and at sites > 100 km post-chalimus stages predominate. A model was fitted to pooled 10-year data time series for sea lice infestation and distance from marine salmon farms to indicate an overall relationship that could be used to support management actions. The average abundance of lice per fish expected very close to farms (1 km) was 50.6. Regression of log-transformed data for individual years showed significant relationships in all years except 1994 and 1999, although substantial variation existed in the data particularly close to farms. Infestations at distances greater than 25 km never reached over 32 lice per fish and were usually much lower. At distances less than 25 km the full range in infestation occurred.

Sea trout have been shown to experience physiological problems and osmoregulatory imbalance at lice levels of approximately 0.7 lice larva/g fish weight. The overall mean size of trout in the present study carrying lice was 79 g, giving an indicative stress level of sea lice infestation of 55 lice/fish. Twenty nine percent of the infested trout had lice levels above this indicative stress level. For fish sampled in bays without farms, 3.4% of the infested trout were above this indicative stress level, while for fish captured in bays with farms this level rose to 30.8%. There was a relationship between the proportion of fish in each sample above 55 lice per fish and distance from salmon farms. There was a significant negative relationship between sea trout marine survival and the level of lice infestation on sea trout in four bays in mid-west Ireland.

A linear model of the relationship between the total number of ovigerous lice produced in two bays between March and mid-May and the average number of sea lice infesting sea trout in nearby rivers showed a significant positive relationship between lice reproductive potential and infestation of trout.

Due to programme timing this contribution had to be presented as a Poster Paper.

The relationships shown in the present study indicate that sea lice from marine salmon farms were a major contributory factor in the sea trout stock collapses observed in aquaculture areas in western Ireland. If recovery of depleted sea trout stocks is to be achieved in this area it is critical to ensure that ovigerous sea lice levels are maintained at near-zero levels on marine salmon farms over the spring period prior to and during sea trout smolt migration. This must be achieved on a consistent annual basis for a successful sea trout recovery.

Introduction

Heavy infestations of the sea louse *Lepeophtheirus salmonis* (Krøyer) on sea trout have been recorded in salmon aquaculture areas in Ireland, Scotland and Norway (Anon., 1997). These sea lice epizootics have followed the development of marine salmon aquaculture in all three countries (Butler, 2002). Currently Ireland produces about 29 000 tonnes of farmed salmon annually, while the Scottish and Norwegian annual production is considerably greater at 120 000 and 400 000 tonnes respectively. In Ireland, a sea trout stock collapse occurred in salmon aquaculture bays in the mid-western area over the 1989/90 period (Anon., 1995). The Connemara district rod catch, which constitutes a large proportion of the mid-western region, fell from an average of 9570 sea trout over the 1974–88 period to 646 sea trout in 1989 and 240 sea trout in 1990. The Connemara sea trout fisheries represented an important resource catering for many visiting anglers and provided an important economic return to the region up to the mid 1980s. Similar serious declines of wild salmon and sea trout occurred in salmon farming areas on the west coast of Scotland in the early 1990s (Walker, 1994).

The history of the mid-Western sea trout stock collapse and subsequent events has been well documented (Whelan, 1993a, b; Poole et al., 1996; Gargan, 2000). Detailed information on the status of sea trout stocks was published by the Sea Trout Working Group over the period 1991–94 (Anon., 1992, 1993, 1994a, 1995) and by the Sea Trout Task Force (Anon., 1994b). In 1989, when sea trout stocks collapsed in western fisheries, sea trout were observed in the lower pools of the Delphi Fishery in Connemara in late May with heavy infestations of juvenile sea lice (*Lepeophtheirus salmonis*). These sea trout were post-smolts, which had returned prematurely from the sea after only 2 or 3 weeks and had little or no sea growth on their scales. Sampling of rivers began in 1990 to determine if this phenomenon was widespread and sea trout post-smolts and some sea trout kelts were recorded in all rivers sampled with infestations of sea lice, predominantly juvenile lice, indicating recent transmission (Tully et al., 1993). The 'sea trout problem' observed in fisheries in the mid-western region can be described as: premature return to fresh water of sea trout post-smolts with heavy sea lice infestation, lice infestation predominated by juvenile lice, a collapse in sea trout rod catches and a change in sea trout population structure (Anon., 1994b).

The scale of the Irish sea trout sampling programme was extended in 1991 and subsequent years to include rivers at varying distances from salmon aquaculture sites.

Sea lice are highly pathogenic to post-smolt salmonids when present in high numbers and significant mortality of fish can occur (Pike & Wadsworth, 2000). Because the observed infestations were high, sea lice have been implicated in the collapse or decline of sea trout populations on the west coast of Ireland. This paper examines the trends in sea lice infestation on sea trout and the relationship with distance to salmon aquaculture sites for a broad geographic range of Irish rivers over the 1992–2001 period. The presence of upstream and downstream traps on four rivers entering bays in the west of Ireland allows determination of sea trout marine survival. The relationship between sea lice infestation on post-smolt sea trout and sea trout finnock marine survival is also examined.

Sea lice (*L. salmonis*) infesting sea trout must originate from either wild or farmed salmonids. Tully & Whelan (1993) have previously shown a relationship between sea lice larval production from farms and the level of infestation that develops on wild sea trout within a bay. This paper explores the trends that exist between lice reproductive potential in two bays and infestations observed on sea trout captured in estuaries in these bays. As in most years the great majority of lice are produced from farmed salmon, examination of these trends is critical for effective management of sea lice infestation on wild fish.

Materials and methods

Sea trout sampling

Sampling sites were restricted to brackish water or freshwater sites close to estuary mouths. Fine-mesh monofilament gill nets (mesh size 38 mm) were used at the majority of sites to capture fish and 72% of the total sample was obtained by gill-netting. Gill nets were fished for a minimum of 2 hours on flooding tide. Fish were also sampled by beach seining or electrofishing or taken in upstream traps from the Invermore and Gowla Rivers, Connemara. In 1992 and 1993 some sea trout were captured by angling and hand-netting at night using strong lamps. Rivers were sampled on a weekly basis where possible from 1 May to 15 June. A target sample size of 20 fish was set for each sampling date although this was not achieved at all locations. All fish were removed from nets or handnets carefully to avoid removing loosely attached lice, placed in individual plastic bags and deep frozen. The samples were then semi-thawed and washed in alcohol to prevent any further decomposition of the lice. The length and weight of each fish was recorded, scale samples were taken and lice were identified to species, life history stage and counted under a dissection microscope.

No sampling took place at sea. Tully & Poole (1999) have shown, however, that infestations are similar at different points along an estuarine gradient in Clew Bay, Ireland. In addition, the majority of fish in the samples analysed here had low infestations. The premise that fish return prematurely to fresh water on acquiring heavy infestations may not therefore hold in all cases. Nevertheless, the adaptive advantage of such behaviour is clear. Premature return of fish to fresh water may stabilize salt and mineral balance and should increase their survival. The samples in the present case, in the light of findings in Clew Bay by Tully & Poole (1999), are therefore regarded as representative of the population of post-smolts generally.

Sea trout kelts were also observed returning to some rivers with heavy lice infestation. A size limit of less than 26 cm or below was used to define post-smolts and larger sea trout captured were generally released. The use of post-smolts further standardized the samples by restricting the size and age of the fish used in the analysis. This size limit was previously confirmed by scale-reading. A field record sheet was completed with information on the date, location, duration of fishing, number of fish retained or released and observations on water flow conditions at the time of sampling. Results from annual surveys were entered on a database each year over the 1992–2001 period. In 1997, concerns were raised as to the efficiency of the sea trout monitoring programme and the robustness of the sea trout/sea lice database (I.G. Cowx, unpublished data). In 1998, a thorough re-examination of the sea lice sampling programme and the sea trout/sea lice database was undertaken in light of the independent reports which identified discrepancies in the original data (Poole et al., 2001). Corrections were made where legitimate discrepancies were identified. The corrected data were added to the database for re-analysis. A comparison between the original database for 1992–95 and the corrected database showed the changes to have little or no effect on the results of the analyses already published (Poole et al., 2001). This corrected database was used in the present analysis. For inclusion in the analysis a 'valid fish' must fall into all of the following criteria: less than 26 cm, captured between 1 May and 15 June and a sample size of three or more valid fish. Table 10.1 shows the number of rivers sampled and the number of valid samples and valid fish.

Table 10.1 Total number of rivers sampled between 1992 and 2001, the number of valid samples ($n \geq 3$) and the number of valid fish in those samples.

Year	Total number of rivers sampled	Number of valid samples	Number of valid fish
1992	15	13	203
1993	52	29	489
1994	35	27	520
1995	40	31	457
1996	30	23	315
1997	31	25	537
1998	30	25	657
1999	30	23	563
2000	26	16	427
2001	15	12	434
Total	304	224	4602

Calculation of sea trout marine survival

Upstream and downstream traps have been operated on a number of fisheries in the west of Ireland (Burrishoole, Erriff, Gowla, Invermore) (Fig. 10.1). Trapping has taken place on the Burrishoole fishery since 1960 and on the Gowla and Invermore systems since 1991 and 1993, respectively, allowing calculation of marine survival. A smolt and kelt trap on the Tawnyard sub-catchment of the Erriff fishery provides an

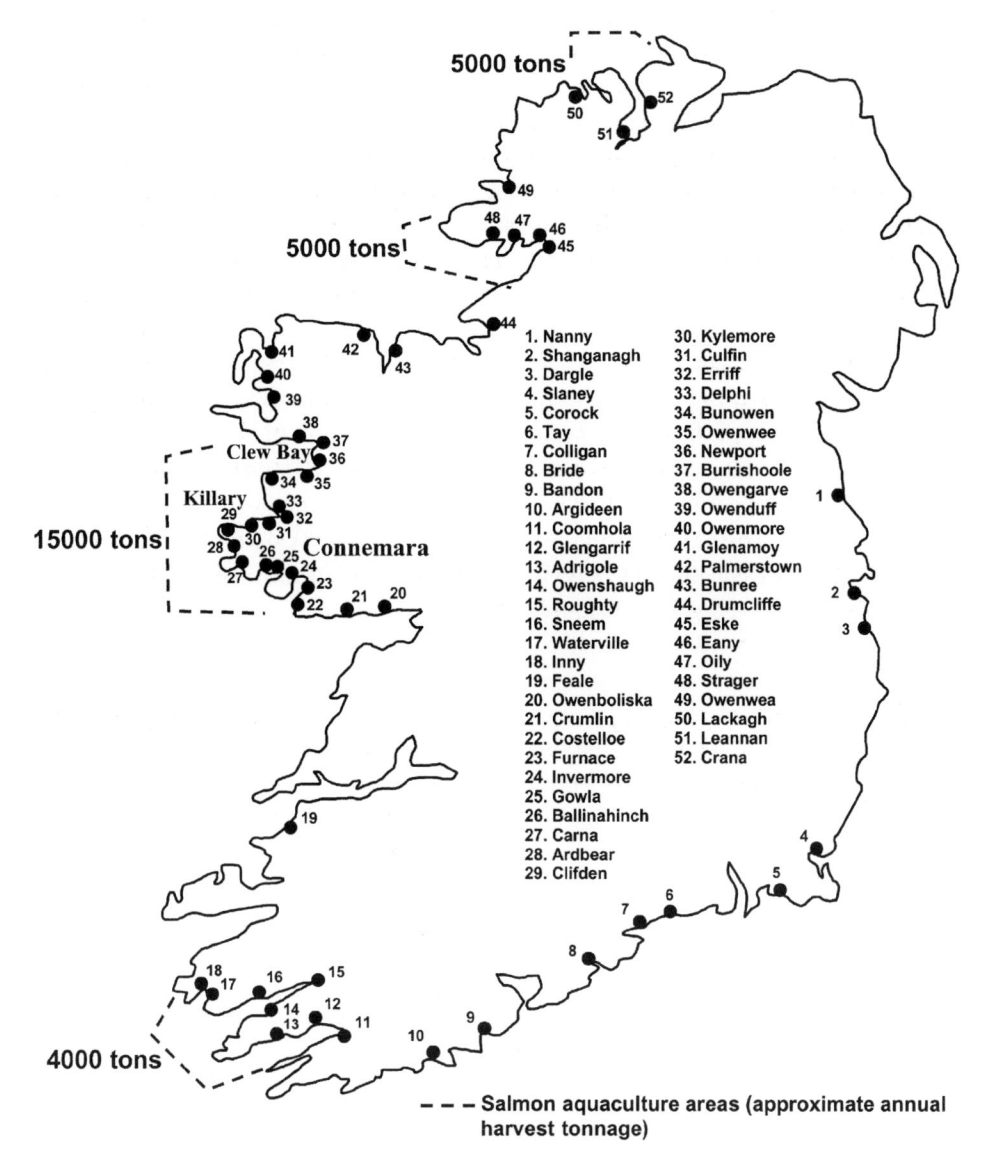

Fig. 10.1 Location of 52 rivers sampled (circles), for sea lice infestation around the Irish coast, 1992–2001.

index of marine survival as all smolts are marked and fin-clipped. Marine survival data presented here return as finnock in the same year as smolt migration. This allows the effect of lice infestation pressure on survival of sea trout post-smolts to be examined.

Calculation of sea lice production

The Irish Marine Institute has undertaken regular monthly monitoring of sea lice levels on all marine salmon farms in Ireland since 1992. Mean ovigerous and total mobile lice levels are published annually (Copley et al., 2001; McCarney et al., 2001, 2002). These data have been used to calculate mean ovigerous lice levels from salmon farms over the March to mid-May period in two bays in the west of Ireland, Killary Harbour and Clew Bay (Fig. 10.1) where long-term data on stock density are available. In order to relate mean lice levels to stock density in spring, the number of 1 and 2 sea-winter salmon present at the end of April was used. These data were obtained from records from Western Regional Fisheries Board inspections of salmon farms, data from the Department of the Marine and Natural Resources, and available records from sea lice monitoring by the Marine Institute. Mean abundance of lice on sea trout in Killary Harbour was obtained by combining data for the Erriff and Delphi Rivers, while Clew Bay data used data averaged for the Owengarve, Burrishoole, Newport, Belclare and Bunowen Rivers.

The numbers of wild salmonids present in Killary Harbour and Clew Bay up to the end of May each year was estimated by adding (1) district commercial salmon catches, (2) counter data (Erriff fishery) and trap data (Burrishoole), (3) salmon and sea trout rod catches to the end of May, and (4) numbers of sea trout in the sampling programme in rivers in each bay. Rod exploitation was estimated at 33% to the end of May and catches raised accordingly. The number of sea trout taken in the annual sampling programme in each bay to 15 June was raised by a factor of ten to estimate the potential total sea trout population present in bays. Wild salmon were sampled from the Killary Harbour draft net fishery in 1995 (Costelloe et al., 1995), providing an estimate of ovigerous lice for wild fish (2.9 lice/fish).

Reproductive potential of sea lice was defined as the total number of ovigerous lice in the bay between March and mid-May infesting both wild and farmed fish. Estimates were calculated as follows: Reproductive potential = (Number of 1 and 2 sea-winter salmon on site at the end of April × Average ovigerous lice per fish) + (Number of wild salmon present to the end of May × An estimate of ovigerous lice on wild fish).

Results

Of the 4602 'valid' fish sampled for sea lice infestation in the national survey over the period 1992–2001 from a range of 52 rivers spread geographically around the Irish coast (Fig. 10.1), 63.8% were infested with one or more lice (total stages) while 59.5%

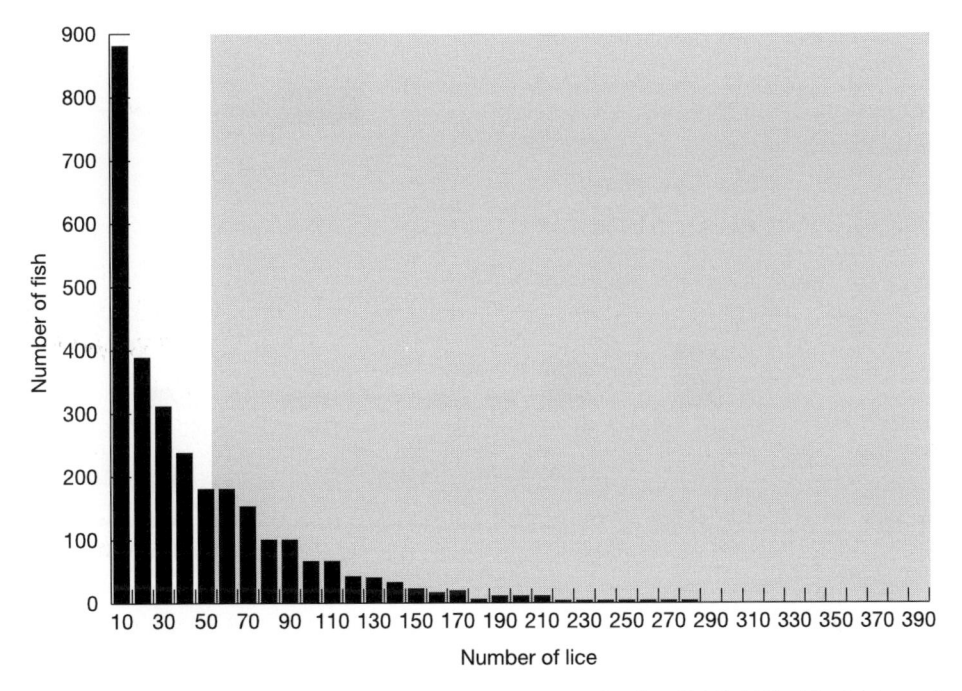

Fig. 10.2 Distribution of lice on fish sampled with one or more lice ($n = 2990$); 36% of the total sample carried zero lice. Hatched area denotes fish with 55 or more lice/fish.

were infested with only chalimus stages. Lice infestation patterns were over-dispersed, with the majority of fish carrying either none (36% of the fish sampled) or low levels of lice (Fig. 10.2).

Relationship between sea lice infestation on sea trout and sea distance to marine salmon farm

The mean lice infestation (chalimus and post-chalimus stages) on sea trout over the period, relative to distance to the nearest salmon farm, is shown in Fig. 10.3. Highest mean levels of total lice intensity (31.5 lice/fish) and chalimus (26.3 lice/fish) were recorded at sites less than 20 km from farms. The mean total lice intensity decreased (20.0 lice/fish) at sites less than 30 km from farms, and beyond 30 km, mean total lice levels were 2.4 lice/fish.

Sea lice population structure on sea trout at distances from marine salmon farms is shown in Fig. 10.4. Chalimus stages predominated at distances of < 20 and < 30 km. At distances < 60 and < 100 km, chalimus and post-chalimus stages were equally represented and at sites > 100 km post-chalimus stages predominated.

A model was fitted to 10 years of pooled data for sea lice infestation (mean abundance) and distance to the nearest marine salmon farm (Fig. 10.5) in order to indicate an overall relationship that could be used to support management actions.

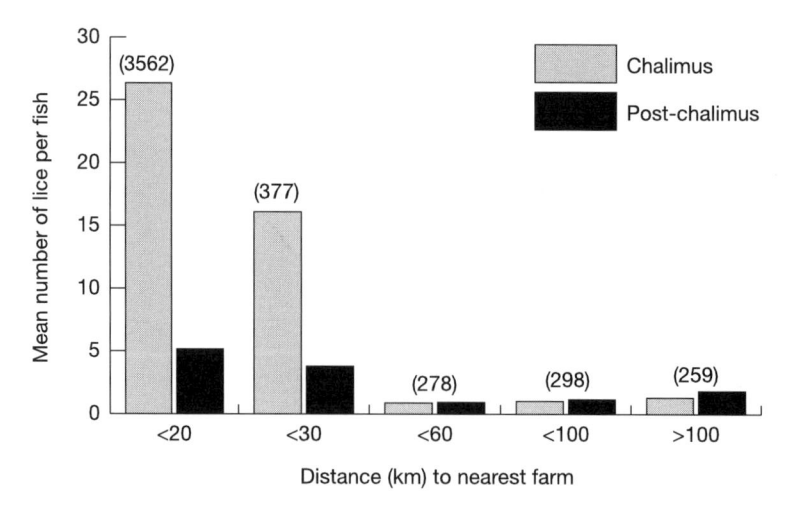

Fig. 10.3 Mean number of lice juveniles (chalimus) and adults (post-chalimus) infesting sea trout smolts in relation to distance categories to the nearest farm. Number of fish in each category is given in parentheses.

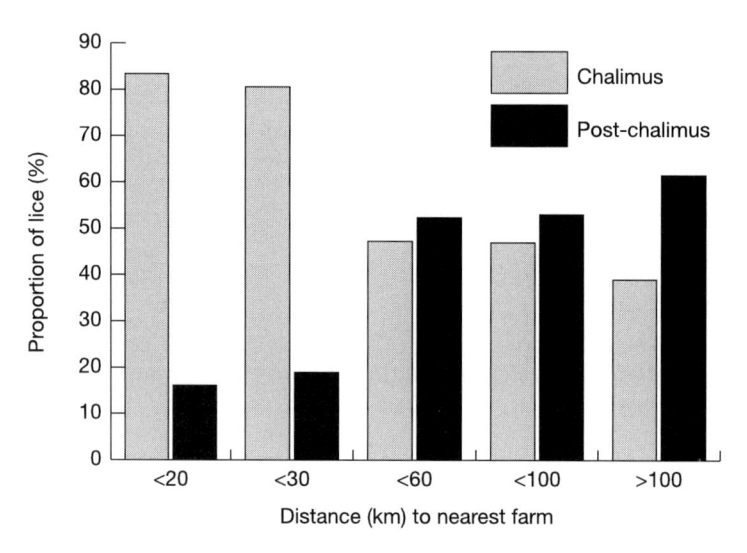

Fig. 10.4 Proportion of chalimus in the lice population infesting sea trout as a function of distance to the nearest farm.

The power function ($y = b1 \times x^{b2}$) fitted to the 10 years of data from 1992–2001 (224 data points) is asymptotic to the y-axis and indicates an intercept of 50.6, which is the average abundance of lice per fish expected close to farms (at 1 km) and, for example, 63.4 at 0.5 km from the nearest farm. Substantial variation exists in the data particularly close to farms. Analysis of the individual years using natural log-transformed linear regression (Table 10.2) showed the negative relationships between lice abundance on sea trout and distance to farms in individual years to be highly significant in

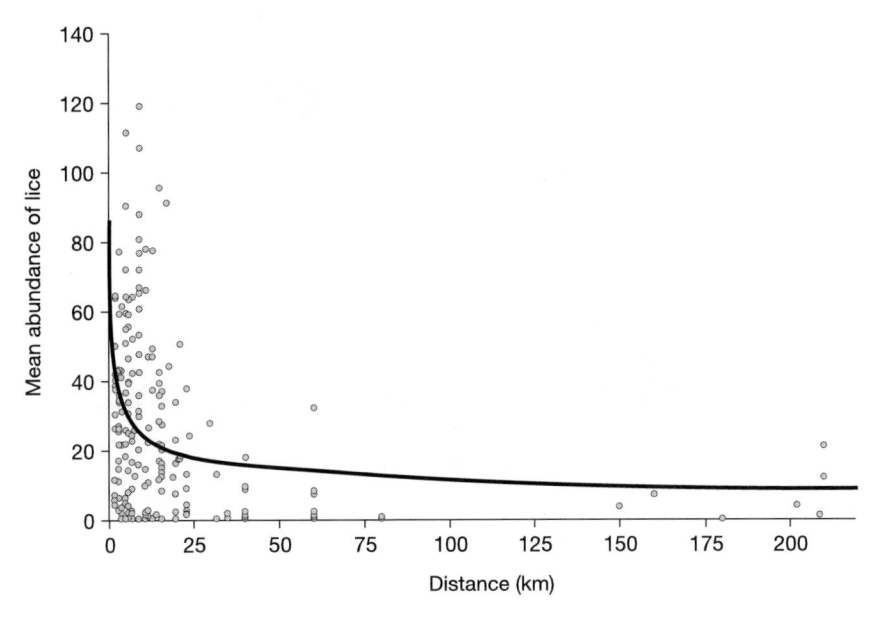

Fig. 10.5 Relationship between sea lice abundance and distance to the nearest salmon farm from 1992–2001. A power function of the form Abundance = 50.6 × Distance^ −0.3256 is fitted to the data.

Table 10.2 Regression parameters for the annual log–log relationships ($y = mx + c$) between mean abundance of lice on sea trout and distance to the nearest salmon farm, 1992–2001.

Year	Number of samples	m	c	F	R^2 (%)	P
1992	13	−1.20	6.01	16.9	60.5	< 0.005
1993	29	−0.57	4.47	14.0	34.1	< 0.001
1994	27	−0.43	2.88	2.25	8.7	= 0.146
1995	31	−0.75	4.28	17.3	37.3	< 0.0005
1996	23	−0.70	4.38	10.0	32.3	< 0.005
1997	25	−0.77	4.03	10.8	31.9	< 0.005
1998	25	−0.71	4.08	12.3	34.8	< 0.005
1999	23	−0.58	3.90	3.78	15.9	= 0.066
2000	16	−0.81	4.03	4.24	23.2	= 0.058
2001	12	−0.64	3.90	6.45	39.2	< 0.05

all years except 1994, 1999 and 2000, while the relationship was approaching significance in 1999 and 2000. The percentage variance explained by the regression was relatively low and ranged from 31.9–60.5% in years with significant relationships. A significant feature of the plot (Fig. 10.5) is the restricted range in lice infestation at distances greater than 25 km from farms. Mean abundance of sea lice on trout at distances greater than 25 km, for 55 samples, never reached over 32 lice/fish [only 14 fish in the total sample ($n = 4602$) carried more than 30 lice at > 25 km from farms]

128 Salmon at the Edge

and was usually much lower. At distances less than 25 km the full range in infestation occurred.

Sea lice infestation of sea trout – risk analysis

Sea trout have been shown to experience physiological problems and osmoregulatory disturbances at lice levels of approximately 0.7 lice larva/g fish weight (Bjorn & Finstad, 1997). The overall mean size of trout in the present study carrying lice was 79 g. Therefore, 28.9% of the trout with lice were above this indicative stress level (≥ 55 lice/fish – shown in Fig. 10.2). Splitting the samples into fish captured in bays with and without marine salmon farms showed two levels of infestation. For fish sampled in bays without farms, 3.4% of the trout with lice were above this indicative stress level (≥ 55 lice/fish), while for fish captured in bays with farms this level rose to 30.8%.

There was a relationship between the percent risk (proportion of fish in each sample above 55 lice/fish) and distance from the nearest salmon farm as demonstrated by the fitting of a power function model (Fig. 10.6). There was a decreasing risk of sea trout carrying levels of lice above that indicative of causing stress with distance from farms. The model indicated a theoretical risk of 35.4% close (1 km) to farms and at distances further than about 30 km the risk dropped to an average of $\sim 6\%$. At 60 km, two samples were above this latter figure; these were two small samples (< 6 fish) with three individuals showing more than 55 lice.

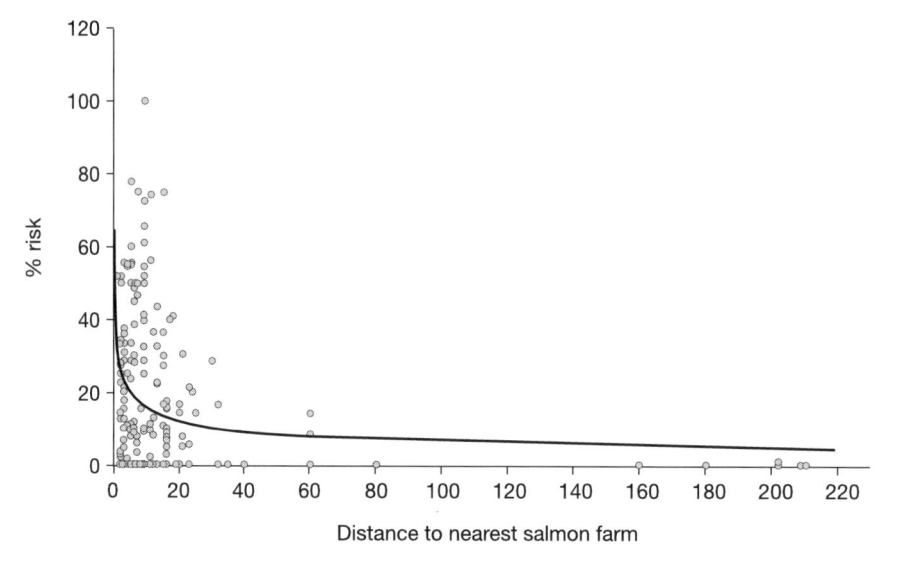

Fig. 10.6 Relationship between the percent risk [number of fish with more than 55 lice/fish/total sample (*n*)] and distance (km) to the nearest marine salmon farm. A power function of the form % Risk = $35.4 \times \text{Distance}^{-0.3670}$ is fitted to the data.

Sea lice infestation and sea trout marine survival

The relationship between sea trout marine survival and the mean abundance of lice infestation in the sea trout population in four bays over the 1992–2001 period is shown in Fig. 10.7. There was a significant negative relationship between sea trout survival and the level of lice infestation taking all four systems together ($y = 6.748 \times e^{-0.0249x}$, $P < 0.01$). Survival during the period ranged from almost 0 to over 16%. Survival was $> 9.5\%$ when lice infestation was < 18 lice/fish ($n = 6$). Survival was consistently $< 7\%$ when infestation was > 30 lice/fish ($n = 13$).

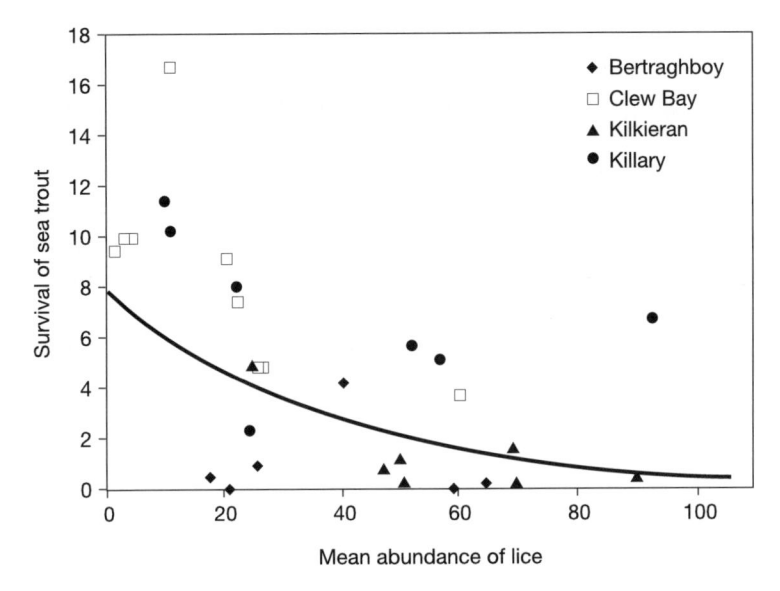

Fig. 10.7 Relationship between mean abundance of sea lice on sea trout (n/sample = > 5) and survival of sea trout smolt to return as finnock.

Relationship between lice reproductive potential in two aquaculture bays and lice infestation of sea trout

The relationships between the total number of ovigerous lice in Killary Harbour and Clew Bay between March and mid-May and the average number of sea lice infesting sea trout in both bays between 1 May and 15 June in each year from 1992–2001 is examined in Fig. 10.8. Linear models ($y = a + bx$) fitted to the data show statistically significant relationships for both bays (Killary, $P < 0.001$; Clew Bay, $P < 0.01$).

Lice from farmed fish produced $> 90\%$ of ovigerous lice in 6 of the 9 years in Killary and all of the 9 years in Clew Bay for years where data were available. Similar findings have been reported on the west coast of Scotland where 1-sea-winter farmed salmon contribute 98% of lice production (Butler, 2002). In Killary Harbour in 1992 only farmed salmon smolts were present in the bay and lice infestation on sea trout was 31.1/fish. This may be due to the model above treating bays as isolated units and

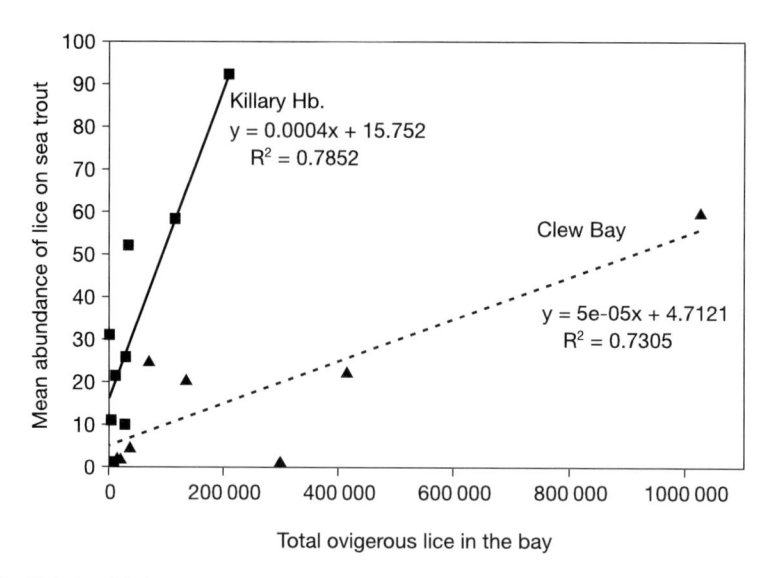

Fig. 10.8 Relationship between number of ovigerous lice (farmed and wild) in Killary Harbour and Clew Bay and mean abundance of lice on sea trout, 1992–2001.

variation around the model may be due to movement of fish out of the bays where different lice infestation may be encountered, and subsequent return to the bay, or the inclusion of sea trout not native to the bay in the sample.

Discussion

Bjorn et al., (2001) have shown very high mean levels of sea lice infestation on sea trout close to salmon farms in Norway in June. This infestation differed significantly from sites distant from salmon farms. Similar high intensities have been recorded on prematurely returned sea trout in other areas in Norway and Scotland with intensive salmon farming (Birkeland, 1996; Birkeland & Jakobsen, 1997; Butler & Watt, this vol.). Similar results were observed from the present study although the infestation levels, while highly over-dispersed, were lower than those reported at sites close to farms in Norway, most likely due to higher densities of farmed fish and greater infestation pressure. A natural background lice abundance of 8 lice/sea trout post-smolt has been calculated by Mackenzie et al. (1998). Tingley et al. (1997) found a low intensity of 5.5 lice/sea trout in south-east England consistent with low recorded lice levels at sites distant from salmon farms in the present study.

Chalimus stages predominated at sites < 30 km from farms in the present study. This was consistent over the 10-year period and also over the 6-week sampling period within each year (data not shown) and indicates recent transmission. Similar results were recorded by Bjorn et al. (2001) in Norway and Butler (2002) in Scotland. This observed dominance of juvenile lice could be due to a number of different processes; heavily infested fish may die soon after lice develop into post-chalimus. In fact the physiological impact of lice infestation increases substantially when chalimus moult to pre-adults (Bjorn & Finstad, 1997). Mortality of lice may also be density dependent and in cases where heavy infestations occur on local areas of the body surface, such as the fins, mortality may be such that few lice survive to post-chalimus stages. In sites with low infestation pressure both fish and lice would be expected to survive. The population of lice would, in this case, mature and post-chalimus stages would come to predominate, as was found at sites distant from farms.

Anon. (1995) demonstrated that there was a relationship between the magnitude of lice infestation on sea trout post-smolts and the proximity of salmon farms in Ireland during the 1992–94 period and developed the physical (dilution and dispersal) and biological (larval behaviour, longevity) rationale behind this distance model. This present study demonstrated a statistical relationship between lice infestation on sea trout and distance to the nearest salmon farm over a 10-year period, with highest infestations and variation in infestation seen close to fish farms. A similar trend has been recorded in Scottish (Mackenzie et al., 1998; Butler & Watt, this vol.) and Norwegian studies (Anon., 1997; Birkeland & Jakobsen, 1997; Bjorn et al., 2001) although the volume of data was lower and no formal statistical relationship was attempted in these studies. The Irish data show a consistency in the relationship between lice infestation and distance. There is a high level of variation in infestations in rivers close to farms. At distances greater than 25 km, however, there were only 0.3% of fish sampled ($n = 4602$) with more than 30 lice/fish. The high level of variation in lice infestation at estuaries close to salmon farms suggests that the process of dispersal of lice from farms, the actual lice production from farms and the mechanism by which these larvae are transmitted to sea trout may differ at each site. This is wholly expected given the different hydrodynamics, topography, salinity variation and behaviour of trout in these systems.

The simple regression model used in relation to distance does not take these variables into account and the relatively low percentage of variance explained by the model in some years at least is likewise not surprising. Nevertheless, the model captures a significant proportion of the variation in the data and is biologically realistic. Larvae from farms disperse from point sources and decline as a function of the volume of water into which they are dispersed. These features are incorporated into the log–log model used. Additional parameters that reflect the limited life span of the larvae would further improve the biological reality of the model. The period of infectivity of the copepodites ranges from 4–6 days and is temperature related (Boxshall, 1976; Wootten et al., 1982). The distance they can travel over this period of time will determine the radius of influence lice production from a farm can have on sea trout infestations and at some point the influence will be zero. Infestation parameters on sea trout may be related to this zone of influence coupled with trout behaviour and movement, host susceptibility or patchy distribution of the parasite in the environment (Tully, 1992). Tully et al. (1999) suggest that there is substantial retention of larvae within the bay in which they are produced by selective tidal stream transport. This idea is consistent with the current findings of localized (< 25 km) effects of larval production on fish migrating into the same system. If larvae were exported efficiently from coastal bays into offshore waters the relationships found here between lice infestation and distance would be less likely.

The feeding activity of sea lice causes mechanical damage such as skin and fin erosion (Bjorn & Finstad, 1998; Dawson et al., 1998), osmotic stress and death (Grimnes & Jakobsen, 1996; Nolan et al., 1999; Finstad et al., 2000). Prematurely returned wild sea trout are emaciated and in osmoregulatory imbalance (Tully et al., 1993). Bjorn & Finstad (1997) have shown that heavily lice-infested sea trout may experience physiological problems a long time before the fish eventually dies. They found a positive correlation between plasma chloride levels and relative densities of late chalimus larvae. This indicates that late chalimus larvae can cause osmoregulatory disturbances in heavily infested sea trout, becoming more serious after the appearance of late adult stages. However, even fish with lower infections experience physiological problems, and osmoregulatory disturbances occurred at 0.7 lice larvae/g of fish weight. The Irish data have shown that at least 28% of the trout sampled with lice had infestations above those previously shown to cause physiological problems (Bjorn & Finstad, 1997). The level of risk from lice infestation was also demonstrated to be reduced to low levels at distances greater than 30 km from salmon farms.

It is difficult to assess the impact of these infections on the whole population, as only a proportion of the entire population is sampled during the early return in the May/June period and an unknown number of fish may also die at sea. Both Anderson & Gordon (1982) and Bjorn et al. (2001) note an excessive mortality of the heaviest infested fish, suggesting that these fish die at sea. In addition, sea lice infestation may result in inhibition of growth and reproductive failure in sea trout (Bjorn et al., 2001). These fish may also have increased susceptibility to secondary infection and may be more easily predated (Bjorn & Finstad, 1997), leading to increased marine mortality. In areas with epizootics, lice have been implicated in the mortality of 32–47% of all migrating sea trout smolts (Bjorn et al., 2001) and 48–86% of wild salmon smolts (Holst & Jakobsen, 1998). It is reasonable to assume, therefore, that infections of sea lice have been a major contributor to increased marine mortality of sea trout observed since the late 1980s in the west of Ireland. Marine survival of Burrishoole sea trout historically (1971–87) ranged from 11.4–32.4% (Poole et al., 1996). Data presented in this study show marine survival to have dropped dramatically below these historical levels. Survival during the period ranged from almost 0 to over 16% and was negatively related to mean lice abundance on trout in the bays.

Previous studies in Ireland (Tully & Whelan, 1993), Scotland (Butler, 2002) and Norway (Heuch & Mo, 2001) have indicated that in spring, the majority of nauplii arise from ovigerous lice infesting farmed salmon. Tully et al. (1999) have demonstrated that the presence of salmon farms significantly increased the level of sea lice infestation on sea trout post-smolts in Ireland. Similar findings have been reported from Norway (Grimnes et al., 2000) and Scotland (Mackenzie et al., 1998; Butler, 2002). The results of the present study are consistent with these findings, that the majority of lice production arose from farmed salmon and sea lice infestation of sea trout was proportional to the level of lice production.

It is clear from the data presented that there is a strong relationship between high infestation of juvenile lice stages on sea trout and proximity to salmon farms and the patterns of infestation and infestation levels change markedly beyond about 25–30 km from salmon farms. There is also a decrease in risk of osmoregulatory imbalance and mortality from sea lice infection at distances greater than 25–30 km from farms. From these relationships we therefore conclude that sea lice from marine salmon farms were a major contributory factor in the sea trout stock collapses observed in salmon aquaculture areas in western Ireland, western Scotland and western Norway (Bjorn et al., 2001; Butler et al., 2001; this study). If recovery of depleted sea trout stocks is to be achieved in these areas, from a management perspective, it is critical to ensure that ovigerous sea lice levels are maintained at near-zero levels on marine salmon farms over the spring period prior to and during sea trout smolt migration. Given the longevity of the sea trout life cycle, this effective control of sea lice on marine salmon farms must be achieved on a consistent annual basis for a successful sea trout recovery.

Acknowledgements

We acknowledge the contribution of the staff of the Regional Fisheries Boards who carried out annual sea trout lice monitoring surveys over the period. Inspector Patrick O'Flaherty, Western Board, and his staff undertook monitoring of traps on the Connemara fisheries, while Roy Peirce and James Stafford monitored the Erriff Trap. Marine Institute staff monitored the traps on the Burrishoole fishery and also undertook river sampling. The Sea Trout Working Group of the Department of the Marine provided considerable expertise and guidance over the 1992–95 period.

References

Anderson, R.M. & Gordon, D.M. (1982) Processes influencing the distribution of parasite numbers within host populations with special emphasis on parasite influenced host mortalities. *Parasitology*, **85**, 373–98.
Anon. (1992) *Report of the Sea Trout Working Group, 1991.* Department of the Marine, Dublin.
Anon. (1993) *Report of the Sea Trout Working Group, 1992.* Department of the Marine, Dublin.
Anon. (1994a) *Report of the Sea Trout Working Group, 1993.* Department of the Marine, Dublin.
Anon. (1994b) *Report of the Sea Trout Task Force.* Department of the Marine, Dublin.
Anon. (1995) *Report of the Sea Trout Working Group, 1994.* Department of the Marine, Dublin.
Anon. (1997) *Report of the Workshop on the Interactions Between Salmon Lice and Salmonids*, Edinburgh, Scotland, 11–15 November 1996. ICES CM 1997/M:4, Ref.:F.
Birkeland, K. (1996) Consequences of premature return by sea trout, *Salmo trutta*, infested with the salmon louse, *Lepeophtheirus salmonis* (Krøyer): migration, growth, and mortality. *Canadian Journal of Fisheries and Aquatic Sciences*, **53**, 2808–13.
Birkeland, K. & Jakobsen, P.J. (1997) Salmon lice, *Lepeophtheirus salmonis*, infestation as a casual agent of premature return to rivers and estuaries by sea trout, *Salmo trutta*, juveniles. *Environmental Biology of Fishes*, **49**, 129–37.
Bjorn, P.A. & Finstad, B. (1997) The physiological effects of salmon lice infection on sea trout post smolts. *Nordic Journal of Freshwater Research*, **73**, 60–72.

Bjorn, P.A., Finstad, B. & Kristoffersen, R. (2001) Salmon lice infestation of wild sea trout and Arctic char in marine and fresh water: the effects of salmon farms. *Aquaculture Research*, **32**, 947–62.

Boxshall, G.A. (1976) The host specificity of *Lepeophtheirus pectoralis* (Muller, 1776) (Copepoda: Caligidae). *Journal of Fish Biology*, **8**, 255–64.

Butler, J.R.A. (2002) Wild salmonids and sea louse infestations on the west coast of Scotland: sources of infection and implications for the management of marine salmon farms. *Pest Management Science*, **58**, 595–608.

Butler, J.R.A., Marshall, S., Watt, J., et al. (2001) *Patterns of Sea Lice Infestation on Scottish West Coast Sea Trout: Survey Results, 1997–2000*. Association of West Coast Fisheries Trusts, Scotland.

Copley, L., McCarney, P., Jackson, D., Hassett, D., Kennedy, S. & Nulty, C. (2001) The occurrence of sea lice (*Lepeophtheirus salmonis* Krøyer) on farmed salmon in Ireland (1995 to 2000). *Marine Resource Series*, **17**, 14 pp. Marine Institute, Dublin.

Costelloe, J., Costelloe, M. & Roche, N. (1995) Variation in sea lice infestation on Atlantic salmon smolts in Killary Harbour, west coast of Ireland. *Aquaculture International*, **3**, 379–93.

Dawson, L.H.J., Pike, A.W., Houlihan, D.F. & McVicar, A.H. (1998) Effects of salmon lice *Lepeophtheirus salmonis* on sea trout *Salmo trutta* at different times after seawater transfer. *Diseases of Aquatic Organisms*, **33**, 179–86.

Finstad, B., Bjorn, P.A., Grimnes, A. & Hvidsten, N.A. (2000) Laboratory and field investigations of salmon lice (*Lepeophtheirus salmonis* Krøyer) infestations on Atlantic salmon (*Salmo salar* L.) post-smolts. *Aquaculture Research*, **31**, 795–803.

Gargan, P.G. (2000) The impact of the salmon louse (*Lepeophtheirus salmonis*) on wild salmonids in Europe and recommendations for effective management of sea lice on marine salmon farms. In: *Aquaculture and the Protection of Wild Salmon* (eds P. Gallaugher & C. Orr). Simon Fraser University, Vancouver, British Columbia, pp. 37–46.

Grimnes, A. & Jakobsen, P.J. (1996) The physiological effects of salmon lice on post smolts of Atlantic salmon. *Journal of Fish Biology*, **48**, 1179–94.

Grimnes, A., Finstad, B. & Bjorn, P.A. (2000) Registrations of salmon lice on Atlantic salmon, sea trout and charr in 1999 (in Norwegian with English Abstract). NINA Oppdragsmelding, **634**, 1–34.

Heuch, P.A. & Mo, T.A. (2001) A model of salmon louse production in Norway: effects of increasing salmon production and public management measures. *Diseases of Aquatic Organisms*, **45**, 145–52.

Holst, J.C. & Jakobsen, P.J. (1998) Dodelighet hos utvandrende postmolt av laks som folge av lakselusinfeksjon (in Norwegian). Fiskets Gang, **8**, 13–15.

Mackenzie, K., Longshaw, M., Begg, G.S. & McVicar, A.H. (1998) Sea lice (Copepoda: Caligidae) on wild sea trout (*Salmo trutta* L.) in Scotland. *ICES Journal of Marine Science*, **55**, 151–62.

McCarney, P., Copley, L., Jackson, D., Nulty, C. & Kennedy, S. (2001) National survey of the sea lice (*Lepeophtheirus salmonis* Krøyer and *Caligus elongatus* Nordmann) on fish farms in Ireland – 2000. Fisheries Leaflet, **180**, 20 pp. Marine Institute, Dublin.

McCarney, P., Copley, L., Kennedy, S., Nulty, C. & Jackson, D. (2002) National survey of the sea lice (*Lepeophtheirus salmonis* Krøyer and *Caligus elongatus* Nordmann) on fish farms in Ireland – 2001. Fisheries Leaflet, **181**, 28 pp. Marine Institute, Dublin.

Nolan, D.E.T., Reilly, P. & Wendelaar Bonga, S.E. (1999) Infection with low numbers of the sea louse *Lepeophtheirus salmonis* (Krøyer) induces stress-related effects in post-smolt Atlantic salmon (*Salmo salar* L.). *Canadian Journal of Fisheries and Aquatic Sciences*, **56**, 1–14.

Pike, A.W. & Wadsworth, S.L. (2000) Sea lice on salmonids: their biology and control. *Advances in Parasitology*, **44**, 233–337.

Poole, W.R., Whelan, K.F., Dillane, M.G., Cooke, D.J. & Matthews, M. (1996) The performance of sea trout, *Salmo trutta* L., stocks from the Burrishoole system, western Ireland, 1970–1994. *Fisheries Management and Ecology*, **3** (1), 73–92.

Poole, W.R., O'Maoileidigh, N., Jackson, D., Gargan, P. & Keatinge, M. (2001) *Review of sea lice monitoring and sea trout/sea lice database*. Report to the Minister for Marine and Natural Resources, Dublin, 24 pp.

Tingley, G.A., Ives, M.J. & Russell, I.C. (1997). The occurrence of lice on sea trout (*Salmo trutta* L.) captured in the sea off the East Anglian coast of England. *ICES Journal of Marine Science*, **54**, 1120–28.

Tully, O. (1992) Predicting infestation parameters and impacts of caligid copepods in wild and cultured fish populations. *Invertebrate Reproduction and Development*, **22** (1–3), 91–102.

Tully, O. & Poole, W.R. (1999) *Parameters and impacts of sea lice infestation of sea trout along a salinity gradient in Clew Bay, Ireland.* Report to the Atlantic Salmon Trust, Pitlochry, and the Salmon Research Agency, Newport, Ireland.

Tully, O. & Whelan, K.F. (1993) Production of nauplii of *Lepeophtheirus salmonis* (Krøyer) (copepoda: caligidae) from farmed and wild Atlantic salmon (*Salmo salar* L.) on the west coast of Ireland during 1991 and its relation to infestation levels on wild sea trout (*Salmo trutta* L.). *Fisheries Research*, **17**, 187–200.

Tully, O., Poole, W.R. & Whelan, K.F. (1993) Infestation parameters for *Lepeophtheirus salmonis* (Krøyer) (Copepoda: Caligidae) parasitic on sea trout (*Salmo trutta* L) post smolts on the west coast of Ireland during 1990 and 1991. *Aquaculture and Fisheries Management*, **24**, 545–57.

Tully, O., Gargan, P., Poole, W.R. & Whelan, K.F. (1999) Spatial and temporal variation in the infestation of sea trout (*Salmo trutta* L.) by the caligid copepod *Lepeophtheirus salmonis* (Krøyer) in relation to sources of infection in Ireland. *Parasitology*, **119**, 41–51.

Walker, A.F. (1994) Sea trout and salmon stocks in the Western Highlands. In: *Problems with Sea Trout and Salmon in the Western Highlands* (ed. R.G. Shelton). Atlantic Salmon Trust, Pitlochry, pp. 6–18.

Whelan, K.F. (1993a) Historic overview of the sea trout collapse in the west of Ireland. In: *Aquaculture in Ireland – Towards Sustainability* (ed. J. Meldon). Proceedings of Conference, Furbo, Co. Galway, 30th April–1 May 1993. An Taisce, Dublin, pp. 51–3.

Whelan, K.F. (1993b) Decline of sea trout in the west of Ireland: an indication of forthcoming marine problems for salmon? In: *Salmon in the Sea and New Enhancement Strategies* (ed. D. Mills). Fishing News Books, Oxford, pp. 171–83.

Wootten, R., Smith, J.W. & Needham, E.A. (1982). Aspects of the biology of the parasite *Lepeophtheirus salmonis* and *Caligus elongatus* on farmed salmonids, and their treatment. *Proceedings of the Royal Society of Edinburgh*, **81B**, 185–97.

Chapter 11
Mortality of Seaward-Migrating Post-Smolts of Atlantic Salmon Due to Salmon Lice Infection in Norwegian Salmon Stocks

J.C. Holst[1], P. Jakobsen[2], F. Nilsen[1], M. Holm[1], L. Asplin[1], and J. Aure[1].

[1] *Institute of Marine Research, PO Box 1870, N-5817 Bergen, Norway*
[2] *University of Bergen, Norway*

Abstract: Since the early 1990s premature returns due to heavy salmon lice infections have been observed in Norwegian sea trout stocks. Following these observations it was hypothesized that the salmon lice could cause serious problems and mortality also in seaward-migrating post-smolts of salmon. However, due to the direct migration into the high seas of this species, evidence was hard to secure. In 1998 the Institute of Marine Research, Bergen, Norway, in co-operation with the University of Bergen initiated fjordic trawl surveys aiming at estimating salmon lice infection and mortality in seaward-migrating post-smolts of western Norwegian salmon stocks. Through the development of a live catching trawl device (Holst & McDonald, 2000), the Ocean-Fish-Lift, it has been possible to secure live samples of post-smolt salmon with their scales and salmon lice infection intact. Fjordic trawling has been carried out yearly during the period 1998–2002, the numbers of fjords sampled varying somewhat between the years. Sampling of post-smolts has also taken place in the open ocean later in the season when the salmon lice have grown to their most aggressive stages and mortality due to salmon lice infection has more or less ceased. It has also been possible to catch live post-smolts infected with sea lice in the fjords and to run a controlled experiment to estimate mortal level of salmon lice infection on wild salmon post-smolts.

The mean infection levels of copepodites and chalimus stages on seaward-migrating post-smolts in western Norway have been observed to vary from 0 to 104 per fish between years and fjords. The hydrographic features of the specific fjord and year appear to be a major factor governing the infection level, with much fresh water being unfavourable for the sea lice. There have also been signs of a general decreasing trend in infection level, possibly related to the efforts by the fish farming industry to lower its mean level of salmon lice on its fish. It is, however, premature to conclude on this matter and a longer time series is necessary for a conclusion. It is also too early to evaluate the limit set on numbers of salmon lice allowed per farmed fish set by the veterinary authorities. At present there are signs that the limit at 0.5 adult females is too high in densely farmed and narrow fjords.

The controlled experiment suggested a mortal level at 11 adult salmon lice on wild post-smolts. This number is in close accordance with the oceanic observations, where none of several thousand post-smolts taken in the Norwegian Sea during a period of 10 years (Holm et al.,

2000) has been observed to carry more than 10 adult salmon lice in July, after the fish mortality has ceased. Oceanic-caught fishes with close to 10 adult salmon lice are observed to be in a bad condition with almost no growth since sea entrance and typical haemoglobin levels below 20. Many of these fishes appear to be in a state where their chances of survival are small.

Based on the observed infection levels in the fjords and a conservative mortal limit at 15 adult salmon lice, estimates varying from 0% up to 95% mortality between years and fjords of western Norwegian post-smolts of salmon due to salmon lice infection have been presented during the period 1998–2002. In general, mortality due to salmon lice infection has been observed to be a major factor regulating stock size in many western Norwegian salmon stocks in all these years except 2002 when conditions improved.

Although the reported mean number of adult females in fish farms in the studied area is down towards the allowed level at 0.5 per fish in the spring, salmonid populations still appear to be negatively affected by salmon lice in many stocks in western Norway. At some stage the present veterinary regulation on salmon lice in fish farms should be evaluated as to whether it is acceptable for securing a sustainable development in local salmon and sea trout stocks. As it seems unrealistic that the salmon lice levels in the fish farms will be further lowered in the near future, additional measures for critically affected rivers could be to treat the running wild smolts with a protective chemical against salmon lice infection or to tow wild-caught smolts out through the salmon lice belt into open waters. Such experiments are currently carried out on the Norwegian west coast.

References

Holm, M., Holst, J.C. & Hansen, L.P. (2000) Spatial and temporal distribution of post-smolts of Atlantic salmon (*Salmo salar* L.) in the Norwegian Sea and adjacent waters. *ICES Journal of Marine Science*, **57**, 955–64.

Holst, J.C. & McDonald, A. (2000) FISH-LIFT: a device for sampling live fish with trawls. *Fisheries Research*, **48**, 87–91.

Chapter 12
A Two-Generation Experiment Comparing the Fitness and Life History Traits of Native, Ranched, Non-Native, Farmed and 'Hybrid' Atlantic Salmon Under Natural Conditions

P. McGinnity[1], A. Ferguson[2], N. Baker[2], D. Cotter[1], T. Cross[3], D. Cooke[1], R. Hynes[2], B. O'Hea[1], N. O'Maoiléidigh[1], P. Prodöhl[2], and G. Rogan[1]

[1] *Salmon Management Services Division, Marine Institute, Newport, Co. Mayo, Ireland*
[2] *School of Biology and Biochemistry, Queen's University, Belfast, BT7 1NN, Northern Ireland*
[3] *National University of Ireland, Department of Zoology and Animal Ecology, Lee Maltings, Cork, Ireland*

Abstract: Since Atlantic salmon from different rivers are genetically different, non-native introductions have the potential to change the genetic make-up and juvenile production (fitness) of the recipient wild populations. While there has been much theoretical discussion on the genetic and ecological impacts on native populations of the deliberate and inadvertent introductions of Atlantic salmon, there have been only a few empirical studies. The development of DNA profiling, involving minisatellites and later microsatellites, has enabled accurate parentage identification and opened the way to direct comparison of stocks from egg stage onwards under realistic natural conditions.

An experiment, comprising three cohorts (1993, 1994 and 1998) of Atlantic salmon was undertaken in the Burrishoole system in western Ireland. This involved multiple families of the following eight groups: native wild (WILD – all cohorts); native ranched (RANCH – 1998 only); non-native from the adjacent Owenmore River (OWEN – 1998 only); farmed (FARM – all cohorts); F_1 wild × farmed (male and female reciprocal groups) (F_1Hy – 1993 and 1994 cohorts); F_2 wild × farmed (F_2Hy – 1998 cohort); B_1 backcrosses to wild (BC_1W – 1998 cohort); and B_1 backcross to farmed (BC_1F – 1998 cohort). The aim of the experiment was to look at genetic differences, without the confusion of behavioural differences.

Survival, growth, migration and maturity characteristics were examined at each life stage and overall lifetime success was estimated. In total, 7033 parr and 1502 smolts from the experimental river together with 1385 returning adults were examined.

In a situation where a river is not at its parr carrying capacity, farmed salmon have a lifetime success equivalent to 3% relative to wild fish. In a river at carrying capacity this increases to 6%. The 'hybrids' show intermediate fitness decreasing in the rank order of: BC_1 wild;

F_1 wild × farm; BC_1 farm; F_2 hybrid (but marine stage not measured for this group); F_1 farm × wild. The Owenmore group showed an overall success of 17 and 20% in the two scenarios. Only the ranched did not show a reduction in success relative to the wild group.

Introduction

In recent years there has been widespread stocking of Atlantic salmon rivers using juvenile salmon derived from non-native broodstock. Stocking has also been undertaken using surplus juveniles from the Atlantic salmon farming industry. In addition, farmed salmon enter rivers due to escapes from smolt-rearing units and from sea cages. The increase of Atlantic salmon *Salmo salar* culture has meant that large-scale escapes are now inevitable and occur frequently. For example, farmed salmon escapes in Scotland in 2000 totalled 491 980 fish (Intrafish, 2000), five times the total wild catch, with escapees potentially outnumbering wild adult salmon returning to many Scottish rivers. In Norway, escapes generally exceed one million fish per year, which is more than four times the natural spawning runs. Overall about one-third of salmon entering rivers are escaped fish with over 80% in some rivers being escapes (NASCO, 2000).

Since Atlantic salmon from different rivers are genetically different (Youngson et al., 2003), these non-native introductions have the potential to change the genetic make-up and juvenile production (fitness) of the recipient wild populations. In addition to genetic differences as a result of their different geographical origins, farmed fish are often genetically distinct from those of the native populations in the rivers that they enter as a result of directional and inadvertent selection, hybridization and genetic drift (Skaala et al., 1990). Introductions may also have ecological effects due to competition and density-dependent mortality. While there has been much theoretical discussion on the genetic and ecological impacts on native populations of the deliberate and inadvertent introductions of Atlantic salmon, there have only been a few empirical studies (McGinnity et al., 1997; Fleming et al., 2000). Such studies of an anadromous fish with a 4+ year life cycle are expensive to carry out and require a river with total juvenile and adult trapping capabilities. In addition, it is necessary to undertake studies for two generations since, in many organisms, first and second generation hybrids often survive and perform quite differently. Thus F_1 hybrids often show intermediate or even enhanced performance compared with their parents (hybrid vigour) with F_2 hybrids showing reduced performance relative to parents (hybrid breakdown or outbreeding depression).

The only currently available way to determine whether adaptive genetic differences are present between different groups of fish is to carry out 'common garden' experiments. That is, fish are reared from egg to adult in a common environmental situation. With the exception of any maternal physiological effects mediated through the egg, any differences found in performance in this common environment are thus a reflection of genetic differences. Up until about 10 years ago, with a few exceptions where suitable allozyme markers were available, it was not possible to do this as it was not possible to identify different groups established at the egg stage. Separate rearing of each group was required to a size where the fish were large enough to be physically tagged. While such physical tagging may be suitable for examining the marine phase of the life cycle, comparison at the early, high-mortality stage is impossible. The development of DNA profiling, involving first minisatellites and later microsatellites, has enabled accurate parentage identification and opened the way to direct comparison of stocks from egg stage onwards under realistic natural conditions (Ferguson et al., 1995).

Materials and methods

The experiment, comprising three cohorts (1993, 1994, 1998) of Atlantic salmon, was undertaken in the Burrishoole system in western Ireland. This system consists of a freshwater lake, connected to the sea by two channels with permanent smolt and adult trapping facilities, and a number of afferent rivers. One of these afferent rivers was equipped with a juvenile and adult trap and was used for the experiment, which involved multiple families of the following eight groups: native wild (WILD – all cohorts); native ranched (RANCH – 98 only); non-native from the adjacent Owenmore River (OWEN – 98 only); farmed (FARM – all cohorts); F_1 wild × farmed (male and female reciprocal groups) (F_1Hy – 93 and 94 cohorts); F_2 wild × farmed (F_2 Hy – 98 cohort); B_1 backcrosses to wild (BC_1W – 98 cohort); and B_1 backcross to farmed (BC_1F – 98 cohort). As the aim of the experiment was to look at genetic differences, without the confusion of behavioural differences, which have been shown between wild and farmed salmon, eggs and milt were stripped from mature adults and artificially fertilized. Fertilized eggs were incubated to the eyed stage in the hatchery with cumulative mortalities being recorded. At the eyed stage, eggs were counted accurately, families and groups mixed and planted out in incubators (Donaghy & Verspoor, 2000) in a river from which natural spawning had been excluded. Samples of parr were electrofished from the river and all emigrant parr and smolts were captured in the downstream traps. Aliquots of each group were maintained communally in the hatchery, smolts from this communal rearing, together with additional smolts from individual tanks for the 98 cohort, were tagged with coded wire tags and released to sea (except F_2 hybrids), and subsequent returning adults sampled from the Irish coastal drift nets and in the Burrishoole system by angling and at the upstream traps. Offspring were assigned to family and group parentage by minisatellite or microsatellite DNA profiling. Survival, growth, migration and maturity characteristics were examined at each stage and overall life-time success was estimated. In all, 7033 parr and 1502 smolts from the experimental river together with 1385 returning adults were examined.

Results and discussion

The highest mortality in the fertilization to eyed stage occurred in the F_2 hybrid group (66%) and this was significantly higher than all other groups. The backcrosses, which used the aliquots of the same eggs as the F_2 hybrids, did not show significantly higher mortality than the wild group, demonstrating that the high mortality in the F_2 hybrids is not due to maternal or egg quality effects. There was also no difference in mortality between the two male groups of families produced by crossing 15 females with two males, indicating that it is not a paternal effect either. This high mortality of the F_2 hybrids probably reflects outbreeding depression.

From the electrofished samples of 0+ parr taken from each of the three cohorts in August it was found that offspring of farmed Atlantic salmon showed the fastest growth rate. Ranking of both length and weight was as follows: FARM, BC_1F, F_2Hy, F_1Hy, BC_1W, WILD. This ranking indicates that growth rate is genetically controlled. All groups, other than ranched, showed a lower relative representation compared to wild in these August 0+ samples. However, prior to the August sampling parr emigration had occurred from the experimental stream. These emigrant parr captured in the downstream trap showed a greater representation of wild parr relative to farmed and hybrid parr. This downstream migration was inversely proportional to parr size and proportional to cohort density, indicating competitive displacement of wild parr by the larger farm and hybrid fish. The Owenmore also showed a substantial early parr migration, but, as these fish were not significantly different in size from the wild and ranched, which were present in the same section of the river, this appears to have been an active migration. This active downstream migration may represent an adaptation to the Owenmore where there is good nursery habitat in the main river, but this river is unsuitable for spawning, this taking place in the small tributaries.

Smolt output (including autumn presmolts, which were largely mature males) was determined at two places. First, a direct count of smolts was possible at the experimental trap. Second, the number of smolts at sea entry, produced by parr emigrating from the experimental river, was estimated assuming that these emigrating parr had a similar survival to those remaining in the experimental river. For the 93 cohort, where emigrating parr were fin-clipped for parentage identification and tagged prior to release, such tagged fish were found to result in smolts at the sea entry traps. Clearly, survival of displaced parr would only occur in a river system below its parr carrying capacity, i.e. with unoccupied downstream juvenile habitat such as freshwater Lough Feeagh in the Burrishoole system which is known to provide such habitat. Due to the displacement of wild parr from the experimental stream, the lower representation of groups relative to wild seen at the 0+ parr stage was no longer present, with the F_2 hybrid group showing a significantly greater number of smolts than the wild. However, taking the estimated smolt output at sea-entry, all groups except ranched, F_2 hybrid and backcross to wild had significantly lower smolt output than wild. Farm salmon, which were present in all three cohorts, consistently showed a significantly lower smolt output relative to wild (34%, 34%, 55%). Thus in a river at carrying capacity, introduction of farm and hybrid juveniles will lower the wild smolt production due to displacement of wild parr by these large fish. In a river with spare juvenile habitat, overall smolt production could be increased under these conditions.

No significant differences in survival among groups were found in the hatchery communally reared controls. However, overall survival was greater than 90%. Growth differences, similar to those in the experimental river, were found. The control groups serve to show that all groups were potentially equally viable and that the differences found in the experimental river were the result of genetic differences among groups.

Adult fish returned from sea after one and two sea winters. In the 1 SW group, all groups except the ranched and backcross to wild group showed a significantly lower return. Out of 9131 farm smolts released, only 27 returning adults were obtained (0.3%) compared with 7.9% return for wild smolts. The F_1 hybrid groups showed significantly more 2 SW returns than other groups. As egg deposition is likely to be the limiting factor in salmon recruitment, taking account of the differential egg production of 1 SW and 2 SW females, it shows that total egg production was significantly lower than wild for all groups except ranched. That is, although 2 SW salmon produce some 2.4 times the number of eggs of 1 SW fish, this increase is not sufficient to make up for the greater mortality that occurs in the 2 SW returning groups.

The product of survival at the different life history stages can be used as a quantitative measure of overall life-time survival which, by taking account of differential egg production, can be equated to fitness. In a situation where a river is not at its parr carrying capacity, farmed salmon have a life-time success equivalent to 3% relative to wild fish. In a river at carrying capacity this increases to 6%. The 'hybrids' showed intermediate fitness decreasing in the rank order of: BC_1 wild; F_1 wild × farm; BC_1 farm; F_2 hybrid (but marine stage not measured for this group); F_1 farm × wild. The Owenmore group showed an overall success of 17 and 20% in the two scenarios. Only the ranched did not show a reduction in success relative to the wild group.

The impact of deliberate and inadvertent introductions of non-native Atlantic salmon into a river will be highly dependent on the density of wild parr in the river. Thus, where the river is below carrying capacity the introduced fish may survive alongside the natives, resulting in an overall increase in smolt and adult production. The production of hybrids will lower the wild production, but in the case of farm hybrids may increase the return of 2 SW salmon and overall the fitness of the population may increase slightly in the first generation. Depending on the extent of hybridization in the second generation, fitness, however, may be reduced to below that prior to the introduction. Such hybridization will also result in a homogenization of genetic variability and the loss of population-specific genotypes, which is likely to have a detrimental effect on the potential of the species to adapt to future environmental challenges. The poor life-time survival of farmed salmon means that deliberate stocking of such fish will have no beneficial effect.

Where a river is already at carrying capacity, introductions can reduce wild smolt production and reduce fitness in the first generation even if there is an increase in the number of 2 SW fish. Deliberate introduction of farm salmon in such situations is particularly damaging due to displacement of wild fish by farm parr with subsequent poor marine survival of farm fish resulting in an overall reduction in adults. Farm escapes entering a river generally result in hybrids rather than pure farm offspring due to differential spawning behaviour of males and females (Fleming et al., 2000). Again, such hybrids will result in displacement of wild fish, but the reduction in fitness of these hybrids will lower the population fitness. Since reduction in fitness is cumulative, repeated introductions in a population on the verge of self-sustainability could result in an extinction vortex. The low success of non-native wild fish means that deliberate introductions of such fish are much more damaging than farm escapes. Thus Owenmore had a lower life-time success than most wild × farm hybrid groups. Hybridization of these introduced fish with wild fish will lower the fitness of the wild. Also, deliberate stocking involves greater numbers of fish than farm escapes and introductions are generally annual rather than periodic ones as typical of farm escapes.

Acknowledgements

This project was funded by the Marine Institute Ireland (SERV/00/SMS/014), the European Commission (AIR1-CT92-0719) and the UK Natural Environmental Research Council (GR9/4786).

References

Donaghy, M.J. & Verspoor, E. (2000) A new design of instream incubator for planting out and monitoring Atlantic salmon eggs. *North American Journal of Fisheries Management*, **20**, 521–7.

Ferguson, A., Taggart, J.B., Prodöhl, P.A., et al. (1995) The application of molecular markers to the study and conservation of fish populations, with special reference to *Salmo. Journal of Fish Biology*, **47** (Suppl. A), 103–12.

Fleming, I.A., Hindar, K., Mjolnerod, I., Jonsson, B., Balstad, T. & Lamberg, A. (2000) Lifetime success and interactions of farm salmon invading a native population. *Proceedings of the Royal Society London B*, **267**, 1517–23.

Intrafish (2000) Almost 500 000 Salmon Escapes in 2000. www.intrafish.com/article.php?articleID=11610.

McGinnity, P., Stone, C., Taggart, J.B., et al. (1997) Genetic impact of escaped farm Atlantic salmon (*Salmo salar* L.) on native populations: use of DNA profiling to assess freshwater performance of wild, farm and hybrid progeny in a natural river environment. *ICES Journal of Marine Science*, **54**, 998–1008.

NASCO (2000). Salmon Escapes in Scotland. www.asf.ca/Nasco/NASCO2000/escapes.html.

Skaala, O., Dahle, G., Jorstad, K.E. & Naevdal, G. (1990) Interactions between natural and farmed fish populations: information from genetic markers. *Journal of Fish Biology*, **36**, 449–60.

Youngson, A.F., Jordan, W.C., Verspoor, E., Cross, T. & Ferguson, A. (2003) Management of salmonid fisheries in the British Isles: towards a practical approach based on population genetics. *Fisheries Research* (in press).

Chapter 13
Finding Resolution to Farmed Salmon Issues in Eastern North America

A. Goode[1] and F. Whoriskey[2]

[1] *Atlantic Salmon Federation, Fort Ardross, Suite 308, 14 Main Street, Brunswick, Maine, USA*
[2] *Atlantic Salmon Federation, PO Box 5200, St. Andrews, New Brunswick, E5B 3S8, Canada*

Abstract: As a result of pressures in both freshwater and marine environments, the number of wild Atlantic salmon returning to North American rivers declined from 1.5 million in 1975 to 350 000 in 2000. The situation is particularly acute in the southern range near the USA–Canada border where many of the populations now number fewer than 100 adult fish.

Aquaculture, once thought to be the saving grace of declining salmon populations, is now accused of being a significant threat to the restoration of wild salmon stocks in eastern North America. The industry's exponential growth and associated growing pains are the subject of a hot public debate.

A strict regulatory approach has failed to safeguard the wild populations while clouding the future of the salmon farming industry. This air of uncertainty has led to direct talks between salmon farmers and the environmental community on common areas of interest such as fish containment and disease. This collaborative approach has yielded more progress in the past year than previously attempted regulatory solutions. At the same time, outstanding issues remain. There is a need for further collaboration on issues including bay management plans and the establishment of exclusion zones near critical salmon rivers.

The ability of these groups to resolve these problems and to gain acceptance in the regulatory arena will require additional sacrifices. The salmon farmers must recognize their possible impacts on wild salmon populations while the environmental groups need to recognize the industry is not going away. Ultimately, a coherent, regulatory structure based on a model of collaborative resolution may well be the only way to safeguard and possibly restore the beleaguered Atlantic salmon stocks of Maine and southern Canada.

Status of Atlantic Salmon in North America

Historical overview

The history of the Atlantic salmon in North America has been an unhappy one, at least from the salmon's point of view. The World Wildlife Fund recently undertook a river-by-river analysis of the status of global Atlantic salmon populations (World Wildlife Fund, 2001). They concluded that of the 50 and 550 Atlantic salmon rivers in the USA and Canada respectively, salmon populations in 84% of the rivers in the

USA are extinct, and all the rest are presently in trouble. There is a clear crisis for salmon populations in the southern portion of their Canadian range. Drastic declines have been observed in salmon abundances in 32 rivers draining into the Bay of Fundy region, and for many of the Southern Uplands rivers that flow from Nova Scotia's east coast into the Atlantic Ocean (DFO, 2000, 2002).

Factors in the decline

Historical declines were clearly driven by habitat destruction. Watt (1988, 1989) concluded that dams and acid precipitation resulting from the burning of fossil fuels were the principal causes of production losses up to about 1970. It is far less clear why since 1970 salmon populations have continued to decline. Commercial fishing has been phased out in North America, but there has not been a rebound to historic salmon abundance levels. Problems in fresh water may be contributing to the problem; however, evidence for big shifts in habitat abundance or suitability since 1970 is lacking. The exception is Nova Scotia's Southern Uplands rivers, where acid rain is continuing to devastate populations (DFO, 2000). There are many other 'unofficial' hypotheses proposed. Suspects include, but are not limited to, predators, changed ocean environmental and ecological conditions that affect food supplies, reduced food supplies due to forage fish fisheries, and possible exposures to chemicals in fresh water that could affect the species; ability to survive at sea. We do not have the necessary evidence at this time to definitively refute any of these hypotheses.

Rise of the salmon farming industry

Evolution of the industry

The industry has continued to grow exponentially since its inception in the 1970s. The most recent statistics available (ICES, 2002, for 2001) suggest that world production of Atlantic salmon has grown to 961 120 metric tonnes (mt), with most of it coming from countries within the natural range of salmon in the North Atlantic area (704 171 mt). Norway produces the lion's share of this (427 000 mt). While small in comparison to world volumes, the East Coast Canada and USA industry production has also rapidly expanded (33 092 and 13 230 mt respectively for Canada and the USA) (Fig. 13.1).

Sociopolitical context

In the East Coast North American areas in which it is practised, the salmon farming industry has had immense local benefits. The annual value of production in Canada exceeds $200 million (Canadian), and in the USA is worth about $60 million (US). In a region characterized by high unemployment, it has provided stable jobs.

Salmon farming also helped to greatly reduce or eliminate interceptory commercial

Fig. 13.1 Salmon farms in Passamaquoddy Bay, New Brunswick, Canada.

fisheries. Most recently, the low prices being paid for commercially caught wild salmon off the coast of Greenland have led to negotiations to buy out that fishery.

Concerns of the Atlantic Salmon Federation

Depressed wild populations

The East Coast salmon farming industry is concentrated in the Bay of Fundy (Canada) and contiguous Cobscook Bay (USA) regions. These areas are also the places where wild salmon populations are most severely depressed. Cause–effect links between the industry and salmon declines in this region remain hypothetical. However, a number of pathways by which salmon farming could harm wild populations have been identified, and are being investigated. These are described below. We regard the salmon farming industry as a legitimate user of coastal resources, but, as with any new industry, problems arise. Our approach is to attempt to identify and implement solutions to these environmental problems so that wild populations and the salmon farming industry can peacefully coexist.

Escapes

The principal concerns about interactions between wild and farmed salmon revolve around interbreeding of escaped farmed fish with wild salmon (genetic introgression), disease transmission and ecological interactions (competition with escapees; waste loading to the environment). Interbreeding and competitive interactions can only result if the farmed fish escape to the wild.

Whoriskey & Carr (2001) reviewed press reports of recent escape events in the East Coast salmon farming industry in the period 1999–2001. They identified six events releasing from a few thousand to more than 100 000 fish in the period between 1999 and 2001. Causes of the escapes include storms, predator attacks (primarily seals), human error and vandalism. Subsequent to their escape, a number of studies have documented the entry to rivers and spawning of the farmed fish with wild Atlantic salmon (e.g., Carr et al., 1997; McGinnity et al., 1997; Fleming et al., 2000).

Genetic dilution

Due to the high degree of homing exhibited by wild salmon, each river develops its own genetically distinct, river-specific population (e.g. King et al., 2001). Farmed fish have undergone an intensive directed selection programme to remove undesirable wild traits (variable growth rates and ages at maturity, aggressiveness) and replace them with domesticated characteristics that are appropriate for the sea cage and market environments (Friars et al., 1997; Gjøen & Bentsen, 1997; Glebe, 1998). The introduction of these domesticated traits back into wild populations by spawning escaped farmed fish could reduce wild fish fitness and result in population declines. The work of Fleming et al. (2000) supported this hypothesis when they found that escaped farmed fish were only 16% as productive as wild fish when spawning in the wild.

The risk of genetic dilution is heightened by the industry's use of European strains. These foreign strains are currently allowed in the USA and prohibited in Canada, though recent work by the Atlantic Salmon Federation (ASF) and Department of Fisheries and Oceans (DFO) researchers has documented the presence of escaped farmed adult and juvenile hybrids or pure-bred European fish in the Magaguadavic River, near St. George, New Brunswick (ICES, 2002).

Disease transmission (parasites, ISA, other diseases)

The two major disease problems for the East Coast North American salmon farming industry are sea lice and infectious salmon anaemia (ISA). ISA is a viral disease that was first recorded in 1984 and subsequently has caused epidemics and severe mortality in the Norwegian, Scottish and Canadian and USA East Coast salmon farming industries. In 1999, ASF researchers found that wild fish and escaped farmed fish entering Canada's Magaguadavic River were positive for ISA.

The salmon farming industry did not 'create' new disease and parasite organisms. Clearly they came from the wild. However, the intensive culture practices that are used in salmon farming have resulted in epidemics, and both ISA and lice are transmitted through water. Wild fish are susceptible to the same diseases, except that they cannot be captured and treated as farmers can do with fish in pens. Epidemics in the salmon farming industry, especially where wild populations are depressed, pose a clear danger to wild salmon.

Genetically modified Atlantic salmon

Transgenic salmon, whose DNA has been altered using molecular biology techniques to insert genes from other species, are now being considered for authorization for use by the US Food and Drug Agency. The ASF believes additional research is needed before genetically modified fish can be approved.

Other species

Cultured haddock and halibut are already being distributed to markets, and production of these species will grow in the near future. It is not yet clear how the culture of these new species is going to be harmonized with the existing salmon industry. If a massive new expansion of biomass is put into the water without reducing salmon volumes, will wastes exceed the carrying capacity of the environment with catastrophic impacts for all, including wild species? Are these new species asymptomatic hosts of diseases that could infect salmon (or vice versa), and hence their culture in proximity is incompatible, again with potential impacts upon wild species? Careful research work needs to be started on these kinds of questions, in advance.

Regulatory oversight of salmon farming

Just 15 years ago, there was hardly a salmon farming industry to regulate. Today, however, there are over 20 million salmon being cultivated in a 100-mile stretch of the USA/Canadian coastline. This exponential growth has outpaced the ability of regulators to keep pace with the industry. The regulatory agencies cannot document the environmental impacts of a given level of production before it has shifted to a significantly higher level.

International Codes

International guidelines intended to shape the growth of aquaculture have been developed by the United Nations Food and Agriculture Organization (FAO) and by the North Atlantic Salmon Conservation Organization (NASCO, 2002).

The FAO's Code of Conduct for Responsible Fisheries (1995) specifically recommends that governments conduct advance evaluations of the effects of aquaculture development on genetic diversity and ecosystem integrity, based on the best available scientific information. In particular, efforts should be undertaken to minimize the harmful effects of introducing non-native species or genetically altered stocks used for aquaculture including culture-based fisheries into waters, especially where there is significant potential for the spread of such non-native species or genetically altered stocks into waters.

NASCO was established under the Convention for the Conservation of Salmon in the North Atlantic Ocean in 1982 with the objective of contributing to the conservation, restoration, enhancement and rational management of salmon stocks in the north Atlantic Ocean. As with other international treaties, the decisions of NASCO are non-binding.

Since 1987, NASCO has recognized that salmon aquaculture can pose a threat to the wild salmon stocks. A working group on salmonid introductions and transfers has drafted (1992) and amended (1994) protocols that are still awaiting adoption by NASCO. These protocols prohibit non-native strains and call for the establishment of exclusion zones around wild rivers.

Both NASCO and FAO have spent a great deal of time articulating and attempting to integrate the precautionary approach to wild salmon management. Both bodies believe that the absence of adequate scientific information should not be used as a reason for postponing or failing to take conservation and management measures to protect wild salmon stocks.

United States laws and regulations

Federal government

The USA has long recognized the economic potential for the aquaculture industry. The National Aquaculture Act of 1980 states that it is *'in the national policy, to encourage the development of aquaculture in the United States'*.

In the USA, an applicant for a salmon farm needs a permit from the Army Corps of Engineers (Corps) under Section 10 of the Rivers and Harbors Act. Also required are a wastewater discharge permit from the Environmental Protection Agency under the Clean Water Act (this authority will be delegated to the State of Maine in 2002) and a site lease permit from the Maine Department of Marine Resources.

Other federal agencies with regulatory oversight include the National Marine Fisheries Service (NMFS), the United States Fish and Wildlife Service (USFWS) and the US Department of Agriculture. As a result of the listing of Maine's Atlantic Salmon under the Endangered Species Act, any federal agency must consult with NMFS and USFWS prior to issuing a federal permit.

State government

At the state level, the Maine Department of Marine Resources has regulatory oversight as well as the Maine Department of Environmental Protection.

150   *Salmon at the Edge*

Canada regulations

Federal government

Canada, through its Federal Aquaculture Strategy, has also embraced aquaculture development, as the following quotes from the Strategy show:

> (p. 2) 'The Federal Aquaculture Development Strategy rests on two pillars: enabling the aquaculture industry to expand and remain competitive, and promoting the preliminary and pre-competitive new species development.'

> (p. 4) 'Aquaculture is recognized as a growing source of employment and offers the possibility for social and economic improvement in communities with limited economic alternatives.'

The federal government, through the DFO, has regulatory oversight for aquaculture. Among its objectives are providing industry and the interested public with reliable scientific basis for the conservation of marine, anadromous and freshwater fishery resources and for the sustainable development of marine aquaculture.

Provincial governments

The New Brunswick and Nova Scotia departments responsible for the salmon farming industry are primarily concerned with development of the industry and its markets. In both cases, however, there are references to the need for environmental responsibility.

Mixed results

Neither the salmon farmers nor the environmental groups are satisfied with current regulations. While new regulations are currently being contemplated, there is clearly room for clarifying and enforcing the regulations already in place. A good start would be to eliminate competing mandates to both regulate and promote the industry seen at the federal, state and provincial levels of government. These competing mandates can lead to inaction or, more typically, one side being favoured over another.

Unfortunately, both Canada and the USA have chosen not to implement resolutions to longstanding problems that have been developed at the international level. Clearly, European strains should not be allowed in North America by either the FAO's Code of Conduct for Responsible Fisheries, NASCO's North Atlantic Commission protocols on introductions and transfers or by the Precautionary Approach.

Another problem can be the lack of clear definitions for terms like 'environmental responsibility' as seen in Nova Scotia's aquaculture strategy. These terms can hinder effective implementation as they are subject to interpretation. Terms such as sustainable, practicable and significant need to be either defined or not used. Furthermore, criteria for evaluating the effectiveness of regulations should be defined.

The lack of enforcement of existing regulations is seen as another problem with current regulations. In the USA, the Corps Standard Permit Conditions for net pens states that: 'No live anadromous Atlantic salmon (*Salmo salar*) whose original source as fertilized eggs or gametes was outside of the North American continent shall be introduced or transported to marine waters within the State of Maine'. Yet the Corps continues to allow the use of European strains. This suggests that better enforcement rather than more regulations may be a part of the solution.

The regulators must do a better job of keeping up with a fast-growing industry. In the early 1990s, the US Environmental Protection Agency decided not to issue wastewater discharge permits as the industry was deemed too small to warrant concern. Since that time, the industry has grown dramatically and a recent court case determined that in fact marine net pens do require discharge permits under the Clean Water Act. This agency oversight will cost the salmon farmers millions of dollars in penalties.

Collaborative approaches

New approach

In North America, ASF's relationship with the salmon farming industry dates back more than 15 years. ASF helped in the early development of the fledgling industry as a participant in a government-financed effort to develop strains of Atlantic salmon particularly well suited for captive breeding. At that time, ASF believed the commercial fishery for wild salmon off Canada and Greenland was the chief factor in the decline of wild stocks. They saw the development of the industry as a way to eliminate the high seas fishery.

As is now well documented, the high seas Atlantic salmon fishery was closed off Canada by 1998 and largely reduced off Greenland in the early 1990s. However, the rebound in the wild stocks did not materialize. At the same time, the salmon farming industry was growing exponentially and reports were coming in from countries like Norway, Scotland and Ireland that the industry had its environmental costs.

In 1999, the ASF Board ratified an aquaculture policy based on the best scientific information to articulate a set of principles meant to protect wild stocks from any potential harm from salmon farming. The policy states that, 'ASF will work with industry, local groups and government to communicate available information, to conduct research, and to help develop strategies to minimize the risk of interactions of wild and cultured salmon'. This directive has since led ASF staff in Canada and the USA to work with salmon farmers on areas of common interest.

In 1999, USA negotiations between the federal agencies, the State of Maine and the salmon farming industry broke down over the issues of escape, marking and European strains. The negotiations were beset not only by different professional opinions, but also years of bad will and conflicting personalities. This stalemate created an opportunity for ASF to begin a dialogue with salmon farmers based on common interests rather than differences.

From these initial dialogues a working group emerged that included representatives from the three large salmon growers in Maine (Heritage Salmon, Atlantic Salmon of Maine, Stolt Sea Farm), the Maine Aquaculture Association and three environmental groups (Atlantic Salmon Federation, Conservation Law Foundation, Trout Unlimited).

The issue of containment seemed to hold the most promise for collaboration. Salmon farmers wanted to keep salmon in their pens as each fish represented a $20 bill swimming away. The wild salmon interests wanted the fish kept in the pens to prevent any interaction with the wild fish. After several meetings, the defined objective of the group was 'for all parties to work towards a predictable and stable regulatory climate for the aquaculture industry that minimizes interactions between sea run salmon and farmed salmon in the waters of the State of Maine'.

Cooperation agreement

After 8 months of discussions, a cooperation agreement was signed between the industry and the environmental groups entitled 'Framework for a Salmon Aquaculture Containment Policy in the State of Maine'. This framework outlined the development of a containment management system (CMS) based on a set of standards designed to minimize interactions between wild and farmed salmon. The use of standards allowed the industry to use its expertise and awareness of evolving technologies and management practices to meet the desired outcomes.

A CMS is a process control system built on seven principles which, when implemented together, form a logical and realistic system for minimizing the escape of farmed salmon. The seven principles are:

(1) The development of a process flow chart describing the salmon farming operation.
(2) An analysis of the process' hazards and risks (hazard analysis).
(3) A determination of the places where a fish could escape, defined as critical control points (CCP).
(4) The development of monitoring schedules and procedures for the CCPs.
(5) The establishment of predetermined corrective actions.
(6) The establishment of an independent auditing system.
(7) The establishment of a record-keeping system and procedures.

A stakeholder group of salmon farmers, state and federal regulators and the environmental community worked for 12 months on the design of a CMS for Maine. Implementation began in August 2002, with complete industry compliance expected by the spring of 2003. This new containment system will standardize for the first time the containment of farmed salmon in Maine. Every farmer will be required to abide by the standards which will be verifiable, enforceable and will require mandatory reporting of any significant known or suspected fish escape. Importantly, a database will be developed to identify trends and allow for adaptive management.

Why collaboration worked

The precedent of having groups that had been at such odds with each other just a year prior should not be under estimated. There were several factors that allowed this initiative to succeed:

- *Direct involvement of companies:* In past negotiations, the salmon farmers were represented by their trade association and legal counsel, with little progress ever achieved. The involvement of senior company management resulted in open, frank discussions, as well as a willingness to find middle ground.
- *Part of the solution:* By allowing the salmon farmers to play a central role in the development of the containment guidelines, they had commitment and ownership to the process. As a result, implementation now has a better chance of success.
- *Non-regulatory arena:* The negotiations between the companies and the environmental groups were simply discussions with no required legal or regulatory outcome clouding the ability of the parties to freely work together. While the industry was facing pressure to make progress in the face of the Endangered Species Act, they could have walked away from the discussions at any time.
- *Outside expertise/funding:* The hiring of an outside expert to facilitate the development of the CMS and a grant to pay for it were both crucial.

Adult stocking programme

A second example of collaboration is the adult stocking programme that released approximately 1500 adult Atlantic salmon in 2000 and 2001 to spawn in several Maine rivers. The adults were raised from river-specific fish stocks by the industry over a 2-year period. This project has generated good publicity for the industry and has established better working relationships between the farmers and regulators. This model of a creative partnership needs to be encouraged as there is a great deal of expertise on both sides that can benefit the wild salmon restoration effort.

Areas in need of collaboration

Bay area planning

The salmon farming industry in North America can benefit from European expertise that has over 30 years of experience dealing with disease, parasites and coastal allocation issues. The fact that the major growers in North America are owned or have operations in Europe only lends further reasoning to this argument.

Recently, environmental groups in North America have begun advocating for the concept of bay area management. This concept recognizes the fact that the coastline logically breaks down into a series of bays. These bays differ in physical size, tidal movements, salinity and a host of other parameters including different human use patterns. These differences indicate the need for bay-specific management schemes for salmon farming. For each bay, the suitability for salmon farming and the appropriate density of pens need to be considered in a comprehensive approach. The current management and regulatory approach treats each farm site on a case-by-case basis and does not allow for a holistic approach.

Coordinated local aquaculture management system (CLAMS)

The need for CLAMS in Ireland arose from growing conflict and frustration between salmon farmers and other historic users along the coastline (exactly the current situation in North America). What CLAMS does is set up a process of communication and information exchange at the local level amongst the farmers and to a lesser, but still important degree, the local community.

CLAMS provides a concise description of the bay in terms of physical characteristics, biology, history, aquaculture operations, future potential and human use. It provides the framework from which a management and development plan for aquaculture in the bay can be drawn.

In the USA, the National Marine Fisheries Service (NMFS) has been working with the Irish Marine Institute on a variety of Atlantic salmon projects including aquaculture development. Based on discussions and site visits to Ireland, NMFS would like to see bay management integrated into Maine's salmon farming industry and has recommended the potential for a joint US/Canadian effort to develop a bay area plan for shared waters (Baum, 2001).

The Maine salmon farmers have been reluctant to adopt the concept of bay area management for fear it is another attempt to curb the growth of their industry. However, the current system of reviewing one farm application at a time is not satisfactory to anyone due to its cost, lengthy review time and, often, bitter debate. The public (citizens groups, environmental groups, other commercial fisheries) feels left out of the process and is forced to use legislative, legal and media arenas to be heard.

The many benefits to bay area planning and management include:

- *Less conflict*: The development of a bay plan with broad stakeholder input will lead to greater acceptance of the salmon farming industry. It will provide a forum for the exchange of ideas and expertise. In the end, it should lead to fewer objections when a new site is proposed in a bay and act as buffer to legal challenges.
- *Streamlined permitting*: The development of a bay-wide plan should reduce the 2-year permitting process and associated costs for the applicant.

- *Regulatory certainty*: By incorporating regulators into the development of bay plans, a farmer will know what to expect in terms of permits and oversight and conversely the regulators will have a greater degree of confidence about how a bay is to be managed.
- *Healthy environment*: The current permitting system looks primarily at the physical and biological suitability of an individual site with little regard for the cumulative impact the site might have on the whole bay. By treating the bay as one functioning ecosystem, scientific criteria can be developed to determine its carrying capacity, appropriate site location and be used to monitor biological or ecological changes.
- *Protection of rare species*: In regards to Atlantic salmon, bay area plans need to consider proximity to wild salmon rivers, migration paths and strain selection. By incorporating wild salmon interests into the development of the plans, wild salmon can be safeguarded and salmon farmers can be assured of less conflict. Currently, wild salmon interests have no assurance on the location or number of pens that will be granted in any of the bays of North America.

Exclusion zones

From a wild salmon perspective, there are areas where salmon farming is appropriate and places where it is not. The North Atlantic Salmon Conservation Organization draft protocols on introductions and transfers recommend salmon farms be sited at least 30 km from the mouths of pristine salmon rivers. Ireland's CLAMS programme calls for excluding pens from segments of bays to protect rare species and other significant natural resources. Even a conservative interpretation of the Precautionary Approach would favour separating salmon farms from the wild stocks where feasible.

In European countries with a longer history of salmon farming, the use of exclusion zones and coastal planning has been recognized as a legitimate and reasonable conservation strategy for protecting wild salmon populations. Iceland has a number of bays and fjords where rearing of fertile salmon is banned. This is an effort to protect Iceland's most prolific wild salmon runs. A government-appointed committee in Norway has recommended the establishment of national salmon rivers and fjords to protect the largest and healthiest salmon stocks. In November 1999, Scotland instituted restrictions on expansion of salmon farming on the north and east coasts where its most abundant wild salmon runs still exist. In Sweden there are about 70 closed areas enforced at river mouths mainly for the protection of wild Atlantic salmon and sea trout (Gudjonsson, 2001).

Exclusion zones are a form of zoning and the salmon farming industry has spoken out against zoning in general. However, zoning has worked on land and should be recognized for its ability to be a proactive measure to avoid conflict and maintain a healthy marine environment rather than the perception that it will be used to zone salmon farming out of business. There needs to be dialogue between the salmon farming industry and the wild salmon interests to explore the use of exclusion zones.

Research

The industry has been accused of being responsible for the decline in the wild stocks yet at least in North America there is little scientific evidence to prove this. Nor in many cases does the industry have any scientific data to refute the allegations. There are many potential areas of cooperative research between salmon farmers and environmental groups.

The joint DFO/ASF smolt tracking programme in the Bay of Fundy has the ability to track smolts as they migrate past the net pens. This programme can answer many questions raised about the impacts of the pens. The tracking of escaped salmon is another area of potential collaborative research between fish farmers and wild salmon interests. This research could help determine what if any interactions are occurring between wild and escaped salmon. It could also provide information on salmon behaviour that farmers could use to recapture lost stock. Joint research to answer these and other questions can benefit both the industry and wild salmon populations.

Towards resolution

The salmon farming industry is centred in the area where the crisis is deepest for North America's wild salmon. The industry may not have been the trigger for the wild salmon declines; however, it does have the potential to impact the restoration of the wild runs. As the industry has expanded so have the concerns of the wild salmon interests.

Wild salmon perspective

From a wild salmon's perspective, there is no set formula for an 'acceptable level' of impact from salmon farming. The necessary science is not available, nor is it likely to be available in the immediate future, to define what an 'acceptable' impact level is. For example, a river with a run of 5000 fish, of which 5% are farmed escapees, may suffer no ecological or genetic impacts. By contrast, a river whose run is reduced to only 20 fish could be devastated by a similar percentage of escapees.

Salmon farming perspective

Regulatory uncertainty is clouding the future of salmon farming in eastern North America. Add in lawsuits, low prices and disease losses and the future is even murkier.

However, closing the industry down in regions where the salmon is in danger of extinction is unrealistic. The industry has not been shown to be cause of the crisis, and there would be serious socio-economic implications. Certainly progress has been made as evidenced by better cage designs and by examples of individual companies voluntarily moving towards higher environmental standards. Atlantic Salmon of Maine has announced plans to become ISO 14001 certified. This rigorous international certification requires that all corporate activities be considered in light of their environmental impact and steps be taken to reduce these impacts.

Shared resolution

The collaborative path to resolving environmental issues in the salmon farming industry is not the easiest route. A series of events including large escapes, the ISA disease epidemic and the expansion of the industry into new bays continue to make salmon farming a hotly debated issue in both the USA and Canada. However, a decade of little progress in resolving these issues mandates the need for a new approach.

Common ground does exist to resolve longstanding environmental disputes. Both the salmon farmers and environmental groups favour clean water, a healthy environment, abundant wild stocks and economic prosperity. The collaborative work on containment should be a model to resolve other differences between the two sides. Only through continued collaboration can the trust be built and relationships established that could allow for more difficult issues to be addressed.

For collaboration to succeed, all sides need credible representatives. Industry associations as well as some environmental groups have not proven effective in resolving environmental conflicts. Just as important is to have credible individuals at the table, as it only takes one person to roll back hard-achieved progress.

The strict regulatory approach to solving problems has not been effective in adequately protecting or restoring wild salmon populations. This 'top down' approach has often been the regulators' only option as the industry has refused to acknowledge the environmental concerns of the resource agencies in charge of the wild stocks. This attitude has begun to change with the direct involvement of company management in negotiations.

For their part, the regulatory agencies need to be more creative in finding resolutions. The use of adaptive management schemes will encourage continual environmental improvement. The use of regulatory standards can give the industry needed flexibility in achieving desired environmental outcomes. Also needed are predetermined criteria for monitoring and evaluating regulations so they can be amended as necessary.

Ultimately, salmon farmers need to participate in the development of regulations that govern their business. By helping to define solutions, the industry will spend less time mounting lobbying and legal challenges to defeat them. This joint development will also ensure that new regulations are actually followed which in the long run will benefit Atlantic salmon most of all.

References

Baum, E. (2001) Atlantic Salmon Unlimited 2001. *US/Ireland Cooperative Progam on Salmon Aquaculture Industry*. Report submitted to NOAA/NMFS, Gloucester, Massachusetts.

Carr, J., Anderson, J.M., Whoriskey, F.G. & Dilworth, T. (1997) The occurrence and spawning of cultured Atlantic salmon (*Salmo salar* L.) in a Canadian river. *ICES Journal of Marine Science*, **54**, 1064–73.

DFO (2000) *The effects of acid rain on the Atlantic salmon of the Southern Uplands of Nova Scotia.* DFO Maritimes Regional Habitat Status Report 2000/2E. Department of Fisheries and Oceans, Ottawa.

DFO (2002) *Atlantic salmon maritime provinces overview for 2001.* DFO Science Stock Status Report D3-14 (2002). Department of Fisheries and Oceans, Ottawa.

Fleming, I.A., Hindar, K., Mjølnerød, I.B., Jonsson, B., Balstad, T. & Lamberg, A. (2000) Lifetime success and interactions of farm salmon invading a native population. *Proceedings of the Royal Society of London,* **B 267**, 1517–23.

Friars, G.W., Bailey, J.K. & O'Flynn, F.M. (1997) A review of gains from selection in Atlantic salmon (*Salmo salar*) in the Salmon Genetics Research Program. *World Aquaculture,* **28** (4), 68–71.

Gjøen, H.M. & Bentsen, H.B. (1997) Past, present and future of genetic improvement in salmon aquaculture. *ICES Journal of Marine Science,* **54**, 1009–114.

Glebe, B. (1998) *East Coast salmon aquaculture breeding programs: history and future.* Canadian Stock Assessment Secretariat Research Document 98/157. Department of Fisheries and Oceans, Ottawa.

Gudjonsson, S. (2001) *Towards Sustainable Salmon Aquaculture – Zoning of the Icelandic Coastline – Abstract.* Institute of Freshwater Fisheries, Iceland.

ICES (2002) *Assess:14.* Report of the Working Group on North Atlantic Salmon, Copenhagen, 3–13 April. ICES CM 2002/ACFM:14.

King, T.L., Kalinkowski, S.T., Schill, W.B., Spidle, A.P. & Lubinski, B.A. (2001) Population structure of Atlantic salmon (*Salmo salar* L.): a range-wide perspective from microsatellite DNA variation. *Molecular Ecology,* **10**, 807–21.

McGinnity, P., Stone, C., Taggart, J.B., et al. (1997) Genetic impact of escaped farmed Atlantic salmon (*Salmo salar* L.) on native populations: use of DNA profiling to assess freshwater performance of wild, farmed, and hybrid progeny in a natural river environment. *ICES Journal of Marine Science,* **54**, 998–1008.

NASCO (2002) *Manual of Resolutions, Agreements and Guidelines.* Sections 1–8. NASCO, Edinburgh.

Watt, W.D. (1988) Major causes and implications of Atlantic salmon habitat losses. In: *Present and Future Atlantic Salmon Management* (ed. R.H. Stroud). Atlantic Salmon Federation (Brunswick, Maine) and National Coalition for Marine Conservation, pp. 101–11.

Watt, W.D. (1989) The impact of habitat damage on Atlantic salmon (*Salmo salar*) catches. In: Proceedings of the National Workshop on Effects of Habitat Alteration on Salmonid Stocks (eds C.D. Levings, L.B. Holtby & M.A. Henderson). *Canadian Special Publication of Fisheries and Aquatic Science,* **105**, 154–63.

Whoriskey, F.G. & Carr, J.W. (2001) Interactions of escaped farmed salmon, and wild salmon in the Bay of Fundy region. In: *Marine Aquaculture and the Environment: A Meeting for Stakeholders in the Northeast* (eds M. Tlusty, D.A. Bengston, H.O. Halvorson, S.D. Oktay, J.B. Pearce & R.B. Rheault Jr.). Cape Cod Press, Falmouth, Massachusetts, pp. 141–9.

World Wildlife Fund (2001) *The Status of Wild Atlantic Salmon: A River by River Approach.* WWF, Washington, DC, 172 pp.

Chapter 14
Delivering the Solutions – The Salmon Farmer's Point of View

Lord Lindsay and G. Rae

Scottish Quality Salmon, Durn, Isla Road, Perth, PH2 7HG, UK

Abstract: The Scottish salmon industry delivers strategic socio-economic solutions around rural Scotland in terms of jobs, communities and local services. It also delivers strategic national economic solutions in terms of primary production, added and retail value and exports. Internationally, it is contributing a solution to the 'Fish Gap'. The prospect of these benefits continuing into the future appears strong. However, the industry is also delivering solutions in terms of our coastline and inshore environment. Scottish Quality Salmon (SQS) has been unique in its proactive and measurable commitment to high-quality, responsible and sustainable husbandry and site management. Independently inspected and certified compliance with externally scrutinized and accredited standards based on best practice and internationally recognized disciplines is mandatory, as is compliance with SQS Codes of Practice and wider responsibilities including collaboration with other national and international stakeholders.

The most potent attribute in terms of site-specific concerns is SQS's mandatory commitment to Environmental Management Systems (EMS) that are independently audited and accredited to ISO 14001. These EMS are based on individual site analysis and monitoring, incorporate periodic reviews of performance against targets, provide a formal vehicle for SQS Codes of Practice, and focus on risk management and the precautionary principle. SQS's EMS regime is facilitating the development of a number of important collaborative programmes such as the Tripartite Initiative.

Against a background of long-term declines in wild salmon stocks throughout their range and their multi-factorial nature, the Tripartite Working Group (TWG) was established in June 1999 with government, the salmon farming industry and wild salmonid interests. A Concordat was signed in June 2000 and sets out terms of reference including measures for the restoration and maintenance of healthy stocks of wild and farmed fish, and measures for the regeneration of wild salmon and sea trout stocks. As a founder member of the TWG, SQS and its members have led the way in the establishment of Area Management Groups and Area Management Agreements designed to contribute to the above objectives. The following text describes progress to date and discusses some of the problems encountered and solutions delivered.

Introduction

The Scottish farmed salmon industry faces many challenges to maintain employment in the Highlands and Islands of Scotland while ensuring that the aspirations of all associated stakeholders are respected. One of the principal stakeholders is the wild salmon industry. In the last few years, a greater understanding and working relationship has developed between these two vital industries, creating solutions to mutual problems and clearly demonstrating that a well-managed marine resource can bring benefits to the people of Scotland while sustaining clean healthy waters which are an essential part of our heritage and the key to future prosperity.

This paper outlines the importance of salmon farming to the people and economy of Scotland; the industry perspective on the specific challenge of establishing sustainable populations of wild salmon in Scottish rivers; solutions already in place to some of these challenges; and finally a consideration of new challenges and solutions being developed for the future.

The salmon farming industry – strategic rural solutions

An independent study commissioned by Highlands and Islands Enterprise (HIE) and Scottish Office: Agriculture, Environment and Fisheries Department (SOAEFD) in 1998 established that the Scottish salmon industry supports around 6500 Scottish jobs. In wages alone, this now equates to a contribution to local families and communities of some £2 million per week – or well over £100 million per year. Add in rents paid to local interests and the figure rises further. Add in the support this generates for local interests such as shops, post offices, petrol stations, businesses and schools and the socio-economic contribution takes a dramatic leap. Indeed, in some areas there are a frightening number of local businesses and services that remain viable solely because of the salmon industry.

Analysis also demonstrates that the jobs being created locally by the industry suit a mixture of age and skills levels, and – of obvious social importance – are especially suited both to young people and to those who have additional interests such as crofting. The industry is therefore helping to retain population and to sustain traditional patterns of economic activity as well as the social infrastructure.

The distribution of these wider benefits is critical. Many of the jobs and opportunities, and much of the wider support, are being generated in areas that are remote, fragile and vulnerable to the decline of more traditional rural industries. Such benefits are of undoubted significance to the Highlands and Islands area; but it should be emphasized that they also extend to the Borders, the south-west, the central belt and the north-east. In the north east, for instance, jobs in the sea-fish processing sector are suffering from Scotland's reducing sea-fish industry. Now, however, the Scottish salmon industry is providing the processing sector with a lifeline, sustaining existing jobs and creating new ones.

Underpinning and enabling this strategic rural solution is a success story for the rural economy. Peter Peacock, when Convenor of the Highland Council, described the Scottish salmon industry as being 'the most important economic development in the Highlands and Islands in the last 30 years'. He was not exaggerating. The Scottish salmon output within the Highlands and Islands area is now worth more than beef and lamb put together. Nationally, the Scottish salmon industry is now delivering primary products valued ex-farm at £300 million per annum. Their value to the

Scottish food industry and our economy does not stop there. With added value, the initial £300 million of produce is worth up to £900 million per annum at the point of retail.

Indeed, the most startling economic achievement of the Scottish salmon industry is that it now accounts for nearly 40% by value of all Scotland's food exports despite its relatively small and dispersed footprint.

If these achievements represent strategic social and economic solutions for rural Scotland, where rural Scotland is both a substantial provider and a substantial beneficiary, what are the prospects for these solutions to continue into the future?

In terms of social and economic factors – two of the three components that comprise sustainability – the fundamental indicators are clear. Valuable social benefits will derive if the Scottish industry is economically viable. Whilst this is not the occasion to analyse the range of crucial issues that will determine future economic viability, such as international competitiveness, there are some key underlying pointers: aquaculture is the world's fastest growing food sector; salmon is probably the most popular, versatile, nutritious and healthy of aquaculture products; and with salmon, Scotland has a proven track record of successfully managing its attributes, advantages and ability to produce and deliver the right product to the right market.

The 'Fish Gap' also provides one further key indicator of future viability and usefully serves to introduce the marine environmental agenda. The 'Fish Gap' anticipates a vital role for aquaculture – whilst consumer demand for fish products is projected to rise by another 30 million tonnes by 2010, 70% of the world's commercial sea-fish stocks are today already either overexploited, fully exploited, recovering from overexploitation or in a state of collapse.

Salmon farming therefore appears to offer Scotland, and in particular rural Scotland, the prospect of continuing social and economic solutions. It is also able to contribute directly and significantly to a marine environmental solution of huge proportions. The fact that 25% by value of all fish products bought by consumers in the UK is farmed salmon and trout represents a substantial easing of the pressure on wild stocks and marine biodiversity.

However, an analysis of the industry's ability to deliver solutions now and in the future rightly requires a more focused assessment of the coastal and inshore environment. SQS has been unique in its proactive commitment to addressing the relevant issues.

Salmon farming and wild fish populations

Much debate centres on the possible association of salmon farming with the continuing decline in wild fish numbers. Proponents on both sides of the argument will no doubt maintain this debate into the future. It is, of course, very clear that wild fish declines have been in evidence for much of the last century – long before salmon farming. It is equally clear that the problem is not confined to the west of Scotland where salmon farming takes place. Other speakers at this symposium will no doubt give us important insights into the many considerable barriers facing our wild salmon including: global climate change; pollution; riverbed erosion; flash floods; commercial fisheries; predators; parasites; and, recently highlighted, the massive by-catch by Norwegian and Russian trawlers of an estimated 900 000 salmon smolts per annum. No wonder the 'King of Fish' has problems.

The challenge for all of us is to ensure that salmon farming is part of the solution by providing quality, healthy fish to demanding consumers – and so remove pressure on wild stocks which could never realistically satisfy this demand.

The salmon farming industry – strategic international solutions

We cannot lose sight of the fact that much of the solution has an international dimension. Global climate change and its marine impact comprise perhaps the most worrying challenge as it is difficult to predict the scale or time span. It is also difficult to see how human resources can achieve meaningful impacts on this issue in the short or medium term. If this causes salmon populations to drift northwards in their range, then we have the prospect of continued downward pressure in many of our rivers.

Control of pollution is another issue with a strong international dimension – but one that we can influence in the short term. The European Water Framework Directive is welcomed by salmon farmers as it takes account of whole river basins and all the discharges from industry, agriculture and forestry. Clearly this is an opportunity for wild fish interests to work with salmon farmers to ensure that this Directive achieves the desired effect.

Genetic modification has been strongly resisted by the Scottish industry for good reason – the consumer. Equally important is the concern of wild fish interests that this technology might not be contained to farmed salmon. This Scottish commitment is of no value unless the international community maintains a united stance on this issue – a continuing challenge for all of us at this symposium.

NASCO is a key player in coordinating the international solution and SQS will continue to play our part in developing and delivering policy.

The salmon farming industry – strategic Scottish solutions

Salmon farming is *already* a leading industry for Scotland, and in order to maintain and enhance that position within a strategic framework for Scotland plc., it needs, and deserves, a ministerially endorsed National Strategy to deliver its maximum potential for the benefit of Scotland. Within such a strategy, it is essential to define the benefits and opportunities for *all* stakeholders and demonstrate how this can best be delivered in a manner that *also* suitably protects the environment.

In June last year, Rhona Brankin MSP, then Deputy Minister for Environment and Rural Development, announced the decision to develop a National Strategy for Aquaculture in Scotland. This move was warmly welcomed by SQS. Ms Brankin's successor, Alan Wilson MSP, has now set up a working group comprising a wide range of interests to develop a draft strategy and we are delighted that SQS has been invited to participate.

SQS believes that the National Strategy for Aquaculture should centre on maximizing Scotland's potential for a quality-driven, sustainable salmon industry that can competitively deliver both premium marketing opportunities and high-quality brands such as Tartan Quality Mark and Label Rouge Scottish salmon. We are seeking to develop a strategy designed to:

- Deliver governance, based on that strategy, which is co-ordinated, coherent and confident.
- Commit to joint government and industry research and development (R&D) as part of a long-term strategic vision, beginning with further studies on carrying capacity.
- Offer a wider range of feasible options, sites and financial support where greater flexibility (e.g. relocation and rotation) would allow particular site issues to be addressed to the satisfaction of all relevant stakeholders.
- Recognize and reward joint stakeholder initiatives such as participation in the Tripartite Area Management Groups, work with non-government observers (NGOs) and other responsible environmental bodies.
- Support those complying with independently accredited and internationally recognized standards leading to a high-quality, sustainable route, and recognize, reward and motivate those investing in a high-quality, long-term future for the Scottish industry and its employees.
- Recognize the vital springboard that salmon farming provides for diversification into other species in terms of its existing investment, technology, R&D, market intelligence and skilled personnel.
- Promote training and educational provision that recognizes the importance of aquaculture to the rural economy.
- Promote the role that salmon farming plays in marine wealth creation in a similar way to other countries that depend on aquaculture as a major strategic industry.

Overall, it is a framework that allows us to deliver the best competitively that we most want.

Regulation

A quality-led industry would be able to compete more effectively in the international arena if underpinned by a national aquaculture strategy and regulatory framework that recognizes quality as a priority and key objective. However, SQS remains highly concerned that good practice is not helped, and is frequently hindered, by the bureaucratic burden of regulation by 10 statutory bodies, 63 pieces of legislation, 43 European Directives, 3 European Regulations and 12 Commission Decisions.

The bureaucratic burden is exacerbated by the unnecessary length of time it takes to achieve consents or permission, given the resource constraints of the public sector offices and the uniquely confusing and repetitive duplication that adds nothing to better decision-making.

In principle, voluntary Codes of Practice are quicker to implement since no legislation is required and present no burden to the taxpayer as the industry in question pays for the appropriate measures to be put in place. However, there are variable qualities of codes of practice and it is important that those that are independently inspected and accredited to recognized international standards, such as ISO 14001, should be merited with greater public sector support than those that are more cosmetic.

SQS Codes of Practice are mandatory for all members, as follows:

- Code of Practice on How to Avoid and Minimize the Impact of Infectious Salmon Anaemia (ISA).
- Code of Practice on the National Treatment Strategy for Control of Sea Lice on Scottish Salmon Farms.
- Code of Practice for Environmental Management Systems.
- Code of Practice on Containment of Farmed Fish.
- Code of Practice on Salmon Farming and Predatory Wildlife.

The lack of government recognition or incentive for industry to strive for standards higher than the legal minimum is counter-productive. Companies currently operating in increasingly competitive, global markets and gaining no public sector advantage in incurring considerable costs for independent inspections and accreditation will feel increasingly driven to the lowest cost, lowest quality route. Consequently, the strategy should consider ways to support work that SQS has established, and continues to develop, on independently accredited and internationally recognized standards (ISO 14001 and EN 45011) leading to a high-quality, sustainable route, and endorse those companies that choose to adopt this route. The lack of formal recognition of this approach within the salmon farming industry is damaging Scotland's long-term requirement for premium standards of production and environmental consideration.

User-friendly and better co-ordinated regulation would expedite matters and be more transparent to all interested parties. Examples in other parts of the world, Norway for example, demonstrate that closer co-ordination between regulatory bodies can improve the efficiency and development of the industry without compromising product safety or environmental concerns.

In fact, SQS worked with the Salmonid Fisheries Forum to produce a joint statement outlining mutually agreeable criteria for improving the regulatory system. Not only was this partnership applauded by politicians of the Scottish Parliament's Transport and Environment Committee but it is a typical example of how, between us, we can deliver the solutions.

Furthermore, SQS welcomes the approach to transfer planning controls from the Crown Estates Commission to local authorities, provided that additional or unnecessary burdens are not placed on the industry and the procedures are simplified and become more streamlined, effective and friendly for all users.

Research and carrying capacity

There is a widely accepted need for continued R&D, particularly on carrying capacity. The industry recognizes that further carrying capacity investigation is a vital and urgent requirement to inform the debate about the development of the industry, although it is important that this is carried out from a holistic viewpoint to mirror the statutory duty expected of the Scottish Environment Protection Agency (SEPA) to balance socio-economic needs and environmental consideration.

SQS supports continued collaboration and consultation between industry and government on initiating and prioritizing research projects. For example, the DEPOMOD modelling technique used by SEPA to determine bio-mass consents, PLGA oral vaccine delivery systems, the development of the ISA Code of Practice and the Containment Code of Practice are all SQS-led initiatives which demonstrate positive progress when industry and government have worked together.

Government and industry should drive R&D together as part of a long-term strategic vision as in Norway where there is a positive climate towards aquaculture supporting a large and varied R&D programme. By encouraging collaborative research, focusing technical and academic expertise on problems facing the aquaculture sector and promoting technology transfer to the industry, research programmes can accelerate, in a responsible way, the commercial development of sustainable aquaculture in Scotland, and open up the means by which aquaculture, in this country, can share in the market opportunity emerging over the next decade.

Sound, objective scientific knowledge is critical in ensuring a sustainable environmental future for the industry. In Scotland, the industry has access to academic excellence, such as the Institute of Aquaculture at the University of Stirling and the Marine Laboratory in Dunstaffnage, where research is conducted into many aspects of salmon farming which have environmental significance. In other countries, such research is considered to be a strategic fundamental contribution to securing long-term prosperity assisted by government.

Environmental sustainability is absolutely crucial to the future of the industry; SQS firmly believes that this is not an optional extra. Scotland has a successful industry because it is an ideal environment in terms of hydrography, temperature and coastal geography. We also now have an enviable skills base.

Progress since the early days of the industry has been immense. Vaccination programmes and the development and use of targeted and increasingly environmentally benign veterinary medicines have aided this progress. The use of antibiotics has declined dramatically, and the use of ad hoc sea lice treatments has been falling since 1998 when the National Treatment Strategy for Sea Lice, developed by SQS, was introduced. This is in addition to the Integrated Management Strategy pioneered by SQS including farming single-year classes of fish and fallowing.

As the leading industry body for salmon farming, SQS believes it can lead the way towards a long-term, sustainable future for Scottish aquaculture. Its commitment includes:

- Independently inspected Product Certification Schemes, accredited to EN 45011 for smolts, fresh, smoked salmon and organic salmon under United Kingdom Register of Organic Farming Standards (UKROFS) accreditation.
- Development and implementation of environmental management systems to internationally recognized standards (ISO 14001).
- Implementing mandatory Codes of Best Practice including National Treatment Strategy for Control of Sea Lice, Containment, Code of Practice on Scottish Salmon Farms and the Control of Predatory Wildlife.
- Participation in Area Management Groups and Agreements.
- Funding of research into fish health, welfare and management strategies in collaboration with internationally recognized academic institutions.
- On-going monitoring for potential threats to the industry, e.g. *Gyrodactylus salaris*, and preparing appropriate contingency plans.
- Collaboration and consultation with the Food Standards Agency, Highlands and Islands Enterprise, Crown Estate, SEPA, Veterinary Medicines Directorate, Royal Society for the Protection from Cruelty to Animals (RSPCA), Scottish Society for the Protection from Cruelty to Animals (SSPCA), wild fish interests and other NGOs and responsible environmental bodies.
- International promotional work, particularly in France where SQS members can market salmon under the prestigious Label Rouge mark.
- Representation in Holyrood, Westminster and Brussels.

SQS members already source key feed ingredients from sustainably managed fisheries and are also considering wider environmental issues such as the substitution of a proportion of plant oils or protein in fish feed. The organization is also developing close working links with the Marine Stewardship Council to address sustainability issues for marine resources.

Scottish Quality Salmon – Tripartite Initiative

SQS's commitment to top-class husbandry and environmental management is perhaps of greatest significance to this symposium in terms of its ability to facilitate and enable collaborative initiatives with other stakeholder interests. A notable example of this is the Tripartite Initiative.

It was recognized by most people that wild and farmed fish interests need to co-exist and work together for their mutual benefit. The Tripartite Working Group was established in June 1999 and is between (1) the Government – represented by the Scottish Executive Environment and Rural Affairs Department (SEERAD), Fisheries Research Services (FRS), SEPA and Scottish Natural Heritage (SNH); (2)

salmon farmers – represented by SQS; and (3) the wild salmonid fisheries interests – represented by the Association of Salmon Fishery Boards (ASFB), Scottish Anglers' National Association (SANA), Association of West Coast Fishery Trusts (AWCFT) and the Atlantic Salmon Trust (AST).

Regular meetings were held over the subsequent months and a Concordat was signed in June 2000 that included the following terms of reference:

'To develop and promote the implementation of measures for the restoration and maintenance of healthy stocks of wild and farmed fish including:

- Environmental standards and husbandry practices.
- The availability and implementation of effective medicinal treatments.
- Fallowing and rotation strategies.
- Location of sites.

To develop and promote the initiation of measures for the regeneration of wild salmon and sea trout stocks including:

- Identification of river systems which are of the highest priority, in terms of imperilled sea trout and salmon stocks.
- The design of procedures to develop restoration projects for these systems.
- Preparation of a broodstock programme to hold stocks from fragile systems until restorative action is possible.

To propose arrangements at a local and national level for taking forward the foregoing and to ensure that the results of this work are reflected in the development of local authority fish farm planning guidelines and framework plans.'

The arrangements proposed were the formation of Area Management Groups (AMGs) to discuss issues of mutual concern that would in turn lead to agreement on the best course of action and the signing of Area Management Agreements (AMAs). AMAs are between salmon farmers and wild fisheries interests and the content of each agreement varies from simple to complex. All, however, have focused on sea lice control.

A template for an AMA is set out in the Concordat and farmers have agreed a set of six headings for action on their part. They are:

(1) Synchronized production
(2) Zero ovigerous lice objective
(3) Furunculosis vaccination
(4) Containment
(5) Relocation
(6) Other Codes of Practice

Synchronized production

The concept of single year-class sites was developed in Scotland in the late 1970s as a means of managing furunculosis, a problem that affects both wild and farmed fish. It involved the use of single year-classes, co-ordinated fallow periods and disease-free smolts. This system worked well and it became clear that all farmers in a biological area had to synchronize their production. Farmers began to sign Management Agreements that covered stocking regimes, immunization, farming practices, monitoring and inspections and communications. This, then, was a good basis for the development of AMAs as required by the TWG.

Zero ovigerous lice objective

The objective of continuously achieving zero ovigerous salmon lice on farmed stocks, especially during the period prior to and during wild smolt migration, has been agreed. Best available methods include the use of single year-classes, fallow periods, the strategic use of veterinary medicines and co-ordination by management agreements, leading on to the development of an integrated sea louse management strategy.

National Treatment Strategy for the Control of Sea Lice

SQS developed a National Treatment Strategy for the Control of Sea Lice after scientists in Marine Harvest found that strategic co-ordinated sea lice treatments between weeks 10 and 20 can have beneficial effects later in the year. Specifically, if gravid females are removed at a time when larval louse survival is low the need for further treatments is reduced.

A Code of Practice defining how the industry should adopt these practices was written in 1998 and the first co-ordinated treatments were also given that year. The initiative has continued and results have improved as new, more efficacious and targeted medicines become available.

Furunculosis vaccination

To reduce the risk to wild and farmed fish from furunculosis in farmed and wild stocks, only smolts vaccinated against this disease should be used within the AMA area. Salmon farmers have been routinely vaccinating their fish against furunculosis for many years and the practice will continue.

This point in the Concordat has been expanded to include control mechanisms for other fish diseases such as *Gyrodactylus salaris*, which everyone agrees must be kept out of the UK.

Containment

To minimize the risk from escaped farmed salmon, the recommendations of the Escapes Working Group should be taken into account by farms within the AMA area. SQS has a Code of Practice on containment that is mandatory for its members.

Relocation

Relocation is an option for current sites that may for one reason or another be better sited elsewhere. The idea has been accepted in principle and work is underway to explore the criteria that would trigger a relocation and the costs involved.

Other Codes of Practice

These have already been covered when we referred to SQS's Environmental Management Systems. Their compliance will be verified independently.

To date, 14 AMGs have been set up and six AMAs have been signed. Progress is monitored by the TWG Management Group consisting of representatives of the three signatories to the TWG Concordat. The workload has grown sufficiently to require the services of a full-time National Development Officer who has now been appointed.

New challenges and solutions

SQS is the Scottish salmon industry's leading quality scheme; our members account for around two-thirds of all production; and – above all else – we endorse without reservation the priority that the industry in Scotland must be sustainable.

The criteria for SQS membership are comprehensive and integrated in every respect. First, membership encompasses all sectors, from feed companies through the freshwater and seawater sectors to smokers and processors. This enables standards and disciplines to be delivered consistently throughout the chain. Second, the mandatory standards and disciplines themselves encompass product quality, product safety, fish health and environmental management.

The determination of SQS to invest in long-term strategic solutions through delivering measurable levels of quality, responsibility, best practice and sustainability has few parallels. The prerequisite for membership is independently inspected and certified compliance with standards that have been scrutinized by recognized experts and accredited by the UK Accreditation Service (UKAS) against internationally recognized disciplines such as EN 45011 and ISO 14001. Additional disciplines include independently verified compliance with mandatory Codes of Practice and a commitment to a range of wider responsibilities including collaboration with fellow stakeholders such as wild salmon interests.

In the context of this Symposium, the most significant attributes of this commitment by SQS relate to good husbandry and environmental management. We should stress that good husbandry and environmental management are commitments that have long been accepted by SQS members as a quality product relies on healthy fish, and healthy fish in turn rely on maintaining a healthy environment. As with their wild cousins, farmed salmon are sensitive to a poor environment and thus so are farm managers, customers, retailers, investors and shareholders.

As with all our standards, we have incorporated the advice of independent experts. Over the last 6 years, we have commissioned ERM, Aspinwalls and URS Dames & Moore as well as consulting extensively with scientists, academics, specialist NGOs and others. We have as a result built up a raft of interlocking disciplines, requirements and practices that collectively comprise the most sophisticated and collaborative salmon farming regime yet developed.

Scottish Quality Salmon – site-specific solutions

One of the most potent elements of the SQS regime when it comes to delivering sustainability at the sharp end is our commitment to Environmental Management Systems (EMSs). All SQS members will have to have an EMS incorporating site-specific criteria that is fully accredited and verified by independent inspectors to the international standard ISO 14001 and subject to UKAS approval.

SQS deliberately chose the most rigorous, verifiable and site-specific environmental management regime after commissioning advice from external consultants. This choice was in preference to less demanding options such as unaccredited or internally inspected EMSs, or less measurable codes of practice.

The mandatory ISO 14001-accredited EMSs that we adopted also incorporate a number of independently inspected disciplines including continuous monitoring, external auditing of the information arising and an annual, or more frequent, review of performance so that the management system can be revised promptly on a rolling and responsive 'Plan – Do – Check – Review' discipline.

Equally significant is SQS's decision to use mandatory ISO 14001 EMSs to guarantee the auditable delivery of other non-statutory requirements. A good example of this are the SQS Codes of Practice and, specifically, the SQS Code of Practice on Containment. This must be translated into each site EMS, address the specific circumstances of each site, and be measured accordingly both for its content and delivery. The SQS Code of Practice on Containment is therefore not optional; nor is it generic; nor is it vulnerable to the one-size-fits-all approach which inevitably involves compromise. It is a mandatory Code of Practice, made to measure site by site, independently inspected and the first of its kind in any salmon-farming industry in the world.

The extent of SQS's commitment to audited, site-specific management and monitoring is a basis for the delivery of sustainability. Furthermore, our commitment is fundamental to addressing the responsibility and obligations posed by the precautionary principle through requiring SQS members to fund real-time data collection that should comprise an effective and verifiable early warning system on each site. The focus, relevance, continuity and quantity of information that can arise from an EMS will deliver more valuable information than public sector agencies could ever hope to be able to afford or achieve.

The salmon farming industry – strategic constraints

Could prospects for our joint interests be further improved? The answer is 'yes'. The Scottish salmon industry needs to be more efficiently regulated. We are currently controlled by more than 10 statutory bodies, 63 pieces of legislation, 43 Directives, 3 Regulations and 12 Commission Decisions. Whilst we can justly claim to be the heaviest regulated aquaculture industry in the world, most believe that there is considerable scope to improve the quality, coherence and co-ordination of regulation. Regulation could also be more intelligent and recognize, motivate and reward good industry practices – at present, it does not.

The volume and strategic capability of R&D underlying Scottish aquaculture needs to be improved. One only has to look at what Norway has achieved to realize how all interests would benefit if the relevant R&D was a greater national priority.

It also has to be said, however, that future prospects for anyone, and least of all for the fish themselves, are probably not being helped by the obsession that salmon farming is a simplistic substitute for a better understanding of demanding multi-factorial complexities. Historic and geographic records over the last 50 years from across the wild salmon's range indicate just how wide ranging the issues are. Despite this broader reality, SQS is fully committed at a national and local level to playing its part in terms of environmental management and collaborative partnerships. Nonetheless, we have to agree with one of our wild fish critics whose view is that with salmon farming comprising just one of 25 factors, 24 are in danger of being neglected because of the preoccupation with just one.

Conclusions

The most critical message for this Symposium is SQS's fundamental commitment to sustainable development. It is our over-riding strategic priority and we repeat that our endorsement of it is without reservation.

This involves achieving commercial viability and delivering socio-economic benefits – indeed, it is important that we continue to make a substantial contribution to jobs, communities, training, investment and the rural and national economies.

But of most relevance to you is the extent of our high-level commitment to sustainable site and loch practices, to good fish husbandry and environmental management, to fellow stakeholders and to wider coastal interests. This is unprecedented in either Scotland or elsewhere.

Gordon Rae was prevented by increasing ill health from delivering this paper and, sadly, he died before its publication. The Atlantic Salmon Trust wishes to recognise his tireless work to encourage co-operation between aquaculture and wild fishery interests for the achievement of environmentally acceptable salmon farming management.

BEACONS OF HOPE

Success Stories

Chapter 15
The Return of Salmon to Cleaner Rivers: A Scottish Perspective

R. Doughty[1] and R. Gardiner[2]

[1] *Scottish Environment Protection Agency, East Kilbride Office, 5 Redwood Crescent, East Kilbride, Glasgow, G74 5PP, UK*
[2] *Fisheries Research Services, Freshwater Laboratory, Faskally, Pitlochry, Perthshire, PH16 5LB, UK*

Abstract: Contemporary accounts indicate that during the 18th century salmon were present in all accessible stretches of major rivers in the central lowlands of Scotland. However, by the end of the century evidence of decline in some rivers was already apparent. This trend continued through the 19th century, so that by 1900 salmon had died out in many rivers where they were formerly abundant. Although several factors contributed to this decline, pollution from industry and the rapidly growing towns and cities was undoubtedly the major cause. Early attempts to control pollution through legislation failed. It was only in the mid-20th century that effective pollution control legislation began to be introduced. Since then, progressive improvements in river quality have allowed salmon to recolonize many of their ancestral rivers. Active management of salmon stocks has until recently been hindered by the absence of fisheries management bodies in the affected catchments. The establishment of fisheries trusts to cover the Clyde and Forth catchments gives some hope for the future, although this must be viewed in the context of threats to salmon populations on a global scale.

Introduction

Before the late 18th century, Atlantic salmon were plentiful in the major rivers of Scotland's Central Lowlands. Over the following 100 years, as this area became increasingly industrialized and the human population moved in large numbers to the towns and cities, many of these rivers became so polluted that viable populations of salmon could no longer survive. By the early years of the 20th century, salmon had effectively become extinct in the River Clyde above Dumbarton and many of the other rivers of the Central Lowlands. Other UK rivers draining heavily populated, industrialized areas were similarly affected.

Restoring salmon to these polluted rivers has taken the best part of another 100 years. Effective water pollution control legislation was first introduced only in the 1950s. It took another 30 years before conditions improved sufficiently to allow salmon to recolonize the Clyde system in significant numbers. Although pollution pressures had been eased, these first returning salmon remained vulnerable because of the absence of an effective fisheries management infrastructure in the catchments concerned.

There have been previous accounts of the demise and subsequent recovery of salmon populations in the rivers of central Scotland (Mackay & Doughty, 1986; Holden & Struthers, 1987; Gardiner & McLaren, 1991). Our aim in this paper is both to provide a synthesis of previously published material and to bring the story up to date, examining the extent to which earlier predictions and management proposals have been realized. Although we have chosen to focus on the Clyde catchment, which has experienced the worst pollution and the most spectacular recovery, we have made reference to other river systems where appropriate.

The pre-industrialization status of salmon in central Scotland

Salmon, along with other species of fish with marine affinities, would have originally colonized the rivers of Scotland as the ice retreated at the end of the last Ice Age, some 10 000 years ago (Maitland, 1977). It seems reasonable to suppose that all waters accessible from the sea were colonized and that until at least the mid to late 18th century, salmon stocks, although commercially exploited, were never seriously threatened (Fig. 15.1).

Certainly, salmon were widespread and common in the Clyde system, being absent only where they were denied access by impassable falls such as Stonebyres Falls, near Lanark. In 1791, it was reported that the Clyde estuary 'abounds with salmon, smelts and trouts which are caught in great plenty'. Salmon were also said to abound in the Black Cart Water, Locher Water and River Gryfe and were reported as being present in the White Cart Water, River Kelvin, North and South Calder Waters, the Avon Water and the Nethan Water (Sinclair, 1791). Aitchison (1964) records that in 1812, as many as 60 boats fished the Clyde at Renfrew during the summer months.

The decline of salmon and its causes

Even as early as the end of the 1790s, there were already indications that all was not well with salmon in the Clyde system. The Statistical Account reported that the fishery at Erskine on the Clyde estuary had been 'rather scanty' for many years and numbers of salmon in the Clyde and Kelvin were said to be in decline (Sinclair, 1791).

By the 1840s, contemporary accounts agreed that although numbers could fluctuate markedly from year to year, salmon were in serious decline in the rivers of the Clyde catchment. Commercial netting continued, but catches were variously estimated at between a tenth and a third of previous levels (Gordon, 1845).

As late as the 1880s, a few salmon were still managing to ascend the Clyde as far as Stonebyres Falls, but it can be assumed that by the early years of the 20th century salmon had been all but eliminated from most of the catchment, with only the River Leven and Loch Lomond and its tributaries managing to sustain a viable population. The River Leven was heavily polluted by bleach and dye works and it was reported

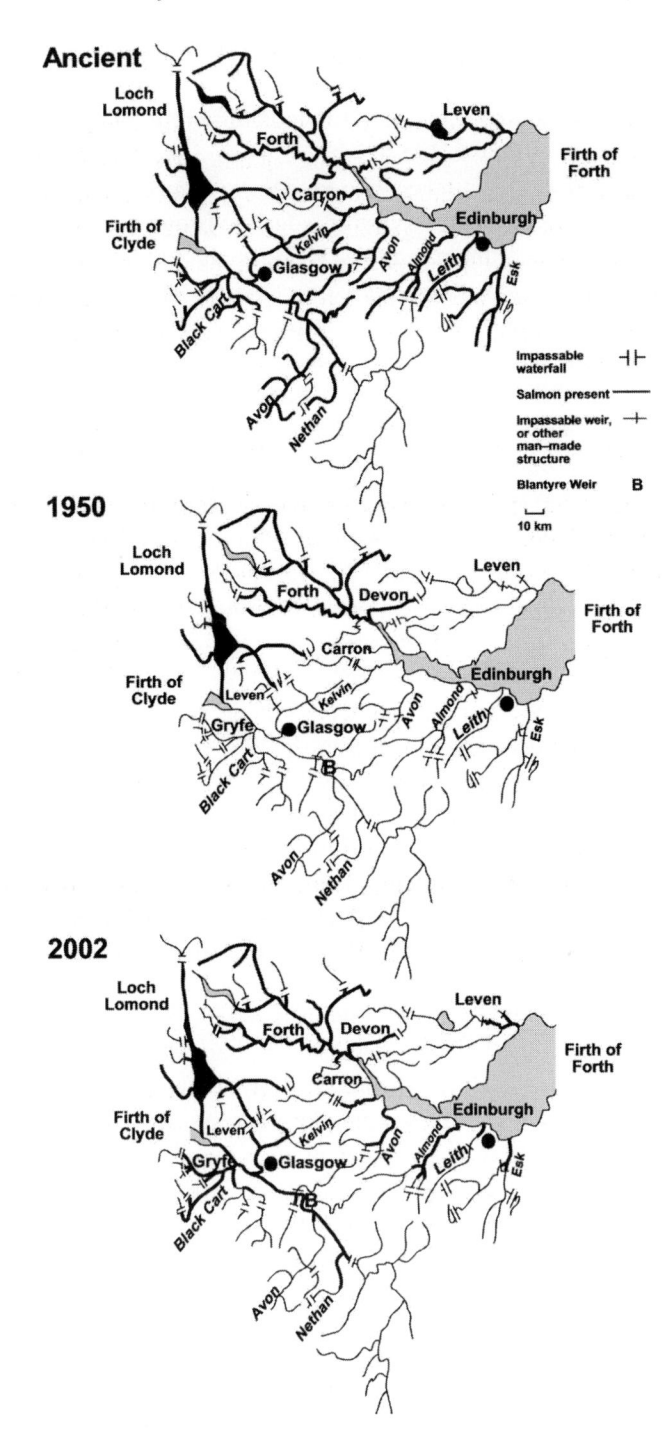

Fig. 15.1 Maps of the Central Belt of Scotland showing the likely ancient (pre-industrialization) distribution of salmon, together with distribution in 1950 and 2002. Only impassable weirs downstream of impassable waterfalls are indicated, and where there are several weirs, only the furthest downstream is shown.

that ascending salmon and sea trout were able to survive only when the river was in flood (Groome, 1884). By 1956 little had changed: it was reported that salmon managed to ascend the Leven 'only with obvious difficulty and frequent mortality' (Clyde River Purification Board, 1957).

It seems clear that several factors combined to drive salmon from the Clyde system. From 1771 onwards, the Clyde estuary was dredged to progressively greater depths to allow the passage of ships of increasing size (Riddell, 1979). At about the same time, weirs were being constructed across the Clyde and its tributaries to supply textile mills. The weir at Blantyre, built in 1785, proved to be such an obstacle that few salmon were able to ascend except in high flows. Similarly, a weir constructed on top of falls at Partick was partly blamed for the disappearance of salmon from the River Kelvin at Kilsyth (Sinclair, 1791).

Around 1800, waterborne sanitation was introduced to Glasgow, with sewers efficiently transporting wastes from industry and the rapidly growing human population directly to the River Clyde. Between 1851 and 1901, the population of Glasgow grew from 330 000 to over 900 000. Probably even more significant was the growth of industry. Dye works, chemical works, bleach works, paraffin oil works, distilleries and tanneries all discharged their wastes untreated into the Clyde and its tributaries. As a result of the dredging, the Clyde estuary retained these pollutants much longer than previously.

By 1872, Glasgow Corporation, in its report to the Rivers Pollution Commission (Rivers Pollution Commission, 1872), described the condition of the Clyde thus:

'Now, through Glasgow downwards, but diminishing below the mouth of the Cart, it is very foul and turbid; in short, a gigantic open sewer, noxious gases being continually evolved, which during summer, are so overpowering as to force the bulk of the passenger traffic from the river to the rail.'

The condition of rivers in the Lothians area was hardly better, although the Rivers Pollution Commission reported that because the various industries tended to be concentrated in particular localities, the nature of the pollution varied between rivers or along the course of a river. For example, the River North Esk and Water of Leith were almost exclusively affected by paper mill waste, while smaller streams draining the City of Edinburgh itself were polluted predominantly by domestic sewage. The River Almond suffered in its upper reaches from the effects of flax-steeping, from distillery discharges further downstream and in its lower reaches from pollution by paraffin oil extracted from oil-bearing shale.

Elsewhere in Scotland the impact of pollution was more localized and the impact on salmon and other fish was limited. In the Tweed catchment, the Gala Water downstream of Galashiels to the Tweed confluence was said to be fishless owing to the effects of sewage and effluents from the woollen industry. Similar blackspots were noted in the Tweed and its tributaries downstream of Hawick, Jedburgh and Kelso. In Ayrshire, the Kilmarnock Water (a tributary of the River Irvine), which until the

1840s had remained virtually unpolluted, was another river to suffer from the effects of the local woollen industry along with sewage from the town of Kilmarnock. The river was said to emit 'a very offensive stench' during the summer months. The Tay catchment remained virtually unpolluted, with the exception of the Dighty Burn, which enters the Firth of Tay beside Dundee and was heavily polluted by linen and jute bleach works. Rivers in the north of Scotland, such as the Dee, were reported to be relatively clean (Rivers Pollution Commission, 1872).

The long road to recovery

The first steps towards reversing the by now appalling state of the Clyde and other polluted rivers in the UK were taken in the 1860s with the appointment of two Royal Commissions. The fourth report of the 1868 Commission dealt with pollution of Scottish rivers (Rivers Pollution Commission, 1872). The deliberations of the 1868 Commission led directly to the Rivers Pollution Act 1876. Unfortunately, the powerful industrial lobby ensured that the Commission's recommendations were so diluted that the Act was severely weakened during its passage through Parliament and its effectiveness was limited.

The Local Government Scotland Act 1887 gave Scottish County Councils the responsibility of administering the 1876 Act, but only Lanark County Council made any serious attempt to enforce its provisions. The Council's efforts, recorded in three reports by the Medical Officer of Health (County of Lanark, 1903, 1909, 1924), resulted in the construction of sewage treatment works and industrial treatment plants, and some prosecutions of polluters. However, these measures were only partially successful and some pollution problems became worse. For example, the coal mining industry expanded to such an extent that by the 1920s, pollution from coal washeries was second only to sewage as the major source of pollution of the Clyde and its tributaries upstream of Glasgow (County of Lanark, 1924).

By the late 1920s, Scottish rivers were in a worse condition than they had been in 1872 (Hammerton, 1994). In 1927, a Scottish Board of Health survey found the Clyde to be the most polluted river basin in Scotland, followed by the Forth and the Tweed. Following this report, the Secretary of State set up the Advisory Committee on Rivers Pollution Prevention, which in its final report in 1936 pointed out the inadequacy of the existing legislation and repeated calls made by the earlier Royal Commission on Sewage Disposal for control of river pollution through catchment-based river boards. No further progress was made until after the Second World War, when further recommendations by a sub-committee of the Scottish Water Advisory Committee culminated in the Rivers (Prevention of Pollution) (Scotland) Act 1951.

Under the 1951 Act, catchment-based River Purification Boards were set up to administer its provisions. It became a criminal offence to discharge sewage or trade effluent without the consent of the river purification authority. The authorities were empowered to refuse consent or to grant consent with conditions attached.

Unfortunately, the 1951 Act applied only to new discharges, whereas most pollution problems were caused by existing discharges. So although the Act was responsible for halting the deterioration in the condition of Scotland's rivers, further legislation was needed to allow the river purification authorities to begin the restoration process. This came with the Rivers (Prevention of Pollution) (Scotland) Act 1965, which extended the powers of the river purification authorities to cover existing discharges.

Over the last 30 years, the control of water pollution in Scotland has been strengthened by further legislation and administrative change which has seen the replacement of the river purification authorities by a single Scottish Environment Protection Agency with powers to control pollution of air and land as well as water, and, in April 2002, the creation of a single water supply and sewage treatment authority for the whole of Scotland. European legislation has become increasingly important and is now the major driving force for pollution control in Scotland and the rest of the UK.

The return of salmon

In the Clyde catchment, salmon managed to survive and reproduce throughout the 19th and 20th centuries only in the River Leven and Loch Lomond system (Fig. 15.1). Then in autumn 1978, salmon were noted ascending the River Gryfe in significant numbers, although occasional salmon had been reported in earlier years (Mills & Graesser, 1981). A further run was noted in 1979 and, since then, salmon have become re-established.

In the Clyde itself, a sea trout and a 4.5 kg salmon were found trapped on cooling water intake screens at a Renfrew power station in January 1978. Five years later, in August 1983, a young angler fishing the Clyde in the centre of Glasgow caught a 3.6 kg salmon. A few weeks later, salmon were seen attempting to jump the weir at Blantyre. The discovery of a 6.4 kg kelt at Crossford provided evidence that at least some fish were successfully ascending the river and were managing to spawn, although it was another 6 years before spawning sites were identified (Mackay & Doughty, 1986; Gardiner & McLaren, 1991).

In 1989 and 1990, salmon were reported in the River Kelvin and the Black and White Cart Waters. Spawning in the Kelvin and Black Cart was confirmed in 1992. At the present time, of those Clyde tributaries that are accessible, only the North, South and Rotten Calder Waters have yet to be recolonized by salmon, at least in part (Fig. 15.1).

Historically, the rivers draining to the Firth of Forth have fared rather better than those on the west coast in that the River Forth and its main tributaries (River Teith, Allan Water and River Devon) have retained their salmon populations. In other rivers draining directly to the Firth of Forth where salmon were eliminated by pollution, they have now started to return.

Following a stocking programme, runs of salmon began to appear in the River

Carron from 1989. However, evidence of successful spawning was to wait until 2000, when sampling by the Fisheries Research Services Freshwater Laboratory showed a healthy population of juvenile salmon to be present in a tributary, the Castlerankine Burn. Populations have become re-established in the Linhouse Water and Murieston Water (two tributaries of the River Almond), and also in the River Esk at Musselburgh and the Fife Leven. Salmon have also been reported in the Avon Water and Water of Leith, and survey work by the Forth Fisheries Foundation in 2002 has confirmed the presence of juvenile salmon in both.

Reasons for the return of salmon

Clearly, the main reason why salmon have managed to recolonize many of the rivers from which they had been eliminated by pollution is the improvement in water quality over the last 30–40 years.

In the case of the Clyde system, the key factor has been the improvement in water quality, particularly in dissolved oxygen concentrations, in the Clyde estuary. Around 1900, three sewage treatment works (STWs) were built to serve the city of Glasgow, all discharging into or at the head of the estuary. While the resulting improvement in water quality was enough to encourage the return of passenger traffic to the river, it was insufficient to allow salmon to return. At the time the Clyde River Purification Board was established in 1956, Shieldhall and Dalmuir STWs provided only primary treatment. Although Dalmarnock STW provided some secondary treatment, its percolating filters were so overloaded that they were usually by-passed (Clyde River Purification Board, 1957).

By 1978, the condition of the upper estuary had improved owing to the upgrading of Dalmarnock STW and the commissioning of a new STW to serve Paisley, together with improvements in the quality of the major tributaries and the decline in industries such as leather tanning and paper making (Mackay et al., 1978). The occurrence of regular salmon runs in the Clyde from 1983 proved that forecasts (Doughty & Best, 1981) that salmon would not be re-established until after the introduction of full biological treatment at Shieldhall (completed 1985) and Dalmuir STW (completed 2001) were unduly pessimistic.

Water quality improvements in fresh waters have subsequently allowed salmon to return to Clyde tributaries such as the Black and White Cart Waters and the River Kelvin.

There has been much speculation about the origins of the salmon returning to the Clyde. Although 6000 salmon fry were stocked into a small burn in Rutherglen in 1980–81, it is most likely that the initial recolonization was natural through fish straying from the adjacent Leven and Loch Lomond system. It is also possible that numbers were supplemented by fish farm escapees, which were common on the west coast of Scotland in the 1980s. Apart from the River Carron, which was actively stocked from 1985, recolonization of east coast rivers also appears to have been natural (Gardiner & McLaren, 1991).

Management of salmon stocks

When salmon returned to the Clyde system it was to a management vacuum. Owing to the absence of salmon for so many years, there was confusion over who owned the salmon fishing rights, although most appeared to be vested in the Crown. Management of the existing trout fisheries was in the hands of several angling associations and there was clearly a potential for conflict between those who wished to maintain the excellent low-cost trout fishing in, for example, the middle reaches of the Clyde, and those who wished to develop salmon fishing. There were understandable concerns that the return of salmon could push up the cost of leases.

In 1983, soon after salmon began to return in numbers, the Clyde River Purification Board facilitated discussions between all the interested parties, including the Crown Estate Commissioners, local authorities, angling associations and the Scottish Landowners' Federation, on the best way of managing the Clyde fisheries. These discussions led in 1984 to the formation of the Clyde Fisheries Management Trust. The Trust was granted leases of the salmon fishing rights by the Crown.

One of the Trust's first actions was to impose a 5-year moratorium on salmon fishing to allow stocks to build up. A limited programme of stocking was also initiated. For example, 8000 salmon parr were introduced to the Avon Water in 1988. In addition, fish passes have been installed at Blantyre Weir on the Clyde (1995), which although not completely impassable was a major obstruction to the upstream passage of salmon, and at the formerly impassable Gavins Mill Weir on the Allander Water in Milngavie (1999). The easing of fish passage at Blantyre Weir has given salmon much improved access to good spawning and nursery areas in the main stem of the Clyde between Blantyre Weir and the Falls of Clyde, and in the Nethan Water, and there is now a marked increase in the number of salmon in the Clyde. The need for further work at both these fish passes is currently under review. Gratifyingly, survey work by the Clyde River Foundation in 2002 has confirmed that salmon are now spawning successfully upstream of Gavins Mill Weir on the Allander Water.

The work of the Trust has been hampered to a degree by a lack of good quality scientific information on which to base management decisions. When salmon were first noted in the catchment, one of the first priorities was to discover where and how successfully salmon were spawning. In 1989–90, the Freshwater Fisheries Laboratory (now Freshwater Research Services Freshwater Laboratory) and Clyde River Purification Board staff carried out some limited exploratory electric fishing surveys of potential spawning sites and found small numbers of salmon parr in the Clyde near Cambusnethan and in the lower Nethan Water at Crossford. Since the Clyde River Purification Board had no official fisheries remit and the Freshwater Fisheries Laboratory had more pressing priorities, more comprehensive surveys could not be justified.

In 1995, the Kelvin Valley Countryside Project commissioned a study of salmonid fish in the River Kelvin and its main tributaries. The work was carried out by the Freshwater Fisheries Laboratory and remains the most detailed such survey carried out to date in the Clyde catchment. The availability of spawning and nursery areas was assessed, together with the availability of holding areas for adult salmon. Juvenile salmon were found in the River Kelvin itself and in the Glazert and Allander Waters. The lower Allander Water was identified as the most productive spawning area, but it is likely that little spawning takes place in the main stem of the Kelvin owing to a lack of suitable habitat. However, the main stem of the Kelvin was found to have plenty of holding areas for adults (Gardiner & Armstrong, 1996).

Following the success of the West Coast Fisheries Trusts and the Tweed Foundation in bringing a more scientific approach to the management of fish stocks in Scotland, the Clyde River Foundation was established as a charitable trust in 1999. The Foundation's aims include collecting information on fish stocks, conserving and enhancing fish populations and running environmental education projects within schools and local communities. The Foundation recently appointed a full-time fishery scientist. The Forth Fisheries Foundation, which has similar aims, was also established in 1999 and has had a fisheries scientist in post since June 2001. It covers the rivers Forth, Teith, Allan, Devon, Leven, Carron, Avon, Almond, Esk and Tyne.

Future prospects

In contrast to the alarming declines in salmon populations observed in some other parts of Scotland, the Clyde stocks appear to be still increasing and now support a significant sport fishery (Table 15.1). However, there is no room for complacency. The Clyde system is largely free of some of the catchment pressures that elsewhere have been blamed for declining salmon stocks (e.g. acidification, fish farming), but it is not immune to the effects of global warming or overfishing at sea.

Table 15.1 Salmon and sea trout rod catches in the River Clyde and selected tributaries in 1999. (Data courtesy of the River Clyde Fisheries Management Trust.)

River	Salmon	Sea trout
Clyde	240	56
Black and White Carts	42	49
Gryfe	31	0
Kelvin	28	87
Avon	0	5

Water quality in the Clyde catchment continues to improve, for example, through the introduction of new sewerage schemes. A prime example of this is the Kelvin Valley Scheme where sewage treatment works serving communities in the Kelvin catchment are in the process of being closed and the sewage diverted to the upgraded Dalmuir STW discharging to the Clyde estuary. The benefits for rivers such as the Allander Water and the lower River Kelvin are already apparent (Fig. 15.2).

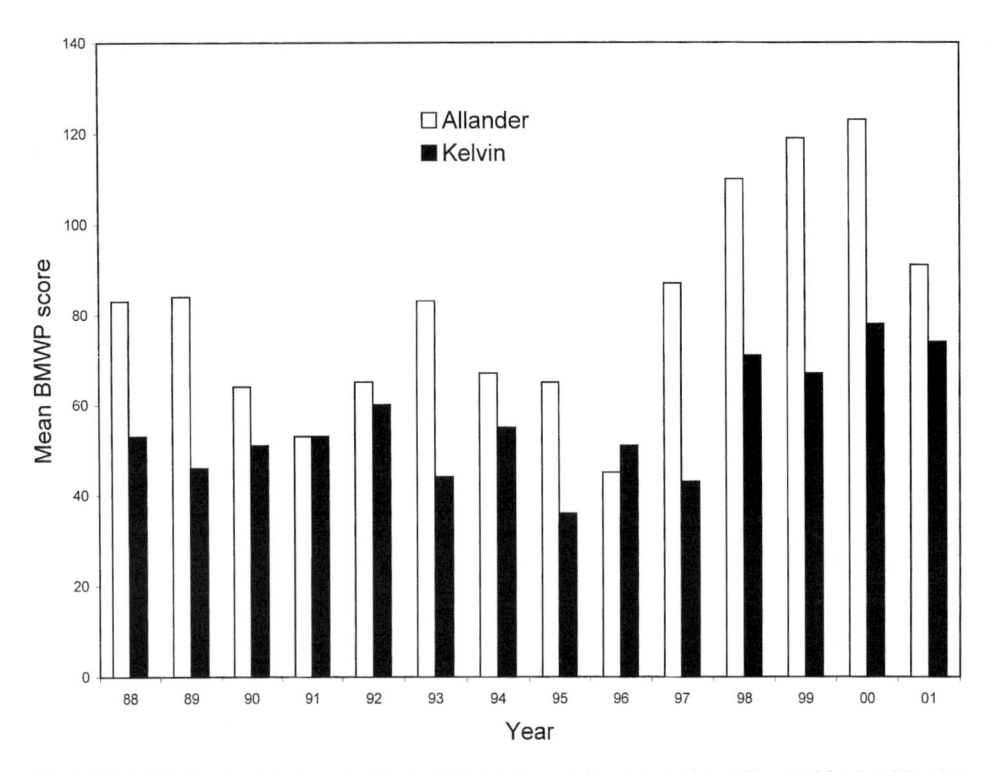

Fig. 15.2 Biological quality [mean Biological Monitoring Working Party (BMWP) score] for the Allander Water above the River Kelvin confluence and the River Kelvin above the River Clyde confluence, showing improvements following closure of Milngavie STW in 1996.

The EC Water Framework Directive, due to be transposed into Scottish legislation by December 2003, will bring further benefits. The Directive will require Member States to achieve 'good status' for surface waters and groundwaters by 2016. The mechanism for achieving good status is through River Basin Management Planning. It is presently unclear how many River Basins will cover Scotland (one to three seems the most likely number) or how management will be implemented at the local, catchment level. However, the Directive should provide the necessary legislative controls to enable problems such as diffuse pollution, river flow regulation and habitat degradation to be tackled much more effectively than at present and will encourage greater partnership working between statutory and non-governmental environmental organizations.

There are further opportunities for removing or overcoming artificial barriers to fish migration. For example, the feasibility of installing passes at falls on the lower River Kelvin and a weir at Ferniegair on the lower Avon Water is being investigated. Physical barriers are also thought to be a critical factor affecting the recovery of salmon stocks in various rivers draining into the Firth of Forth, including the rivers Almond, Leven and Esk. Decisions on whether to extend this policy to natural barriers, thus opening up areas to which salmon never previously had access, but which may be in better condition to support juvenile salmon populations, need to be carefully considered. Such management activities, together with habitat enhancement schemes and fish population monitoring, depend on continued funding of bodies such as the Clyde River Foundation. At present, continued funding cannot be guaranteed.

It may be concluded that while improvements in the water quality of the recovering rivers of central Scotland may be expected to continue as a result of existing and new legislation, the effects of larger-scale impacts such as climate change, together with the precarious existence of non-statutory bodies such as the Clyde River Foundation mean that the long-term future of salmon in these rivers is by no means secure.

References

Aitchison, J. (1964) Salmon in the Clyde. *Scotland's Magazine*, March 1964, 12–14.
Clyde River Purification Board (1957) *First Annual Report 1956–57*. Clyde River Purification Board, Glasgow.
County of Lanark (1903) *Report on the Administration of the Rivers Pollution Prevention Acts*. County Medical Officer, Lanark.
County of Lanark (1909) *Second Report on the Administration of the Rivers Pollution Prevention Acts*. County Medical Officer, Lanark.
County of Lanark (1924) *Third Report on the Administration of the Rivers Pollution Prevention Acts*. County Medical Officer, Lanark.
Doughty, C.R. & Best, G.A. (1981) *Impact of pollution on salmonid fish populations in western Scotland*. Proceedings of the 2nd British Freshwater Fisheries Conference, University of Liverpool, 13–15 April, pp. 240–47.
Gardiner, R. & Armstrong, J. (1996) *Migratory Salmonids in the River Kelvin*. Fisheries Research Services Report 4/96. Scottish Office: Agriculture, Fisheries and Environment Department Freshwater Fisheries Laboratory, Pitlochry.
Gardiner, R. & McLaren, I. (1991) Decline and recovery of salmon in the central belt of Scotland. In: *Strategies for the Rehabilitation of Salmon Rivers* (ed. D. Mills). Atlantic Salmon Trust, Institute of Fisheries Management, Linnean Society, London, pp. 187–93.
Gordon, J. (1845) *The New Statistical Account of Scotland*. Blackwood, Edinburgh.
Groome, F.M. (1884) *Ordnance Gazetteer of Scotland. A Survey of Scottish Topography: Statistical, Biographical and Historical*. Jack, Edinburgh.
Hammerton, D. (1994) Domestic and industrial pollution. In: *The Fresh Waters of Scotland. A National Resource of International Significance* (eds P.S. Maitland, P.J. Boon & D.S. McLusky). Wiley, Chichester.
Holden, A. & Struthers, G. (eds) (1987) *The return of salmon to the Clyde*. Proceedings of Conference of Institute of Fisheries Management Scottish Branch, University of Strathclyde, 15 November 1986.
Mackay, D.W. & Doughty, C.R. (1986) Migratory salmonids of the estuary and Firth of Clyde. *Proceedings of the Royal Society of Edinburgh*, **90B**, 479–90.
Mackay, D.W., Tayler, W.K. & Henderson, A.R. (1978) The recovery of the polluted Clyde estuary. *Proceedings of the Royal Society of Edinburgh*, **76B**, 135–52.
Maitland, P.S. (1977) Freshwater fish in Scotland in the 18th, 19th and 20th centuries. *Biological Conservation*, **12**, 265–78.
Mills, D. & Graesser, N. (1981) *The Salmon Rivers of Scotland*. Cassell, London.
Riddell, J.F. (1979) *Clyde Navigation: A History of the Development and Deepening of the River Clyde*. Donald, Edinburgh.
Rivers Pollution Commission (1872) *Fourth Report. Pollution of Rivers of Scotland*. HMSO, London.
Sinclair, J. (1791) *The Statistical Account of Scotland*. Creech, Edinburgh.

Chapter 16
The Return of Salmon to Cleaner Rivers – England and Wales

G.W. Mawle[1] and N.J. Milner[2]

[1] *Environment Agency, Waterside Drive, Aztec West, Almondsbury, Bristol, BS12 4UD, UK*
[2] *National Salmon and Trout Fisheries Centre, Environment Agency, 29 Newport Road, Cardiff, CF24 0TP, UK*

Abstract: Industrial and urban development during the 19th and early 20th centuries brought about the extinction of salmon from many rivers. Over the last 40 years, increasingly strong regulation of discharges from industry and sewerage systems, combined with reductions in heavy industry and large-scale investment in water treatment, have resulted in major improvements in water quality. As a consequence, salmon are returning to an increasing number of rivers including the Thames, the Taff, the Tyne and, most recently, the Mersey. In some rivers, stocks have recovered sufficiently to provide fisheries. The Tyne now boasts the highest rod catch for salmon of any river in England or Wales at 2513 in 2001. This reflects not only the improved catch in the Tyne but also the decline of some other rivers; 42% of rivers have exhibited a statistically significant decline in rod catch over the period 1974–98, compared to 17% showing an increase. Recovering rivers now contribute about 25% of the annual rod catch of salmon, due largely to improvements in water quality aided by constraints on exploitation and, on some rivers, to stocking and to the removal of obstructions to migration. The impact on water quality of point discharges in urban areas has been, and continues to be, progressively reduced, but pollution in rural areas, particularly from diffuse sources, is an increasing concern. Despite major reductions in sulphur emissions and changes in forestry practices, acidification remains an issue in some upland areas, though mitigation by liming has proved locally successful in restoring salmon stocks where circumstances permit. The severity of some types of pollution associated with the intensification of agriculture has been reduced, notably from slurry and silage, but other aspects particularly from silt and pesticides, and compounded by flow regime changes, remain a challenge.

Introduction

Increasing urban and industrial development, particularly during the 19th and 20th centuries, led to the gross pollution of rivers and their estuaries. When combined with increasing obstruction to migration, abstraction and overexploitation, this degradation of water quality resulted in the loss of salmon from many rivers across England and Wales.

This paper documents the loss and the subsequent return of salmon (*Salmo salar* L.) to these rivers following the reduction of pollution over the past 40 years.

Decline and loss

Prior to the industrial revolution, most of the rivers of England and Wales supported salmon populations, excepting the low-gradient rivers of East Anglia. Netboy (1968) describes the previous abundance and subsequent decline of salmon stocks and their associated fisheries, including those of the Trent, Yorkshire Ouse, Thames, Mersey and Medway.

The losses were gradual. The Thames, which flows through London, was probably the first major river to lose its stock: the last salmon for over a century was recorded in 1833. The Mersey also ceased to produce salmon in the middle of the 19th century (Harris, 1980).

Evidence to the 1860 Commission into Salmon Fisheries highlights pollution as a key factor in the decline, and, in some cases, the loss of some salmon stocks (Anon., 1860). For example, many of the productive salmon rivers of south Wales, including the Taff, were affected by pollution from coal mining, metallurgical industries and sewage from the expanding population (Mawle et al., 1985).

In 'The Salmon Rivers of England and Wales', Augustus Grimble (1913) wrote:

'The Bristol Avon, the Mersey, the Calder, the Weaver, the Taff, Ogmore, Loughor, Neath and Ebbw are ... absolutely ruined (by pollution) and the Ribble and other fine rivers are only too sure to follow suit unless the law steps in.'

In the north-east, the salmon fishery on the River Tees became defunct in the 1930s, whilst those of the Tyne and Ouse finally ceased in the 1950s (Axford et al., unpublished). Champion (1991) describes 1959 as the 'nadir' for the Tyne, a description that would probably apply generally to water quality in salmon rivers across England and Wales. A national survey of water quality in 1958, classified 28% of non-tidal rivers and 59% of tidal rivers as being 'doubtful; poor; or grossly polluted' (National Rivers Authority, 1991).

At this time salmon stocks had apparently been lost from about 30% of the catchment area of England and Wales (Fig. 16.1). On the east coast the main losses were the Tyne, Wear and Tees in the north; the large Ouse and Trent systems which share the Humber estuary; and the Thames and Medway. The biggest losses on the west coast were the rivers of the South Wales coalfield and the Mersey in the north.

Reduction of pollution

Stronger legal measures to control and reduce pollution were implemented in the latter half of the 20th century, including the Rivers (Prevention of Pollution) Acts 1951 to 1961 and the Control of Pollution Act 1974. Compliance with this legislation has required massive investment, both public and private, in wastewater treatment.

As well as improved treatment of effluents, there was a reduction in the need to treat effluents as some heavy industries declined. For example, UK steel production dropped by a third from the 1960s to the 1990s (Office for National Statistics, 1997).

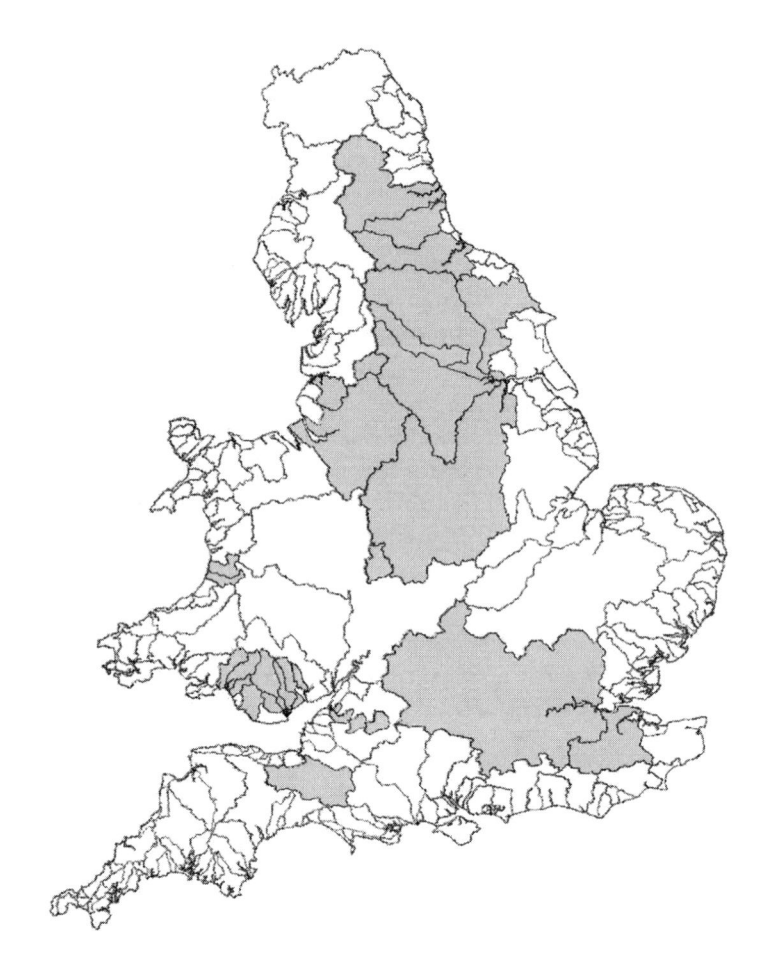

Fig. 16.1 Catchments in England and Wales from which salmon stocks were either lost or severely depleted in the 20th century.

More marked has been the decline in coal production which has fallen by about 80% over the past 40 years; coke production has fallen by 71% (National Statistics, 2000). The consumption of coal to produce coal gas fell from 23 000 tonnes in 1960 to nothing by 1977, as natural gas became available. The coal industry, including the associated production of coke and coal gas, has polluted rivers with suspended solids, ammonia, phenols, thiosulphates and thiocyanates (Edwards, 1981).

Water quality has improved markedly in many rivers (Environment Agency, 1998). The proportion of non-tidal rivers classed as grossly polluted or bad fell from 6.4% in 1958 to 0.4% in 2000 (Fig. 16.2). There have been two main periods of improvement. The first was during the 1960s and 1970s. The second followed privatization of the water industry in 1989, since when there has been stronger regulation, by the National Rivers Authority and subsequently the Environment Agency, as well as increased

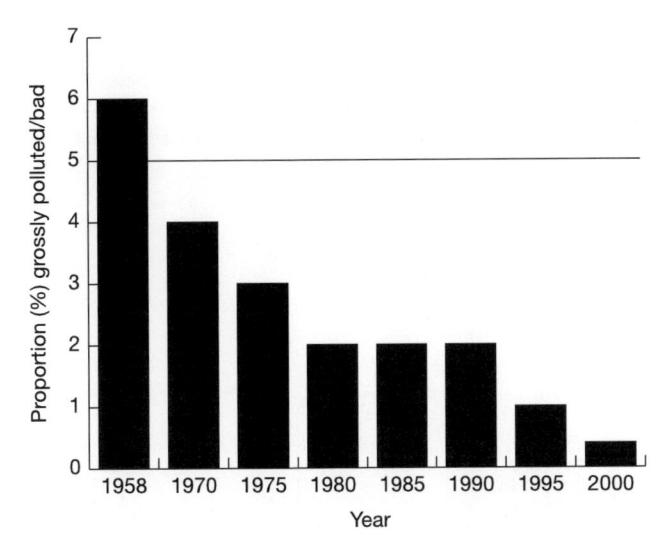

Fig. 16.2 The proportion (%) of non-tidal river length surveyed in England and Wales that was grossly polluted or bad from 1958–2000 (NRA, 1991; Environment Agency, 2002a).

investment by the water industry, equivalent to billions of pounds each year at 1999 prices (Fig. 16.3). Under the industry's current Asset Management Plan, investment will continue at this level until 2005 at least with further benefits for water quality.

Water quality in estuaries has shown similar improvements. A notable example is that of the River Tyne, which Grimble (1913) described as having been the 'first salmon river in England'. Up to 1980 much of the estuary was still 'grossly polluted'. Even in 1990 water quality was poor for about 30% of its length. Since 1995 all of it has been classed as of 'good' or 'fair' quality. The improvement was due to industrial

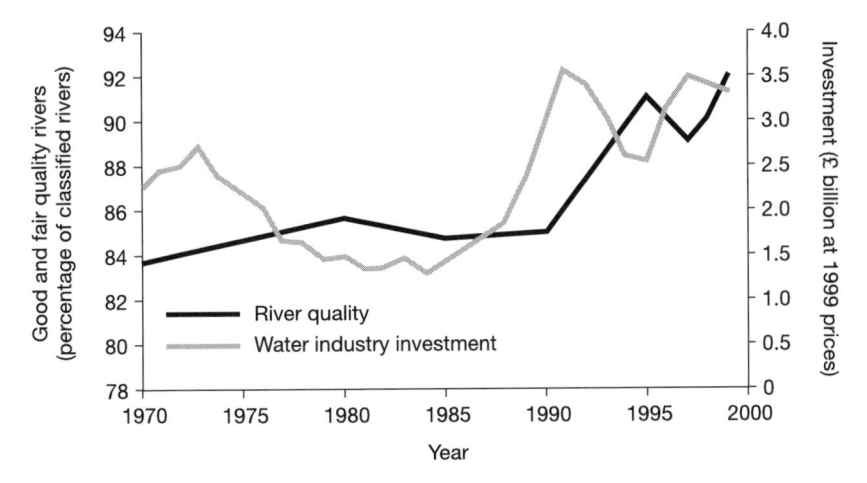

Fig. 16.3 Investment in the water industry in relation to the proportion of river length classed as good or fair in England and Wales from 1970–96. [Updated from Environment Agency (1998).]

decline (particularly the closure of gas works and their associated coke ovens) coupled with major investment in sewerage systems which led to greatly improved estuarine water quality. Similar improvements occurred in the nearby Rivers Tees and Wear. The reduction in ammonia levels in the Wear is shown in Fig. 16.4.

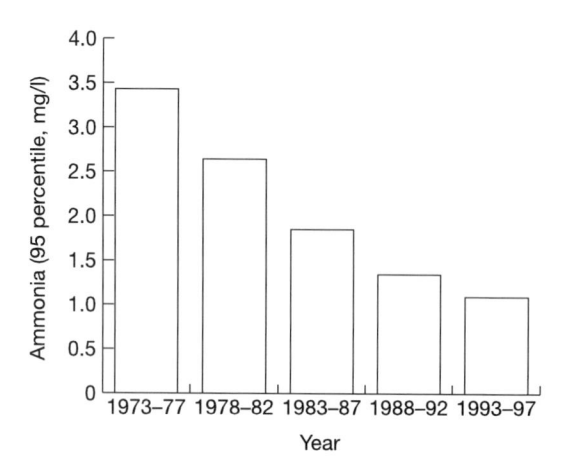

Fig. 16.4 Ammonia concentrations (95 percentile) in the River Wear at Lamb Bridge from 1973–97.

Return of the salmon

Since the 1960s, recovery has been dramatic in many rivers around the country. The principal index of this has been catch returns declared by licensed salmon anglers to the Environment Agency and its predecessors (Russell et al., 1995; Environment Agency, 2000). Within the last 3 years, salmon have been recorded either through such declared catches or by other means (including Environment Agency fisheries surveys and accidental captures by anglers) from almost all the catchments from which they had been lost (Fig. 16.5). The recovering stocks, mostly in the north-east and south Wales, have contributed 20% of the declared rod catch of salmon in England and Wales for the last 3 years (Table 16.1). These changes have been all the more impressive because they have occurred against a wide-scale decline in salmon stocks in many areas of the north Atlantic over recent decades, attributed to reduced marine survival (Potter & Crozier, 2000).

East coast rivers

In the north-east, the Tyne, Tees and Wear all now support significant rod fisheries (Fig. 16.6). The most impressive example of recovery is the Tyne which, with a catch of 2513 salmon in 2001, is now the most productive rod fishery in England and Wales. All these rivers have been stocked with juvenile salmon to aid recovery, particularly the Tyne where 160 000 or more young salmon have been stocked annually as mitigation for the impact of Kielder reservoir.

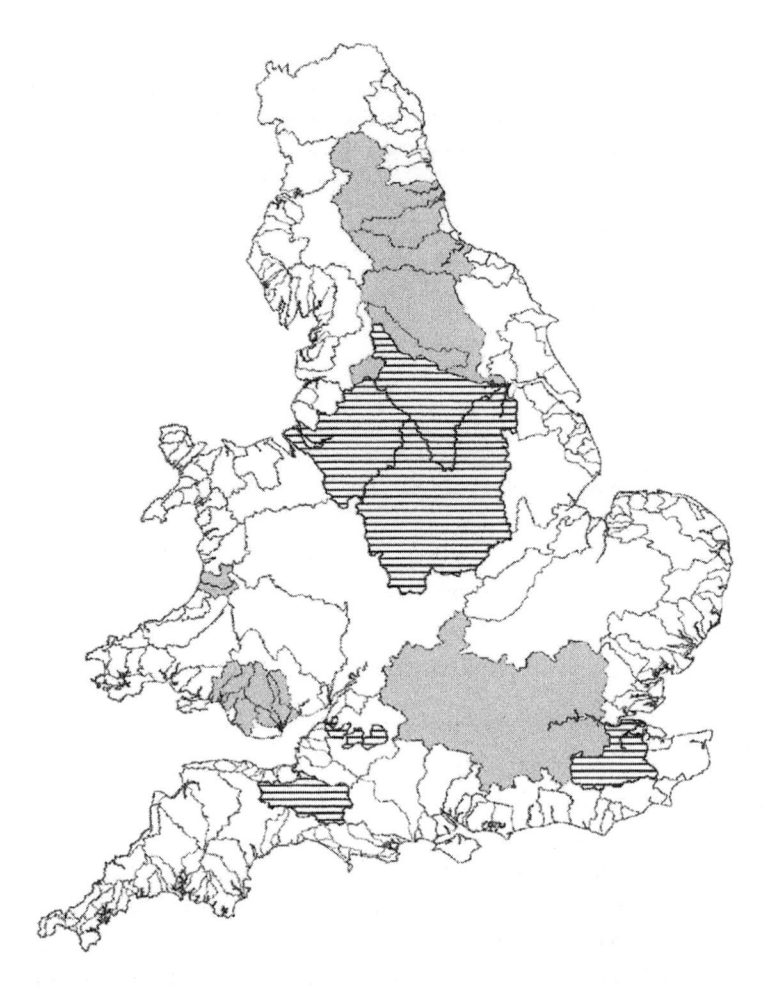

Fig. 16.5 Catchments in England and Wales from which salmon have been recorded in the last 3 years (1999–2001), stocks having been lost in the 20th century. Dark shading indicates catchments from which licensed salmon anglers have declared rod-caught salmon. Hatching indicates that although no licensed rod catch has been declared, salmon have been recorded by other means.

Since 1978, salmon are known to have been breeding again in the River Ure, a tributary of the Yorkshire Ouse, following a stocking programme (Axford, 1991). Despite this, few salmon have, as yet, been reported caught by licensed salmon anglers from the Ouse which feeds the Humber Estuary; only 17 salmon over the past 10 years. Nonetheless, stock abundance now seems to be increasing. In the winter of 2000/2001 about 300 redds were observed in the Ure, where high densities of juveniles have also been recorded in the Environment Agency's routine electrofishing surveys (Axford et al., unpublished). In March 2001, a salmon kelt was accidentally caught by a coarse angler in the River Don, which enters the Ouse just upstream of the Humber Estuary. There have also been reports from the Don of salmon being seen jumping in the centre of Sheffield, and in the adjacent River Aire in 2001 (Axford et al., unpublished).

Table 16.1 Mean annual rod catch declared by licensed salmon anglers for 1999–2001 from rivers in England and Wales in which salmon stocks had previously been lost or severely depleted.

River	Mean annual rod catch of salmon (1999–2001)
Tyne	2236
Tees	412
Wear	98
Ouse	7
Thames	<1
Ebbw	<1
Rhymney	<1
Taff	43
Ely	<1
Ogmore	75
Afan	8
Neath	60
Tawe	77
Loughor	13
Ystwyth	13
Rheidol	27

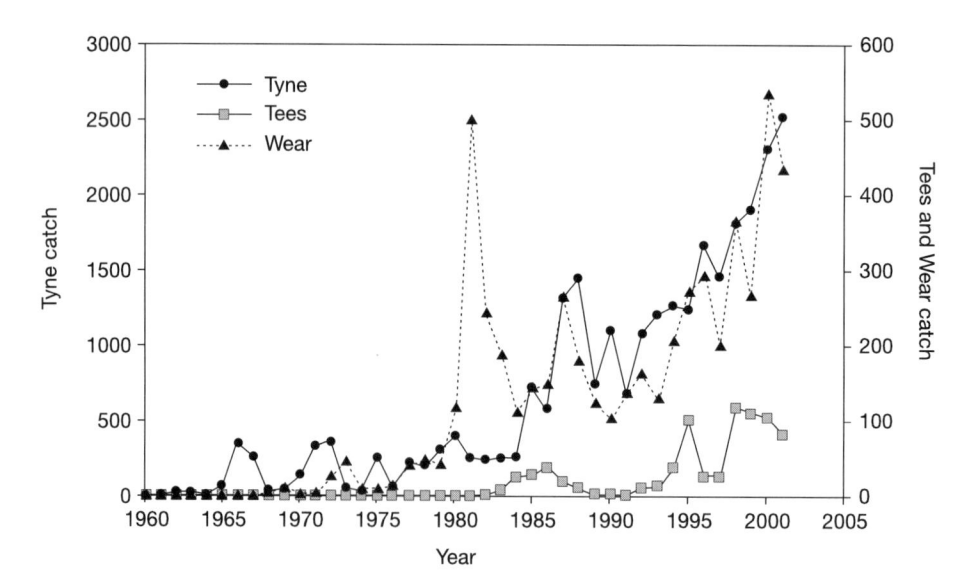

Fig. 16.6 Recovery of the salmon rod fisheries in the north-east of England as indicated by the annual declared rod catch of salmon for the rivers Tyne, Tees and Wear from 1960–2001.

The Return of Salmon to Cleaner Rivers – England and Wales

Between 1967 and 1993, there were over a hundred authenticated reports of salmon in the Trent, the other major river entering the Humber, including 20 authenticated sightings or captures in 1982 (Axford et al., unpublished). A tributary, the Dove, was stocked in 1998 with 150 000 salmon fry of Tyne origin. Other stockings were made in 2000 and 2001 and a Trent Salmon Trust has been set up to fund work to restore a salmon population. In 2001, several salmon were seen leaping at weirs on the Dove, the first authenticated sightings of salmon in this tributary since one was caught by an angler at Egginton in 1932; redds were also observed.

There is a Thames Salmon Trust which has led the funding of work to restore a salmon population to that river. Stocking has helped stimulate runs of several hundred salmon a year in the early 1990s, mostly captured by trap (Fig. 16.7). Up to 34 salmon a year, in 1996, have been reported caught by anglers (Environment Agency, 2002b). These fish are thought to have been wanderers from other rivers or derived from stocking. A chain of fish passes has just been completed on 37 weirs up the Thames and its tributary, the Kennet. For the first time in over 150 years adult salmon can now potentially reach somewhere to spawn in the Thames catchment. The recent poor returns may be related to poor estuarine water quality during the summer and perhaps to changes in stocking practice. On the adjacent River Medway, returns have so far been confined to the occasional fish captured or observed in the lower reaches.

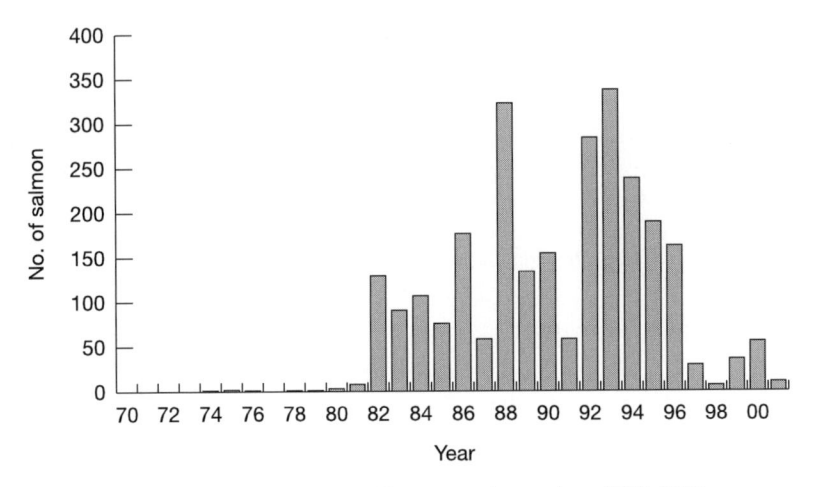

Fig. 16.7 Number of salmon recorded in the Thames each year from 1970–2001.

West coast rivers

Salmon are once again returning to all the rivers in south Wales, which are also supporting stocks of juvenile fish, as recorded in electrofishing surveys, though in several rivers the numbers are still small. Mawle et al. (1985) relate the initial return of salmon to the Taff to improvements in water quality. The declared rod catch on this river reached a peak of 114 salmon in 1988; many were tagged fish derived from stocking as smolts. The future expansion of the stock on this river depends on the provision of access to significant spawning grounds, as well as the successful passage of fish over an estuarine barrage, recently constructed, and through its impoundment (Mawle, 1991). Salmon have been attempting to enter the adjacent River Ely since 1991 but were stopped by a weir at the head of tide. That weir has been removed, but the river is now affected by the same barrage as the Taff. The extent to which the barrage obstructs migration is unclear as yet, but significant numbers of adult salmon and sea trout are passing it. Further west, the Rivers Ogmore, Neath, Afan and Tawe are now all supporting significant rod fisheries for salmon (Table 16.1, Fig. 16.8) and more particularly sea trout (*Salmo trutta* L.). The Tawe has also been impounded in its estuary.

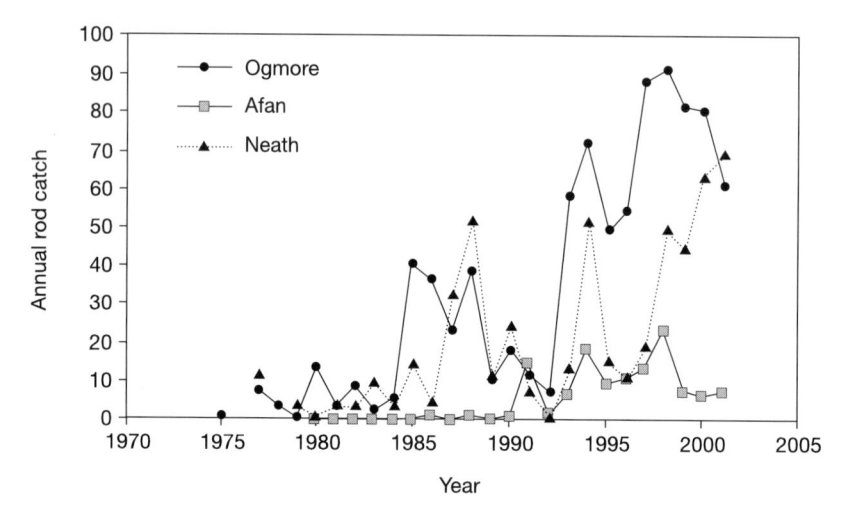

Fig. 16.8 Recovery of the salmon rod fisheries in south Wales, as indicated by the annual declared rod catch of salmon for three rivers, the Ogmore, Afan and Neath, from 1970–2001.

The Stour, a tributary of the Severn, was once grossly polluted by effluents from the industrial West Midlands, known as the 'Black Country', and largely fishless. A fisheries survey in 2000 found salmon in the river for the first time in more than a century (Environment Agency, 2002b).

Further north, salmon have also recently returned to the Mersey. More than £1 billion has been spent on cleaning up the effluents to this river. Three salmon were trapped at Woolston weir on the Mersey in 2001, though as yet there is no evidence that they can reach potential spawning areas.

The future

The process of recovery is still continuing, but for recovery programmes to work well all the conditions for salmon survival and production have to be in place. These include safe passage through clean estuaries, access to silt-free spawning areas, good environmental quality and flow regimes in rearing areas and finally access back to the sea for smolts. If just one of these requirements is missing, even intermittently, then the life cycle is broken and recovery is stalled or at best very slow. The slow progress with the return of salmon to the Thames is indicative of the difficulties faced on some rivers.

As yet the large catchments of the Ouse, Trent, Thames and Mersey are not able to support a salmon fishery of any consequence, though this may happen with time. Nonetheless, the return of salmon to these rivers highlights the major improvements in environmental quality that have been achieved. Regardless of the interests of fishermen, the presence of salmon would seem to be valued by the general public. A household survey in the Thames catchment indicated that residents would be willing to pay about £12 million a year to establish a breeding population of salmon in the river (Spurgeon et al., 2001).

The decline of stocks in rural rivers

Where it has occurred, recovery of stocks in rivers affected by urban and industrial pollution has been noticeable because the affected rivers started from a very degraded position and contrasts with the wide-scale decline of salmon in the north Atlantic (Potter & Crozier, 2000).

In England and Wales, analysis of trends in anglers' declared rod catches has shown that they had significantly declined ($P < 0.05$) in 42% of rivers over the period 1974–98. Superimposed upon the overall decline in salmon is a complex mosaic of change in catch patterns across England and Wales resulting from the influence of local factors (Milner, 2000). Included in the declining stock group are rivers designated as Special Areas of Conservation, such as the Itchen, Hampshire Avon, Camel, Wye, Usk, Teifi, Tywi and Dee (Fig. 16.9). These are located in the most rural and notionally pristine habitats in the country. Decline in these rivers is partly due to factors affecting survival at sea, but there is concern over the cumulative influence of several freshwater environmental factors. These are clearly not related to direct influence of point industrial or domestic pollution, but represent a suite of issues derived from diffuse pollution and river flow changes. Issues include subtle impacts of diffuse pollution from acidification, pesticides and other organic chemicals and silt loading.

Acidification has been a well-known environmental problem in Wales since the early 1980s (Stoner & Gee, 1985; Milner & Varallo, 1990). The mechanisms have been well described and there has been some success in reversing low pH using lime dosing and other methods (Scarr et al. 2002). Although the deposition of SO_2 and NO_x has decreased through emission controls (National Expert Group on Transboundary Air Pollution, 2001), acidification is still a serious issue in the headwaters of some major catchments such as the Wye and other rivers draining the upland regions of Wales and north-west England.

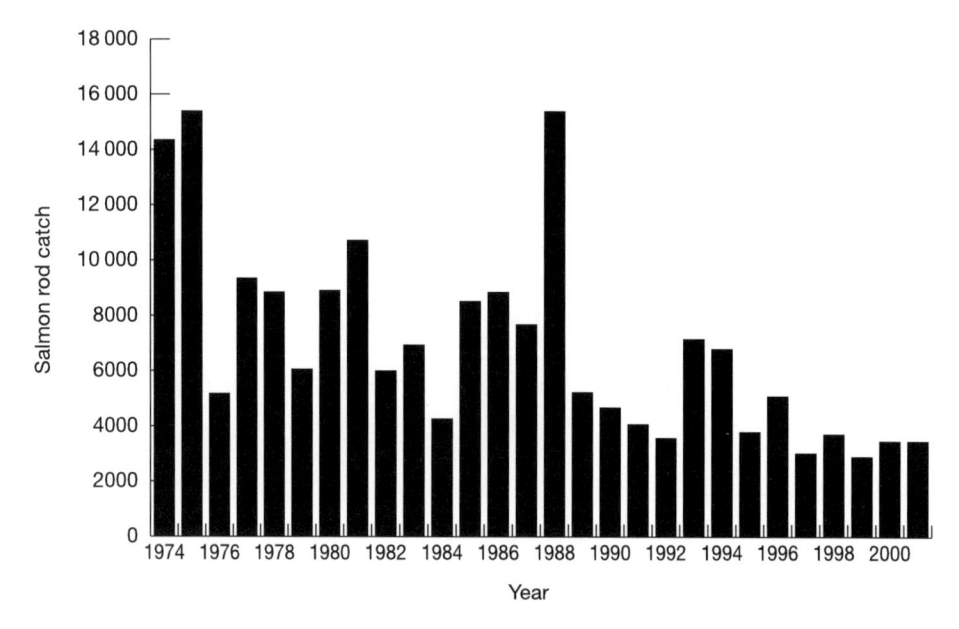

Fig. 16.9 Declared rod catch of salmon from 1974–2001 for eight rivers in England and Wales designated as Special Areas of Conservation for salmon (Rivers Itchen, Hampshire Avon, Camel, Wye, Usk, Teifi, Tywi, and Dee).

Intensive agriculture is thought to have a major impact on the freshwater environment through a variety of processes (Environment Agency, 2000). Intensive stock grazing and changes in stock management (e.g. over-wintering of cattle and pigs in open fields; increased maize production for forage) cause field and stream-bank erosion leading to siltation of river beds. Such changes directly affect salmon by preventing successful egg incubation, trapping alevins in gravel and smothering fry habitat (Theurer et al., 1998; Acornley & Sear, 1999). In addition, siltation may be having an indirect effect on fish production by causing loss of invertebrate production and reducing food availability. Similar concerns have been expressed for salmon habitat in the River Bush, Northern Ireland (Potter & Crozier, 2000).

The use and disposal of pesticides in the process of agriculture and forestry is also regarded as a serious issue (Fairchild et al., 1999). A survey of Welsh streams in 1997 showed that 95% of sites surveyed had pesticide residues present and 26% of farms were a 'high' pollution risk (Fig. 16.10). By 1999, following intensive farm campaigns by the Environment Agency to propagate best practice, the proportion of farms at high risk had reduced to 3%, but pesticides were still present in 67% of sites. The concern is that high levels can be anticipated in any areas where intensive sheep farming occurs. Sheep headage has increased significantly in many rural areas. Between 1980 and 1993 the total number of sheep increased by almost 40% to just under 44 million (Sansom, 1999). Sublethal effects of pesticides on salmon reproductive function have been demonstrated at concentrations well below their nominal Maximum Acceptable Concentration; a standard based on direct toxicity tests, upon

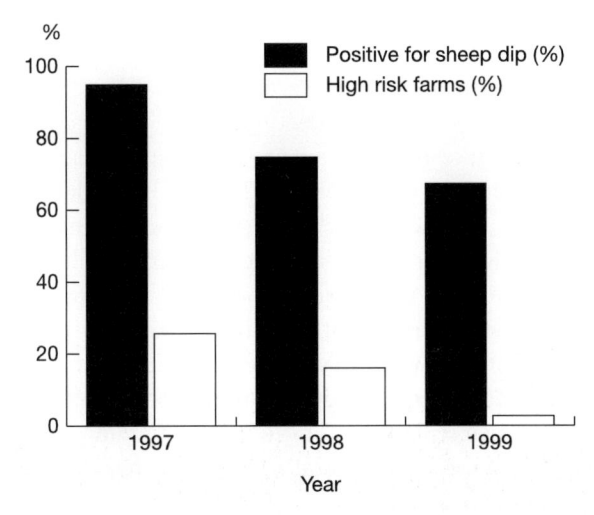

Fig. 16.10 Results of a survey of sheep farming areas in Wales, showing the proportions (%) of (1) farms classified as high risk and (2) stream sites where sheep dip pesticides were recorded.

which environmental standards are set (e.g. Moore & Waring, 1998). In addition to direct effects on fish, pesticides can be expected to have effects on insects, for that is their intended function. Organophosphate pesticides have been recognized as being injurious to human health and have been partially replaced by synthetic pyrethroid pesticides. However, synthetic pyrethroids are far more toxic to invertebrates than organophosphate pesticides and several serious pollution incidents have occurred in which aquatic invertebrates were killed over long lengths of rivers. Establishing cause and effect in complex aquatic ecosystems is difficult, but recent reports of declining invertebrate numbers and diversity (Frake & Hayes, 2002) may be attributable to the cumulative effects of low-level organic contaminants as well as other environmental changes.

Changes to flow patterns are thought to have arisen through a combination of causes varying in importance between catchments. For example, rod catches from the southern chalk rivers (Hampshire Avon, Itchen, Test, Frome and Piddle) show a common pattern of rapid decline following successive droughts in the period 1989–92, exacerbated by substantial abstraction. In contrast, on rain-fed rivers, the more recent increase in winter precipitation linked to climate change (McKenzie Hedger et al., 2000) combined with compacted soils or enhanced drainage will enhance run-off, and can be expected to increase soil erosion, siltation and egg washout. Overgrazing in upland areas has been highlighted as a cause of increased run-off in recent years (Sansom, 1999).

Research on the River Dee (Davidson, pers. comm.) indicates that salmon growth pattern in fresh water may be changing, linked to a warmer climate, leading to production of younger, smaller smolts. Marine survival may be altered as it has been observed to increase with smolt age (Salminen, 1997).

Summary

The overall picture of salmon stock change around England and Wales is therefore a complex one of a broad-scale change, reflecting marine factors, on which is superimposed variable performance in rivers around the country each experiencing different circumstances. There are now more catchments in England and Wales with salmon than at any time for the last 150 years. The recovering rivers represent a significant victory in the long-term campaign against gross pollution that has come about through a combination of massive investment and fortuitous decline of extractive and heavy industries, combined with effective regulation. However, salmon rivers, including those that lie in wholly rural areas, are subject to a complex cocktail of damaging environmental changes, mainly resulting from intensive land use. Thus one set of problems has been replaced by another, which requires new understanding to establish the underlying mechanisms and new approaches to policy and management in order to find solutions.

Acknowledgements

We thank all the various Environment Agency staff for the information provided on the status of individual rivers, including Roger Inverarity, Miran Aprahamian, Roy Sedgwick, Gary Cyster, Darryl Clifton-Dey, Peter Gough, Steve Thomas and Jim Gregory. We are particularly grateful to Rob Evans for his help in the preparation of the figures.

References

Acornley, R.M. & Sear, D.A. (1999) Sediment transport and siltation of brown trout (*Salmo trutta* L.) spawning gravels in chalk streams. *Hydrological Processes*, **23**, 447–58.

Anon. (1860) *Commission into salmon fisheries in England and Wales*. Report and minutes of evidence. Chairman W. Jardine. HMSO, London.

Axford, S.N. (1991) Re-introduction of salmon to the Yorkshire Ouse. In: *Strategies for the Rehabilitation of Salmon Rivers* (ed. D. Mills). Linnean Society, Institute of Fisheries Management and Atlantic Salmon Trust, London, pp. 179–86.

Champion, A.S. (1991) Managing a recovering river – the River Tyne. In: *Strategies for the Rehabilitation of Salmon Rivers* (ed. D. Mills). Linnean Society, Institute of Fisheries Management and Atlantic Salmon Trust, London, pp. 63–72.

Edwards, R.W. (1981) *The impact of coal mining on river ecology*. In: Mining and Water Pollution, Proceedings of Symposium, Institute of Water Engineers and Scientists, Paper 3, pp. 1–8.

Environment Agency (1998) *The State of the Environment of England and Wales: Fresh Waters*. Stationery Office, London.

Environment Agency (2000) *Agriculture and the Environment*. An impact statement prepared by the Environment Agency. EA, Cardiff.

Environment Agency (2002a) *Rivers and Estuaries – A Decade of Improvement. General Quality Assessment of Rivers and Classification of Estuaries in England and Wales*. EA, Cardiff.

Environment Agency (2002b) *Salmon and Freshwater Fisheries Statistics for England and Wales, 2001*. National Salmon and Trout Fisheries Centre of the Environment Agency. EA, Cardiff.

Fairchild, W.L., Swansburg, E.O., Arsenault, J.T. & Brown, S.B. (1999) Does an association between pesticide use and subsequent declines in catch of Atlantic salmon (*Salmo salar*) represent a case of endocrine disruption? *Environmental Health Perspectives*, **107** (5), 349–57.

Frake, A. & Hayes, P. (2002) *Report on the Millenium Chalk Streams Fly Trends Study*. Environment Agency report. EA, Blandford Forum.
Grimble, A. (1913) *The Salmon Rivers of England and Wales*. Kegan Paul, Trench, Trubner & Co., London.
Harris, G.S. (1980) Ecological constraints on future salmon stocks in England and Wales. In: *Atlantic Salmon: Its Future* (ed. A.E.J. Went). Fishing News Books, Farnham, pp. 82–97.
Mawle, G.W. (1991) Restoration of the River Taff, Wales. In: *Strategies for the Rehabilitation of Salmon Rivers* (ed. D. Mills). Linnean Society, Institute of Fisheries Management and Atlantic Salmon Trust, London, pp. 109–121.
Mawle, G.W., Winstone, A.S. & Brooker, M.P. (1985) Salmon and sea trout in the Taff – past, present and future. *Nature in Wales*, New Series, **4** (1 & 2), 36–45.
McKenzie Hedger, M., Gawith, M., Brown, I., Connell, R. & Downing, T.E. (eds) (2000) *Climate Change: Assessing the Impacts – Identifying Responses. The First Three Years of the UK Climate Impacts Programme*. UKCIP Technical Report, UKIP and DETR, Oxford.
Milner, N.J. (2000) Changes in migratory fish populations. Vol. 7, paper no. 10. *Coastal Futures, Proceedings of Coastal Management for Sustainability – Review and Future Trends 2000* (ed. R. Earll). Coastal Management for Sustainability, Kempley, pp. 47–52.
Milner, N.J. & Varallo, P.V. (1990) Effects of acidification on fish and fisheries in Welsh rivers and lakes. In: *Acid Waters in Wales* (eds J.H. Stoner, A.S. Gee & R.W. Edwards). Junk, The Hague.
Moore, A. & Waring, C.P. (1998) Mechanistic effects of a triazine pesticide on reproductive function in mature male Atlantic salmon (*Salmo salar* L.) parr. *Pesticide Biochemistry and Physiology*, **62**, 41–50.
National Expert Group on Transboundary Air Pollution (2001) *Transboundary Air Pollution: Acidification, Eutrophication and Ground Level Ozone in the UK*. Report prepared by NEGTAP, DEFRA contract EPG 1/3/153, 314 pp. NEGTAP, Centre for Ecology and Hydrology, Penicuik, Scotland.
National Rivers Authority (1991) *The Quality of Rivers, Canals and Estuaries in England and Wales: Report of the 1990 Survey*. National Rivers Authority Water Quality Series no. 4. NRA, Bristol.
National Statistics (2000) *Digest of United Kingdom Energy Statistics 2000*. Stationery Office, London.
Netboy, A. (1968) *The Atlantic Salmon, a Vanishing Species*. Faber & Faber, London.
Office for National Statistics (1997) Motor vehicle production and steel production and consumption. Table 4.3 in *Economic Trends, Annual Supplement 1997 Edition*. Stationery Office, London.
Potter, E.C.E. & Crozier, W.W. (2000) A perspective on the ocean life of Atlantic salmon. In: *The Ocean Life of Atlantic Salmon: Environmental and Biological Factors Influencing Survival* (ed. D. Mills). Fishing News Books, Oxford, pp. 19–36.
Russell, I.C., Ives, M.J., Potter, E.C.E., Buckley, A.A. & Duckett, L. (1995) *Salmon and Migratory Trout Statistics for England and Wales, 1951–1990*. Data Report 38, MAFF Directorate of Fisheries Research, Lowestoft, 252 pp.
Salminen, M. (1997) Relationships between smolt size, post smolt growth and sea age at maturity in Atlantic salmon ranched in the Baltic Sea. *Journal of Applied Ichthyology*, **13**, 121–30.
Sansom, A.L. (1999) Upland vegetation management: the impacts of overstocking. *Water Science Technology*, **39** (12), 85–92.
Scarr, A., Edwards, P. & Bishop, M. (2002) *Acid Waters Remediation and Long Term Monitoring in Wales*. Environment Agency R&D Technical Report P2-246/TR. Environment Agency R&D Dissemination Centre, c/o WRc, Swindon, Wiltshire.
Spurgeon, J., Colarullo, G., Radford, A.F. & Tingley, D. (2001) *Economic Evaluation of Inland Fisheries*. Environment Agency R&D Project W2-039/PR/2 (Module B).
Stoner, J.H. & Gee, A.S. (1985) Effects of forestry on water quality and fish in Welsh rivers and lakes. *Journal of Water Engineers and Scientists*, **39**, 27–45.
Theurer, F.D., Harrod, T.R. & Theurer, M. (1998) *Sedimentation and Salmonids in England and Wales*. Environment Agency R&D Technical Report P194 for R&D Project P2-103. Environment Agency R&D Dissemination Centre, c/o WRc, Swindon, Wiltshire.

Chapter 17
Opening Up New Habitat: Atlantic Salmon (*Salmo salar* L.) Enhancement in Newfoundland

C.C. Mullins[1], C.E. Bourgeois[2] and T.R. Porter[2]

[1] *Department of Fisheries and Oceans, Joseph Roberts Smallwood Building, 1 Regent Square, Corner Brook, Newfoundland, A2H 7K6, Canada*
[2] *Department of Fisheries and Oceans, Northwest Atlantic Fisheries Centre, PO Box 5667, St. John's, Newfoundland, A1C 5X1, Canada*

Abstract: Many Newfoundland rivers are characterized by natural obstructions that are impassable to upstream migrating Atlantic salmon (*Salmo salar* L.). Vast amounts of excellent juvenile Atlantic salmon spawning and rearing habitat occur upstream of the obstructions in many of these watersheds. A programme to expand the range of Atlantic salmon was initiated in the mid-1900s. Fishways were constructed on some rivers and colonization of Atlantic salmon occurred using several techniques: natural straying, adult transfer or stocking with unfed fry. The relative success of enhancement on four rivers (Exploits River, Rocky River, Terra Nova and Torrent River) representing the three colonization methods is evaluated. In all cases, Atlantic salmon successfully colonized the new habitat. All stocks improved, with the largest growth increments being achieved within the first three generations. Differences in the rate of growth of the populations using the different enhancement methods are examined. In contrast to enhanced stocks, selected other Newfoundland rivers showed no change in returns of small salmon (< 63 cm) and returns of large salmon (> 63 cm) declined over the same period. Compared to unenhanced stocks, stocking in new habitat contributed to improved sport fishery catches. The results are discussed with respect to habitat, species diversity/interactions, the 1992 moratorium on commercial salmon fishing and the Canadian Biodiversity Strategy.

Introduction

Many Newfoundland rivers are characterized by natural obstructions that are impassable to upstream migrating Atlantic salmon (*Salmo salar* L.). Vast amounts of excellent juvenile Atlantic salmon rearing habitat occur upstream of the obstructions in many of these watersheds. An enhancement programme to increase production of Atlantic salmon through range extension into inaccessible areas was implemented in the 1940s and continued until the mid-1990s. Initially, the primary objective of the enhancement was to increase socio-economic benefits from the commercial Atlantic salmon fishery and more recently from the recreational fishery. The enhancement programme involved opening up new habitat by constructing fishways and colonization by either one or a combination of three methods (Farwell & Porter, 1976): natural straying of adult salmon; stocking of adult salmon; and stocking of unfed salmon fry (using artificial spawning channel and/or upwelling incubation boxes). This paper provides an account of the enhancement programme and an evaluation of the success of the three enhancement techniques. Relative success of the enhancement techniques is assessed based on changes in colonized Atlantic salmon populations on Great Rattling Brook (a tributary of the lower Exploits River), Middle Exploits River, Rocky River, Terra Nova River and Torrent River (Table 17.1, Fig. 17.1). Enhanced populations are compared with unenhanced populations on Gander River, Middle Brook, Northeast Brook (Trepassey), Northeast River (Placentia), Lomond River and Western Arm Brook (Table 17.1, Fig. 17.1).

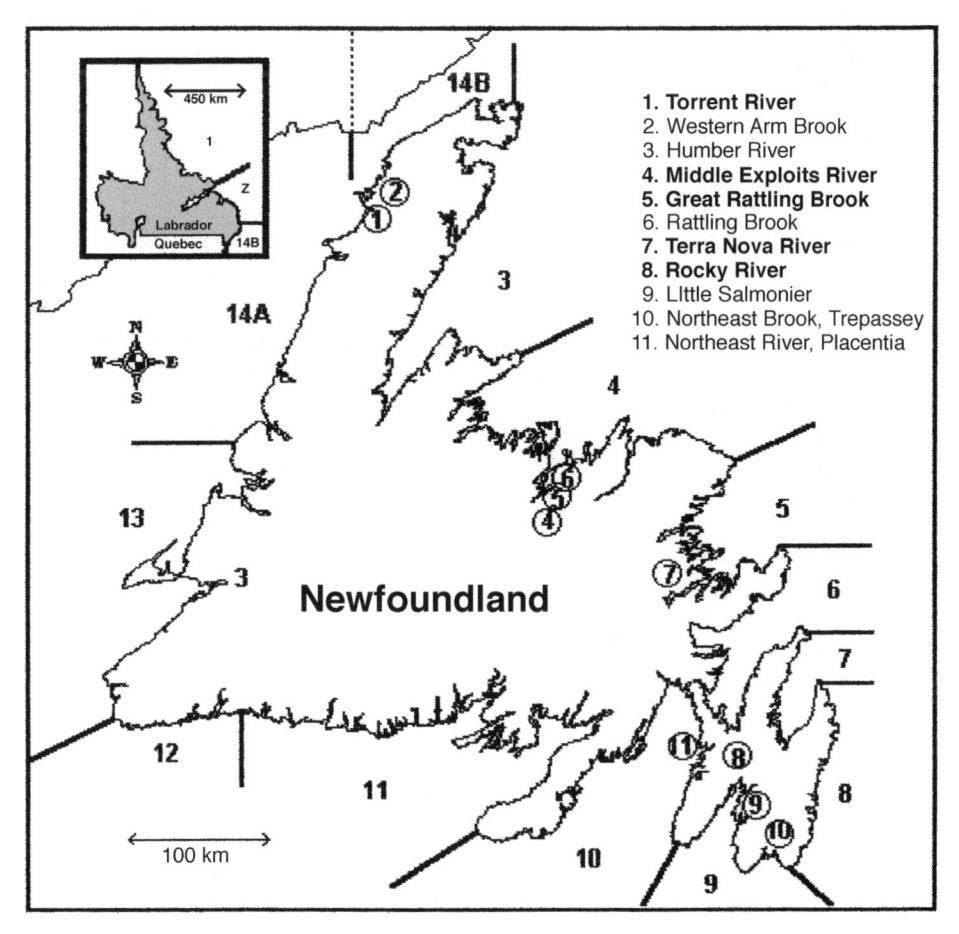

Fig. 17.1 Map of Newfoundland and Labrador showing salmon fishing areas and location of enhanced rivers and donor streams.

Table 17.1 Stocking history and monitoring of rivers with enhanced and unenhanced Atlantic salmon stocks.

River	Type of obstruction	Enhancement method	Year of fishway construction	Stocking method	Stocking period	Years of stocking	Monitoring period
Enhanced stocks							
Great Rattling Brook	Natural, complete	Fishway	1960	Adult transfer and natural straying	1957–1964	8	1960–2001
				Fry stocking[1]	1987–1992		
Middle Exploits River	Natural, complete	Fishway	1974	Fry stocking and natural straying	1970–1992	22	1970–2001
Upper Terra Nova River	Natural, complete	Fishway	1955	Natural straying	1955–2001		1955–2001
Rocky River	Natural, complete	Fishway	1987	Fry stocking	1984–1987	4	1987–2001
				Adult transfer	1987	1	
				Fry stocking	1995–1996	2	
Torrent River	Natural, complete	Fishway	1965	Adult transfer and natural straying	1972–1976	5	1966–2001
Unenhanced stocks							
Gander River							1989–2001
Middle Brook							1956–2001
Northeast Brook, Trepassey							1984–2001
Northeast River, Placentia							1968–2001
Lomond River							1961–2001
Western Arm Brook							1971–2001

[1] Fry were stocked at a rate equivalent to 20% natural egg-fry survival to compensate for brood removal for Middle Exploits

Table 17.2 Physical description of rivers with enhanced and unenhanced Atlantic salmon stocks. SFA refers to Salmon Fishing Area as shown in Fig. 17.1; N/A = not applicable.

River	SFA	Total drainage area[1] (km^2)	Approximate % of habitat opened	Fluvial area (1000 m^2)	Lacustrine area (100 ha)	Ratio lacustrine: fluvial	Stream order (at mouth)
Enhanced stocks							
Great Rattling Brook	4	1 351	90	492	62	125:1	3
Middle Exploits River	4	11 272	100	34 800	338	10:1	6
Upper Terra Nova River	5	1 883	40	3 260	174	53:1	5
Rocky River	9	296	100	1 080	22	20:1	4
Torrent River	14A	619	95	517	23	45:1	4
Unenhanced stocks							
Gander River	4	6 398	N/A	15 956	215	13:1	4
Middle Brook	5	276	N/A	264	46	176:1	2
Northeast Brook, Trepassey	9	21	100	56	0	5:1	2
Northeast River, Placentia	10	94	N/A	14	10	741:1	2
Lomond River	14A	470	100	216	16	73:1	4
Western Arm Brook	14A	149	N/A	290	20	70:1	3

[1] Drainage area data from Porter et al. (1974)

Description of rivers

The drainage areas of the five study rivers range from 296 km^2 for Rocky River to 11 272 km^2 for Middle Exploits River with stream orders ranging from 3 to 6 (Table 17.2). The watersheds are comprised of both fluvial and lacustrine Atlantic salmon parr rearing habitat (Table 17.2). In Newfoundland, juvenile salmon extensively use lacustrine habitat for rearing and over-wintering (Pepper, 1976; Chadwick & Green, 1985; Ryan, 1986; O'Connell et al., 1990). Great Rattling Brook is a relatively large tributary of the Exploits River, and a fishway was constructed at a falls, approximately 8 km upstream from its confluence with the Exploits River, in 1959. A fishway was constructed on the Exploits River, at Grand Falls (up-river from Great Rattling Brook tributary and 25 km from the estuary) in 1974. A fishway was constructed on the Upper Terra Nova River, 14 km from the estuary, in 1955; on Rocky River, in the estuary, in 1987; and on Torrent River, 2 km from its estuary, in 1965. The fishways opened between 40 and 100% of the habitat currently available to salmon on the five rivers (Table 17.2). Prior to the introduction of anadromous Atlantic salmon, these rivers supported mainly brook trout (*Salvelinus fontinalis* Mitchill), American eels (*Anguilla rostrata* Lesueur) and sticklebacks (*Gasterostidae*). Brown trout (*Salmo trutta* L.) occurred in Rocky River and Arctic charr (*Salvelinus alpinus* L.) in Exploits and Terra Nova Rivers. Landlocked Atlantic salmon occurred in all watersheds except Torrent River.

Methods

Enhancement activities

Colonization was usually spread over at least 5 years, one complete generation, in order to establish all year-classes (modal smolt age of 3+ years) of Atlantic salmon known to occur within each river system. Broodstock used for fry stocking was removed from donor streams systematically throughout the run. Transferred adult salmon were introduced into selected headwater tributaries of Great Rattling Brook and Torrent River from nearby donor streams that were known to have healthy salmon stocks. Table 17.3 shows the average numbers of adult salmon transferred and used as broodstock for fry stocking for each generation or 5-year period in each of the five enhanced rivers. The average number of natural spawners as well as total spawners, small salmon recruits and the ratio of recruits:spawners are also given.

Great Rattling Brook

Rattling Brook, a river adjacent to the mouth of the Exploits River, was the donor stream for Great Rattling Brook, a tributary of Exploits River. O'Connell & Bourgeois (1987) describe the adult stocking on Great Rattling Brook. As compensation for a hydroelectric development in 1956 that created an impassable obstruction at the

Opening Up New Habitat 205

Table 17.3 Results of enhancement of five Atlantic salmon stocks in Newfoundland. The column heading 'Generation i' refers to 5-year periods of salmon growth from hatch to returning adult salmon. SFA refers to Salmon Fishing Area as shown in Fig. 17.1.

Generation	Years	Transfers/ broodstock (Gen. i)	Natural spawners (Gen. i)	Total spawners (Gen. i)	Small recruits	Ratio recruits (Gen. $i+1$):Spawners (Gen. i)
Great Rattling Brook, SFA 4 (adult transfer and natural straying)						
1	1957–1961	532	95	627	744	—
2	1962–1966	71	1131	1173	2335	3.72
3	1967–1971	—	1358	1358	2225	1.90
4	1972–1976[1]	—	2445	2445	6880	5.06
5	1977–1981	—	3634	3634	8731	3.57
6	1982–1986	—	4226	4226	9272	2.55
7	1987–1991[2]	1061	877	1938	4153	0.98
8	1992–1996	—	2929	2929	3997	2.06
	Mean					2.84
Middle Exploits River SFA 4 (fry stocking and natural straying)						
1	1970–1974	466	7	473	164	—
2	1975–1979	1087	23	1110	628	1.33
3	1980–1984	1280	1997	3277	6942	6.25
4	1985–1989	3313	613	3926	6475	1.98
5	1990–1994	1942	2566	3732	5193	1.32
6	1995–1999	—	9942	9942	11569	3.10
7	2000–2001	—	7589	7589	8515	0.86
	Mean					2.47
Upper Terra Nova River, SFA 5 (natural straying)						
1	1955–1959	—	74	74	75	—
2	1960–1964	—	224	224	346	4.68
3	1965–1969	—	394	394	560	2.50
4	1970–1974	—	508	508	616	1.56
5	1975–1979	—	547	547	1378	2.71
6	1980–1984	—	707	707	1477	2.70
7	1985–1989	—	794	794	1493	2.11
8	1990–1994[3]	—	1170	1170	1261	1.59
9	1995–1999	—	1350	1350	1368	1.17
10	2000–2001	—	1681	1681	1568	1.16
	Mean					2.24
Rocky River, SFA 9 (fry stocking and adult transfer)						
1	1984–1987	263	0	263	263	—
2	1990–1993	—	285	285	436	1.66
3	1994–1997[4]	69	281	350	533	1.87
4	1998–2001	—	408	408	353	1.01
	Mean					1.51
Torrent River, SFA 14A (adult transfer and natural straying)						
1	1966–1971	—	40	40	197	—
2	1972–1976	131	164	295	464	11.60
3	1977–1981	—	1377	1377	3136	10.63
4	1982–1986	—	2333	2333	4987	3.62
5	1987–1991	—	1997	1997	4854	2.08
6	1992–1996	—	4902	4902	4882	2.44
7	1997–2001	—	4298	4298	4135	0.84
	Mean					5.20

[1] Means are for only 3 years due to missing counts in 1973 and 1974
[2] Fry were stocked into Great Rattling
[3] Means are for only 3 years due to missing data in 1990 and 1991
[4] Broodstock used in 1995 and 1996 were from Rocky River

mouth of Rattling Brook, adult salmon were transferred out of Rattling Brook and into Great Rattling Brook to an area upstream of a natural impassable obstruction. An average of 532 adult salmon per year were transferred from Rattling Brook into Great Rattling Brook starting in 1957–61 and an average of 71 adult salmon in 1962–64. A fishway was constructed at the natural obstruction on Great Rattling Brook in 1960 to provide passage for returning adult salmon produced by the transferred adults. Salmon passed through the fishway in the first year of operation in 1960 (prior to any returns from adult transfers), indicating that natural straying from below the fishway also occurred.

Middle Exploits River

Initial broodstock for fry stocking on the Middle Exploits River came from Adies Stream, a tributary of Humber River, with subsequent brood from Great Rattling Brook and finally from the Middle Exploits River itself (O'Connell & Bourgeois, 1987). Fry stocking into the Middle Exploits River is described in O'Connell et al. (1983) and O'Connell & Bourgeois (1987). Fry produced from controlled-flow artificial spawning channel and deep-substrate incubation boxes were stocked into a portion of the watershed between Grand Falls and a storage dam at Red Indian Lake (both complete obstructions). Fry were produced from an average of 1589 broodstock per year beginning in 1970 and continuing until 1992. Fry were distributed by helicopter and stocking was limited to areas characterized by parr rearing habitat as described by Elson (1957).

Terra Nova River

A fishway was constructed in 1955 to provide fish passage above a complete obstruction on the upper section of the river. No other enhancement activity was carried out. Between 10 and 140 Atlantic salmon adults passed through the fishway each year in the first 5 years of operation.

Rocky River

Rocky River was inaccessible to Atlantic salmon because of a complete obstruction at the mouth of the river. A fishway was constructed in 1940 but was not made operational until 1987. During 1984–87 an average of 263 salmon (Table 17.3) were used to produce 333 980 unfed fry per year that were stocked in to Rocky River. The donor stream for the broodstock was Little Salmonier River (Fig. 17.1). Fry were stocked throughout the watershed in selected riverine habitat at an average stocking density of 51 (range 45–60) fry/100 m^2. There was also supplementary fry stocking in 1995–96. Adult salmon were transferred to Rocky River in 1 year only (1987), the fifth year of the enhancement programme. In 1987, 124 female salmon (and 16 males) were also transferred into the watershed to spawn naturally.

These are included in the 1984–87 average. Bourgeois (1998) describes the adult transfer into Rocky River.

Torrent River

Between 56 and 213 adult Atlantic salmon were annually transferred to Torrent River from Western Arm Brook, located about 80 km north of Torrent River, from 1972–76. The donor stock is a grilse population with a female to male sex ratio of approximately 4:1 (Mullins et al., 1999). For the initial 6 years (1965–71) following construction of the fishway (Fig. 17.2), 23–58 Atlantic salmon annually passed upstream through the fishway. Presumably, these were strays from the small indigenous population of salmon in the lower section of the river. During the period of stocking, 41–388 salmon passed through the fishway. The fishway on Torrent River and adult transfer from Western Arm Brook, 1966–73, are described by Porter & Davis (1974) and Anon. (1978).

Fig. 17.2 Fishway on the Torrent River, Newfoundland, constructed in 1965 to provide anadromous Atlantic salmon with access to new habitat upstream of the 10-m-high falls located 2 km from the mouth of the river.

Evaluation

The evaluation of the success of the enhancement methods is based on changes in annual and average recruitment and total spawning escapements of small plus large Atlantic salmon; recreational catches and effort; and the ratio of small ($<$ 63 cm) recruits:total spawners. Changes in production from the enhanced rivers were compared to estimates of production of Atlantic salmon in Newfoundland obtained from Anon. (2002). Changes in annual recruitment of small salmon from enhanced rivers were also compared with those in nearby rivers that had not undergone enhancement (Table 17.2).

Annual counts of both small and large Atlantic salmon are available for most years since fishways were constructed on all five rivers. Recreational catch statistics are available for all rivers from 1953–2001 except Rocky River, where recreational fishing has not been permitted. The fishery was also closed in some years on the other rivers until stocks became established in the enhanced areas.

Total annual returns to each river (TRR) were derived separately for small and large salmon by combining counts of salmon at the fishway with retained and 10% of hooked-and-released catches of salmon in the recreational fishery downstream of the fishway:

$$TRR = CNT + RET_{d/s} + HRM_{d/s}$$

where:
CNT = count of small or large salmon at the fishway
$RET_{d/s}$ = number of salmon retained downstream
$HRM_{d/s}$ = hook-and-release mortality downstream

$$HRM_{d/s} = REL_{d/s} \times 0.1$$

where:
$REL_{d/s}$ = number of salmon hooked and released downstream

The mortality rate of 0.1 for hooked-and-released Atlantic salmon is an assumed value based on consultations with anglers. This value is similar to the 0.082 value observed in an experiment on Conne River, Newfoundland (Dempson et al., 2001). No adjustments were made for any other unrecorded mortality.

Recreational salmon fishery catches of Great Rattling Brook salmon were derived from the total angling catch on the Exploits River based on the proportion of returns to Great Rattling Brook fishway counted at Bishop Falls fishway. Similarly, recreational salmon fishery catches of Middle Exploits salmon were based on the proportion of returns to Middle Exploits (Grand Falls) counted at Bishop Falls fishway. Recreational catches of Upper Terra Nova River salmon were based on the total angling catch in the Lower Terra Nova River adjusted for the proportion of returns to Upper Terra Nova River fishway counted at the Lower Terra Nova fishway.

Natural spawning escapements (NSE) were calculated separately for small and large salmon for the area upstream from the fishways by subtracting retained catches, mortality due to catch and release and other known removals (primarily broodstock for fry production) upstream of the fishway from total counts of salmon at the fishway:

$$NSE = CNT - RET_{u/s} - HRM_{u/s} - OTH_{u/s}$$

where:
  CNT = count of small or large salmon at the fishway
  $RET_{u/s}$ = number of salmon retained upstream
  $HRM_{u/s}$ = hook-and-release mortality upstream
  $OTH_{u/s}$ = other known removals upstream

Total spawning escapements to areas upstream from the fishway include natural spawners plus transferred adult Atlantic salmon or broodstock used to produce the fry stocked in the enhanced area.

In years when commercial fisheries were in place, total small salmon recruits were determined by adjusting the total returns to the river of small salmon for exploitation in the Newfoundland–Labrador commercial fishery. Commercial exploitation rates (*er*) for small salmon were taken from Dempson et al. (2001):

$$Recruits_{small} = TRR_{small}/er$$

The percentage of large salmon in Newfoundland rivers is low (5–6 % on average) and they are predominantly repeat spawners. Apportioning large salmon to their appropriate year-classes based on smolt age distribution would not have added substantially to the analysis. Small salmon in Newfoundland also have a small repeat spawner component, but these are treated the same as virgin recruits. Because the data are summarized by generation, an overestimation of the number of small salmon recruits would cause an underestimation of the recruits per spawner (R/S) ratio. However, in this case the difference would not be significant.

Analyses were conducted by comparing the average annual recruits and spawners per generation for each river. A generation stratum corresponds to the number of years from egg to small Atlantic salmon spawner. Generation stratum on Exploits River, Terra Nova River, Rocky River and Torrent River is 5 years. Rocky River produces a high number of 2-year-old smolts, but the modal age is 3 years as on the other rivers.

Results

Anadromous Atlantic salmon stocks were successfully established on all five rivers where new habitat was opened up – including on Terra Nova River where no stocking was carried out (Table 17.3, Fig. 17.3). Stocking programmes were succeeded by

210 *Salmon at the Edge*

Great Rattling Brook
(Adult transfer and natural straying)

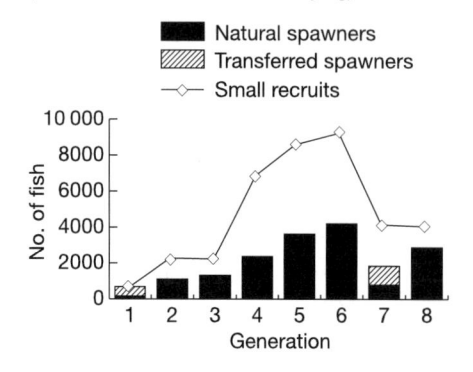

Upper Terra Nova River
(Natural straying)

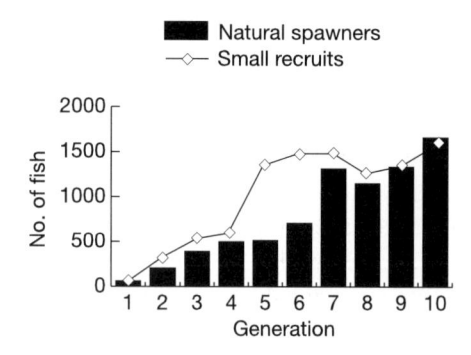

Middle Exploits River
(Fry stocking and natural straying)

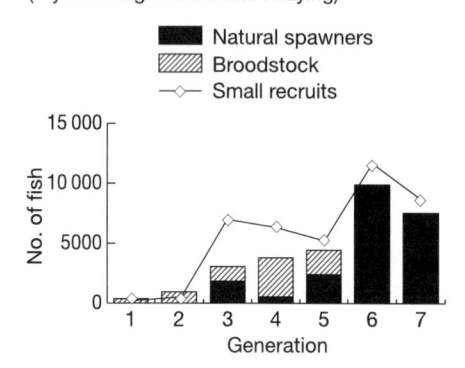

Rocky River
(Fry stocking and adult transfer)

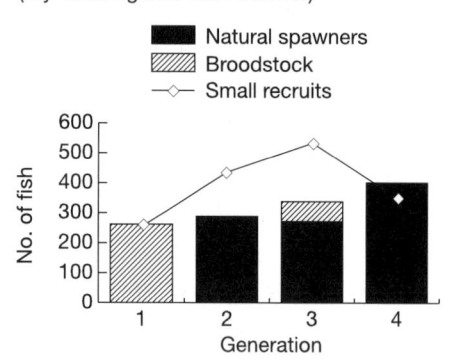

Torrent River
(Adult transfer and natural straying)

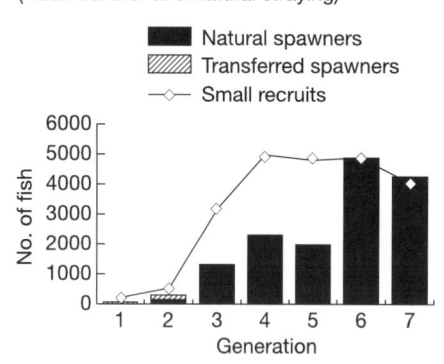

Fig. 17.3 Stocking history of Atlantic salmon and number of small salmon produced by the five enhanced stocks in Newfoundland.

seven generations on Great Rattling Brook, two on Middle Exploits River, three on Rocky River and five on Torrent River (Fig. 17.3).

With some fluctuation, the number of recruits and the number of spawners increased with successive generations after stocking on all rivers (Table 17.3, Fig. 17.3). Monitoring of the stock on Great Rattling Brook stopped in 1996 so the performance of the stock past generation 8 is unknown.

The number of recruits increased in all five stocks in generation 2 (the first generation after stocking) and in four out of the five stocks in generation 3 (Fig. 17.3). Three stocks continued to increase in generation 4 and two into generation 5.

The total number of recruits leveled off or declined after generation 3 on Rocky River, generation 4 on Torrent River, generation 5 on Upper Terra Nova River and in generation 6 on Great Rattling Brook and Middle Exploits River (Fig. 17.3). The increase in numbers of recruits in generation 6 on the Middle Exploits River coincides with an increase in recruits/spawner (Figs 17.3, 17.4). The decrease in recruits on Great Rattling Brook in generation 7 coincides with a large decrease in recruits/spawner (Figs 17.3, 17.4).

Analysis of trends in recruitment of small salmon indicated that the Middle Exploits River had the strongest overall growth based on the slope of the regression line (Fig. 17.5). Torrent River and Great Rattling Brook showed similar growth trends (Fig. 17.5), but Great Rattling Brook declined in generation 7 (Figs 17.3, 17.4). The reason for this decline is not readily apparent. Terra Nova River and Rocky River had the lowest growth (Fig. 17.5). Terra Nova River had the lowest annual variation based on R^2 values, possibly indicating greater stock stability. After only four generations, Rocky River had the weakest growth trend.

The highest population growth rates, measured as the number of recruits per spawner (R/S), ranged from 1.33 to 11.60 for the five rivers and usually occurred within the first three generations of the start of the enhancement programme (Table 17.3, Fig. 17.4). The highest R/S ratio was for generation 1 spawners on Torrent River and Terra Nova River, while for Middle Exploits River and Rocky River, the highest R/S ratio was for generation 2 spawners and generation 3 spawners on Great Rattling Brook (Fig. 17.4). Generation 2 recruits on Torrent River and Terra Nova River were actually the progeny of spawning by strays into the new habitat in generation 1 (Table 17.3). The progeny of transferred adult salmon on Torrent River returned to the river as recruits in generation 3. In all cases, the R/S ratio decreased over time (Fig. 17.4) but with greater fluctuation on Great Rattling Brook and Middle Exploits River. Generally, the R/S ratio was above 2.0 for Great Rattling Brook, Upper Terra Nova River and Torrent River. On the Middle Exploits River only in two of six generations was the R/S ratio above 2.0 and in no years on Rocky River.

The highest average population growth rate (average R/S ratio) for the five rivers was on Torrent River (5.20) and the lowest (1.51) was on Rocky (Table 17.3).

The average number of recruits in the first four generations on all five rivers was above estimates of their cohorts [i.e. the R/S ratio was greater than 1.0 (Table 17.3, Fig. 17.6).

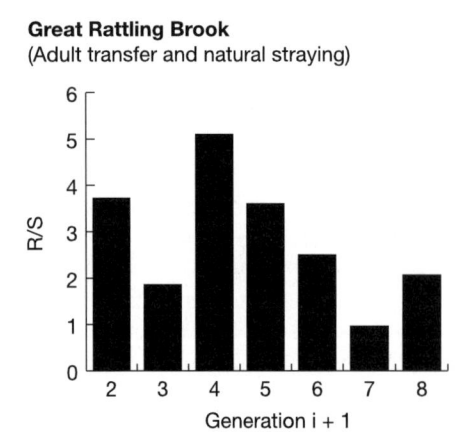

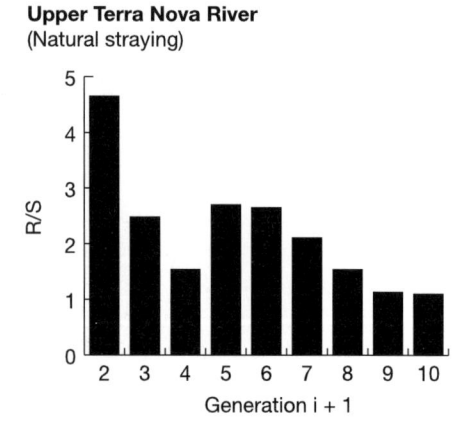

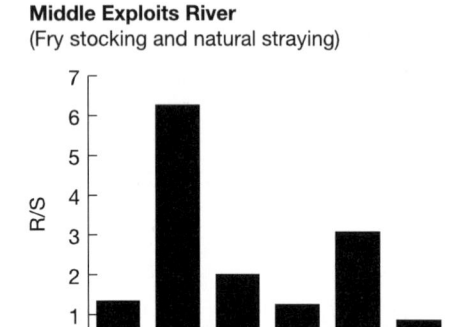

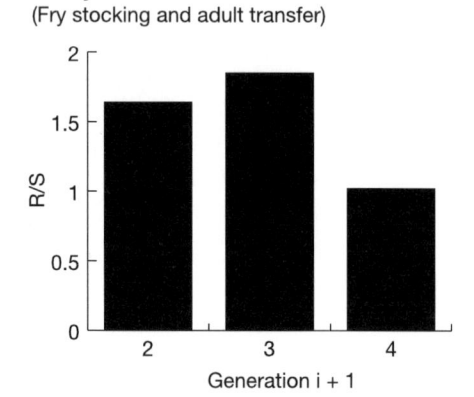

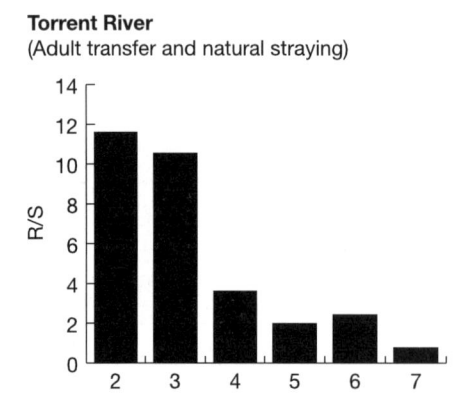

Fig. 17.4 Average number of small Atlantic salmon recruits (R) produced in generation $i+1$ per spawner (S) in generation i on five enhanced rivers in Newfoundland.

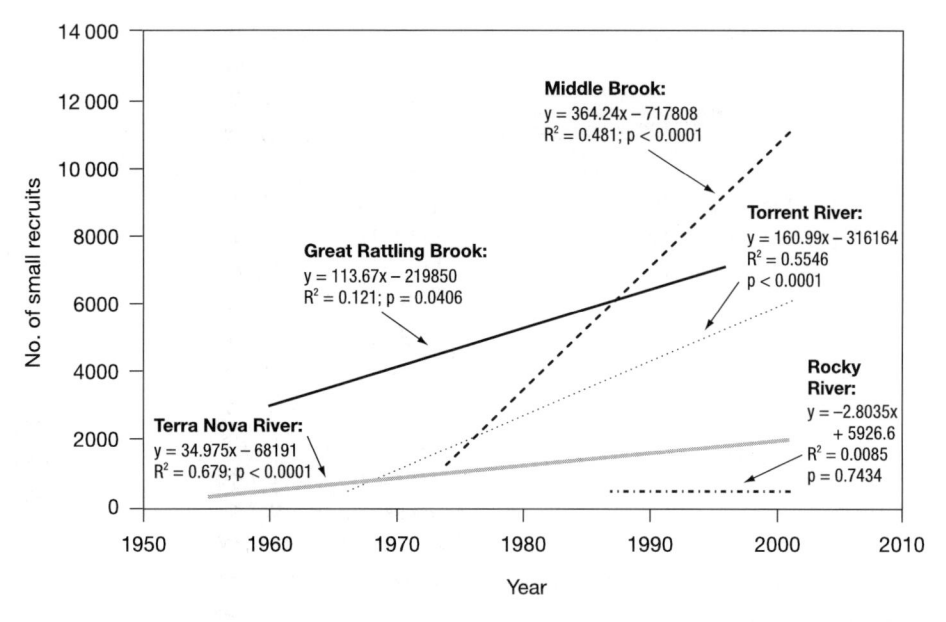

Fig. 17.5 Trends in small recruits of Atlantic salmon produced by enhancement on five Newfoundland rivers.

In contrast to the enhanced stocks, the salmon population in insular Newfoundland has declined (Figs 17.3, 17.7). The total small recruits of four of the five enhanced stocks were highly positively correlated with year ($P < 0.05$) indicating overall growth during the time series (Table 17.4). The total salmon population in insular Newfoundland was highly negatively correlated with year, indicating a general stock decline both before and after the commercial salmon fishery moratorium in 1992 (Table 17.4). Three of the six unenhanced stocks examined (Middle, Northeast Placentia and Lomond) also had a significant positive correlation with year (Table 17.4). These are rivers where fishways were constructed to improve existing runs but no enhancement was carried out. Three of the unenhanced stocks (Gander, Northeast Trepassey and Western Arm) were negatively correlated with year, similar to insular Newfoundland as a whole. Counts of salmon on these rivers were from counting fences – since no fishways were constructed.

The period of greatest decline in insular Newfoundland was during the late 1980s (Fig. 17.7). This decline resulted in a moratorium on commercial salmon fishing in 1992. Enhanced stocks also experienced a decline during this period (Fig. 17.8). The commercial salmon fishery moratorium resulted in only limited improvement to either insular Newfoundland as a whole (Fig. 17.7) or to enhanced stocks (Fig. 17.8). Of the five enhanced stocks, only Middle Exploits showed a large increase in population after 1992, indicating that it is still in a growth phase but other stocks may be at capacity. Since 1992, the five enhanced stocks contributed approximately 13% (min.

214   Salmon at the Edge

Great Rattling Brook
(Adult transfer and natural straying)

Upper Terra Nova River
(Natural straying)

Middle Exploits River
(Fry stocking and natural straying)

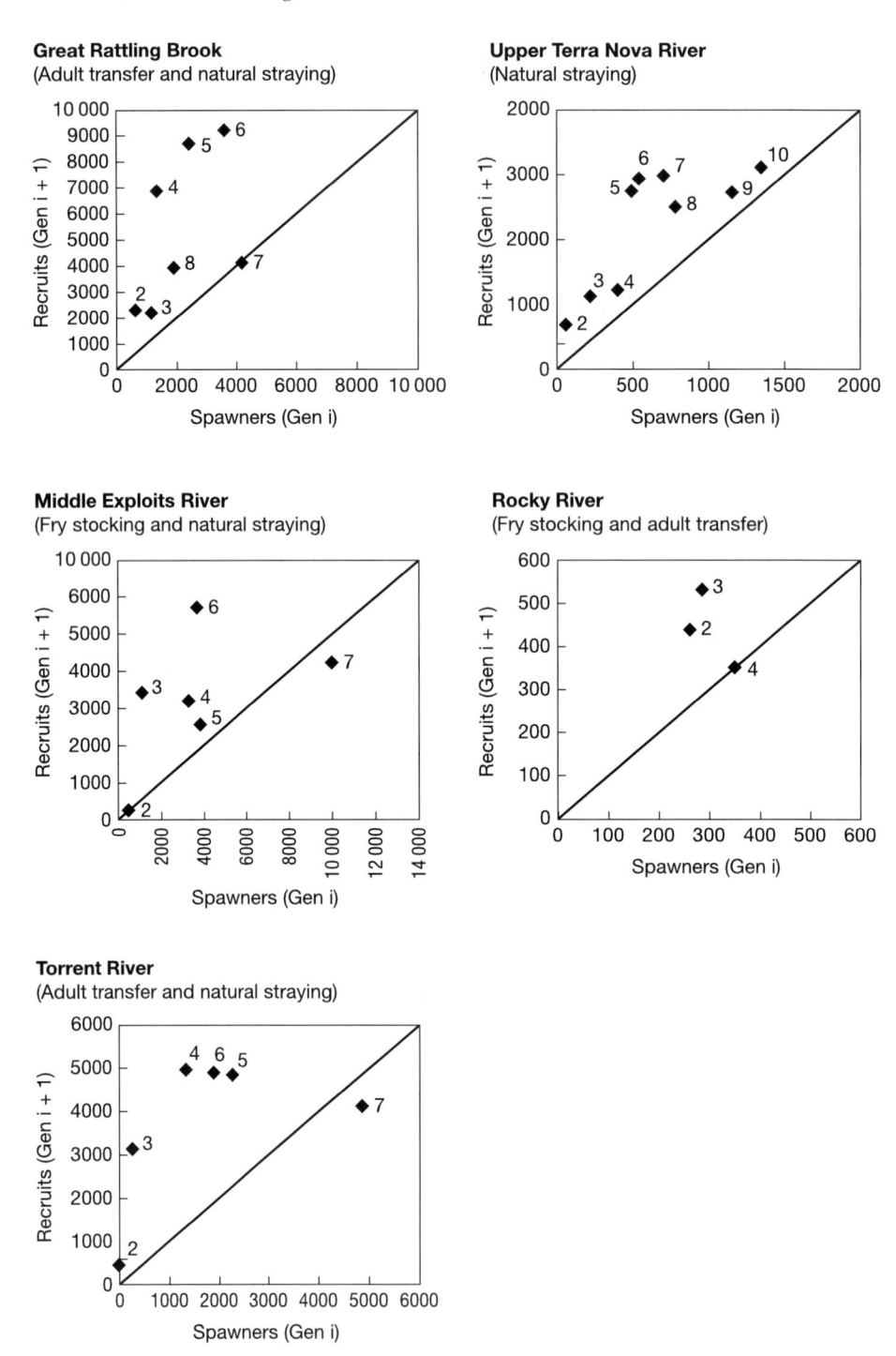

Rocky River
(Fry stocking and adult transfer)

Torrent River
(Adult transfer and natural straying)

Fig. 17.6 Atlantic salmon spawners in generation i and small salmon recruits in generation $i+1$.

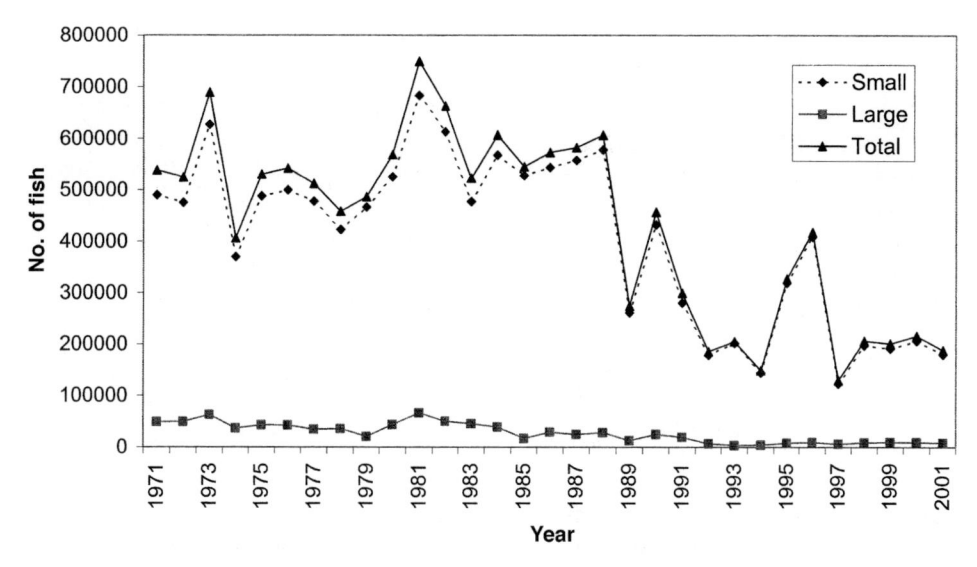

Fig. 17.7 Estimated total annual production of small, large and total Atlantic salmon recruits in insular Newfoundland, 1971–2001.

Table 17.4 Correlation analysis of Atlantic salmon recruitment with year on enhanced and unenhanced rivers. Includes data to 1991. Values in italics indicate rivers where recruitment had a significant positive correlation with year at the 0.05 level. Underlined values were significant at the 0.01 level.

River	r value
Enhanced stocks	1.00
Great Rattling Brook	*0.48*
Middle Exploits River	0.49
Terra Nova River	0.87
Rocky River	0.11
Torrent River	*0.81*
Total enhanced	0.79
Average enhanced	0.74
Unenhanced stocks	
Gander River	−0.95
Middle Brook	*0.59*
Northeast Brook, Trepassey	−0.64
Northeast River, Placentia	0.70
Lomond River	0.75
Western Arm Brook	−0.30
Total unenhanced	0.82
Average unenhanced	0.72
Total insular Newfoundland	−0.44
n	39

10%, max. 19%) of the estimated total production of Atlantic salmon in insular Newfoundland (Fig. 17.8).

Angling catches increased in all rivers compared to pre-enhancement levels (Fig. 17.9). Angling catches prior to 1992 increased with year on Torrent River ($Y = 3.85x - 7489.34$; $R^2 = 0.35$; $P = 0.0002$), Exploits River ($Y = 42.27x - 82309.5$; $R^2 = 0.47$; $P \leq 0.0001$) and Terra Nova River ($Y = 14.5x - 28111.8$; $R^2 = 0.65$; $P \leq 0.0001$). Effort also increased significantly on Exploits River ($Y = 203.28x - 396948$; $R^2 = 0.63$; $P \leq 0.0001$) and Terra Nova River ($Y = 14.5x - 28111.8$; $R^2 = 0.25$; $P = 0.0013$) but not on Torrent River. Catch per unit effort (CPUE) measured as catch per rod day increased on Torrent River ($Y = 0.0062x - 11.9$; $R^2 = 0.53$; $P \leq 0.0001$) but declined on Exploits River. Catch rate increased initially on Terra Nova River but remained stable overall (Fig. 17.9).

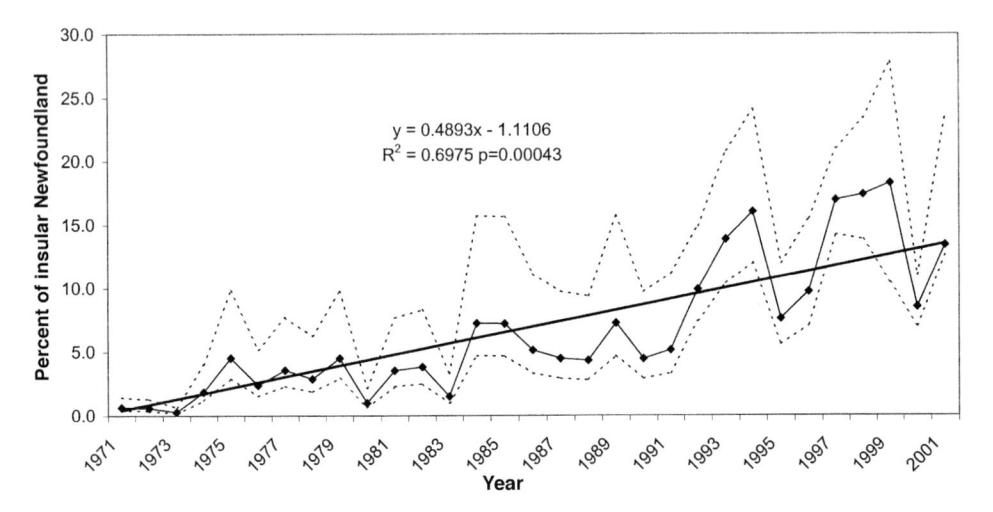

Fig. 17.8 Contribution of Atlantic salmon from five enhanced stocks to the total production of small salmon in insular Newfoundland. Dashed lines represent minimum and maximum percentages and the solid line is the regression line.

Regression of catch rate on population size was significant for Torrent River ($Y = 74.87xe^{-6}x + 0.1444$; $R^2 = 0.63$; $P \leq 0.0001$) and for insular Newfoundland in general ($Y = 954.07xe^{-9}x + 0.1371$; $R^2 = 0.64$; $P \leq 0.0001$) but not for Great Rattling Brook, Middle Exploits River or Terra Nova River. Angling data for the Exploits River is for the entire system including Great Rattling Brook and the Middle Exploits. Hence, it was not possible to make a true comparison between catch rate and number of fish on either Great Rattling Brook or Middle Exploits River. The decline in catch rate with increased returns on Terra Nova River was not significant at the 5% level. Catch rates in some rivers may be dependent on factors other than the number of fish, such as water discharge, water temperature and fisheries management plans. Recreational Atlantic salmon fishery management plans introduced in Newfoundland in 1978, 1984 and 1992 resulted in shorter seasons and reduced bag limits

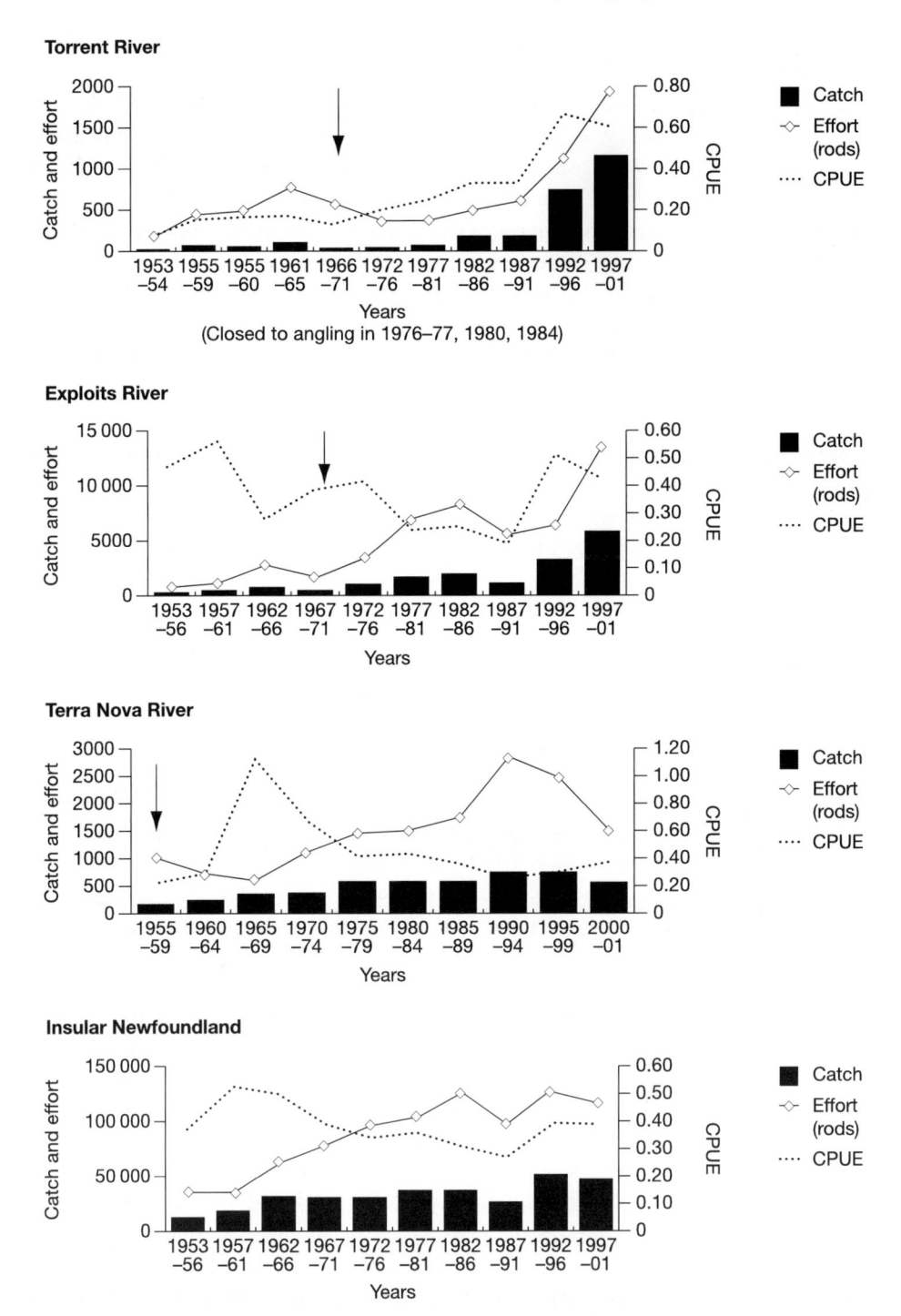

Fig. 17.9 Recreational Atlantic salmon fishery catch, effort and catch-per-unit-effort (CPUE) by 5-year means for three enhanced rivers and insular Newfoundland. Arrows indicate start of enhancement programme.

which would have affected the amount of angling effort and catch rates on some rivers depending on angling patterns. In addition, the recreational salmon fishery was closed completely on Torrent River in 1976–77, 1980 and 1984.

Discussion

All of the enhancement programmes were successful in establishing anadromous Atlantic salmon populations. The size of the populations of all five established stocks increased during a period when the overall population of Atlantic salmon in insular Newfoundland remained relatively constant or declined (Figs 17.3, 17.7, 17.8, Table 17.3). The enhanced populations also increased during a period when salmon populations in three of the six unenhanced stocks and in insular Newfoundland in general declined (Table 17.4). These observations suggest that the natural survival rates (probably juvenile in-river) for these stocks were generally higher than for other Newfoundland stocks. The three unenhanced stocks examined that also showed an increase in salmon populations had fishways constructed to improve fish passage for existing runs, suggesting a benefit from enhancement even in the form of improved access to spawning and rearing habitat. The rate of increase in annual number of recruits, and the rate of growth of the enhanced populations, as measured by R/S ratio, were variable among enhanced rivers (Table 17.3, and Figs 17.3, 17.4, 17.5). There are likely many factors that would influence this variability, such as: (1) number of fish stocked; (2) inter- and intra-specific competition; (3) distribution of spawning and rearing habitat; (4) amount of lacustrine habitat in relation to spawning habitat; (5) quality of habitat; and (6) genetic fitness of introduced fish.

In all rivers the R/S ratio was highest in either the first, second or third generation (Table 17.3, Fig. 17.4). Low intra-specific competition was likely a contributing factor to this observation. The average R/S ratios were higher for Torrent River (5.20) and Great Rattling Brook (2.84), which were stocked with adult Atlantic salmon, than either of the other three rivers (Table 17.3). The average R/S ratio for Rocky River (1.51) was the lowest of all enhanced rivers (Table 17.3). The average R/S ratio (2.47) for the Middle Exploits River was slightly higher than that (2.24) for Upper Terra Nova River (Table 17.3). However, only two of six generations for the Middle Exploits River stock had an R/S ratio > 2.0 compared to five out of nine for Upper Terra Nova River (Table 17.3). Thus, it would appear, for these five rivers that naturally spawning Atlantic salmon, either stocked or strays, provided more recruits per spawner than fry stocking. However, there are insufficient data to determine the extent that some of the factors, referenced in the paragraph above, may have contributed to the low R/S ratios on the Middle Exploits and Rocky Rivers. Inter-specific competition with brown trout is believed to be one of the causes of the low R/S ratio for Rocky River.

The rate of increase in total population size appears to be influenced by the numbers of spawners. The Middle Exploits River had the highest increase in population growth of the four rivers, followed by Great Rattling Brook and Torrent River, even though Torrent River had the highest R/S ratio (Fig. 17.4, Table 17.3). Natural straying was the most cost-effective enhancement technique; however, the population growth is slow. Adult transfers and fry stockings appear to produce higher population growth in a shorter time span than natural straying without any form of stocking.

The natural survival rate of Atlantic salmon at sea declined in the late 1980s and early 1990s and has remained low, causing the decline in overall population size observed in Fig. 17.7 (Anon., 2002). This low survival is a contributing factor for the lower R/S ratio observed in the enhanced stocks in recent generations (Table 17.3, Fig. 17.4) and places constraints on interpretation. However, with only two exceptions, R/S ratios in recent generations were all > 1.0 (Table 17.3). This suggests that these stocks were in a growth mode (Fig. 17.6). However, growth of the spawning population in some generations was minimal or declined. In a stock with a healthy spawning population it is suggested that some annual fluctuation in the spawner:spawner relationship is to be expected, with points falling both above and below the replacement line in a 50:50 distribution.

Natural survival in fresh water of juvenile fish in particular can also influence the R/S ratio. Both density-independent and density-dependent factors can affect freshwater survival. Density-independent factors include changes in habitat quality, annual variation in water discharge and water temperature. Density-dependent factors can include increased competition for food and space as population size increases.

In spite of declining salmon stocks in insular Newfoundland, Great Rattling Brook, Middle Exploits River, Upper Terra Nova River, Rocky River and Torrent River contributed an average of 13% of the total salmon population in 1992–2001 compared to only 2.5% in 1972–81 (Fig. 17.9). In addition, there are numerous other rivers where fishways were constructed to improve existing runs that also contribute to the overall population. Thus, increasing the range of Atlantic salmon by opening up new habitat has been a successful technique in improving Atlantic salmon production in Newfoundland even at times when natural survival is low.

Stocking unfed Atlantic salmon fry, while not the most commonly used approach throughout the range of Atlantic salmon, has proven to be successful in Newfoundland (O'Connell & Bourgeois, 1987; Bourgeois, 1998). This success may be attributed to a number of factors: (1) fry were stocked into habitat not being utilized by anadromous Atlantic salmon; (2) eggs were incubated under gravity-fed water from donor rivers and water temperatures similar to recipient river, hence timing of fry emergence (swim-up) was similar to those in the river; (3) fry incubated in substrate incubators were significantly larger than those incubated in spawning channels (Bourgeois, unpublished data); (4) fry were evenly distributed throughout pre-selected under-yearling habitat at densities of 75 fry/unit; (5) transportation time for fry was usually less than 1 hour, resulting in minimal stress to the fry.

A number of factors have influenced the evaluation of the success of the enhancement. The use of a constant commercial exploitation rate to estimate the total number of recruits prior to 1992 is problematic. Commercial exploitation can be highly variable depending on environmental conditions. However, the error surrounding the estimates was reduced by using estimates of commercial exploitation derived for the same or nearby river. The estimated numbers of small and large Atlantic salmon include some repeat-spawning fish. Therefore the number of recruits produced by each generation is overestimated. There are no data to make the necessary adjustment each year although it is believed that previous spawners comprise less than 10% of the runs.

A number of fisheries management changes impacted positively on returns and spawning escapements to all rivers in Newfoundland. The biggest changes were the closure of the commercial salmon fishery in 1992 and subsequent reductions in bag limit in the recreational fishery. For the most part, enhanced stocks responded to these changes similar to other rivers of insular Newfoundland. With the exception of Middle Exploits, enhancement activities had ceased for several generations by 1992 and stocks were beginning to show signs of stabilizing.

In Canada, future range extension of Atlantic salmon through opening up of new habitat, previously inaccessible to Atlantic salmon, will receive considerably more scrutiny than it did in the past. Today there is much more concern about the impacts that introduction of a new species, intentional or unintentional, will have on the biodiversity and the integrity of the ecosystem. Canada has adopted a biodiversity policy 'to conserve bio-diversity ... by development and implementation of measures ... to prevent alien and living modified organisms from adversely affecting biodiversity' (Anon., 1995). The enhancement of the five rivers referenced in this paper would be unlikely to get approval, if proposed today. Such enhancement would contravene Canada's biodiversity policy. For example, Atlantic salmon are known to out-compete brook trout from some habitats (Gibson, 1973), and almost all rivers in Newfoundland contain brook trout populations. Also, providing fish passage at waterfalls for Atlantic salmon may also provide passage for unintentional introductions.

All proposed introductions in Canada are subject to Canada's National Code on Introductions and Transfers of Aquatic Organisms, adopted in January 2002. Proposals also undergo a risk assessment analysis to evaluate the potential of harmful alterations of natural ecosystems; risks of deleterious genetic changes in indigenous fish populations; and, risks to fish health from the potential spread of pathogens and parasites.

References

Anon. (1978) *Atlantic Salmon Project, Torrent River and Western Arm Brook, 1965–1977*. Department of Environment, Freshwater and Anadromous Fisheries Management, St. John's, Newfoundland.

Anon. (1995) *Canadian Biodiversity Strategy. Canada's Response to the Convention on Biological Diversity*. Minister of Supply and Services, Ottawa, 80 pp.

Anon. (2002) *Report of the Working Group on North Atlantic Salmon*. ICES CM 2002/ACFM:14, 305 pp.

Bourgeois, C.E. (1998) Assessment of the introduction of Atlantic salmon, *Salmo salar*, into Rocky River. In: *Stocking and Introduction of Fish* (ed. I.G. Cowx). Fishing News Books, Oxford.

Chadwick, E.M.P. & Green, J.M. (1985) Atlantic salmon (*Salmo salar* L.) production in a largely lacustrine Newfoundland watershed. *Verhandlungen Internationalis Verein Limnologae*, **22**, 2509–15.

Dempson, J.B., Schwarz, C.J., Reddin, D.G., O'Connell, M.F., Mullins, C.C. & Bourgeois, C.E. (2001) Estimation of marine exploitation rates on Atlantic salmon (*Salmo salar* L.) stocks in Newfoundland, Canada. *ICES Journal of Marine Science*, **58**, 331–41.

Elson, P.F. (1957) Using hatchery reared Atlantic salmon to best advantage. *Canadian Fish Culturist*, **21**, 7–17.

Farwell, W.K. & Porter, T.R. (1976) *Atlantic salmon enhancement techniques in Newfoundland.* In: Proceedings FAO Technical Conference on Aquaculture, Kyoto, Japan, 26 May–2 June, FIR:AQ/Conf/76/E.31 February 1976.

Gibson, R.J. (1973) Interactions of juvenile Atlantic salmon (*Salmon salar*) and brook trout (*Salvelinus fontinalis*). *International Atlantic Salmon Foundation, Spec Publ Ser*, **4** (1), 181–202.

Mullins, C.C., Caines, D. & Lowe, S.L. (1999) *Status of Atlantic Salmon (*Salmo salar *L.) Stocks of Three Selected Rivers in Salmon Fishing Area 14A, 1998.* CSAS Res Doc 99/101, 38 pp.

O'Connell, M.F. & Bourgeois, C.E. (1987) Atlantic salmon enhancement in the Exploits River, Newfoundland, 1957–1984. *North American Journal of Fisheries Management*, **7**, 207–14.

O'Connell, M.F., Davis, J.P. & Scott, D.C. (1983) An assessment of the stocking of Atlantic salmon (*Salmo salar* L.) fry in the tributaries of the Middle Exploits River, Newfoundland. *Canadian Technical Report of Fisheries and Aquatic Sciences*, **1225**.

O'Connell, M.F., Dempson, J.B. & Porter, T.R. (1990) *Spatial and temporal distribution of salmonids in Junction Pond, Northeast River, Placentia, and Conne Pond, Conne River, Newfoundland.* CAFSAC Res Doc 90/77, pp. 27–88.

Pepper, V.A. (1976) Lacustrine nursery areas for Atlantic salmon in insular Newfoundland. *Fisheries and Marine Service Resource Development and Technical Report*, **671**, 61 pp.

Porter, T.R. & Davis, J.P. (1974) *Fishway and counting fence data – 1973.* Data report series no. NEW/D-74-6. Fisheries and Marine Service (Newfoundland Region), Resource Development Branch, 53 pp.

Porter, T.R., Riche, L.G. & Traverse, G.R. (1974) *Catalogue of rivers in insular Newfoundland.* Data Report Series no. NEW/D-74-9. Resource Development Branch, Newfoundland region.

Ryan, P.M. (1986) Lake use by wild anadromous Atlantic salmon *Salmo salar*, as an index of subsequent adult abundance. *Canadian Journal of Fisheries and Aquatic Sciences*, **43**, 2531–4.

Chapter 18
Optimizing Wild Salmon Production

F.G. Whoriskey

Atlantic Salmon Federation, PO Box 5200, St. Andrews, New Brunswick, 5EB 3S8, Canada

Abstract: A key component of salmon optimization involves marshalling scarce resources in order to generate better knowledge to permit more informed decision-making. In this regard the Atlantic Salmon Federation (ASF) has been involved in three key programmes targeted at providing new resources so that salmon could be better managed. The ASF/J.D. Irving Ltd. Salmon Optimization programme involved working with an industrial partner to provide biological information on salmon in two key North American watersheds. The New Brunswick Smolt Wheel programme involved finding the money necessary to put equipment into the hands of local watershed groups and government biologists to generate important new knowledge. The third programme involved working with the salmon farming industry to replace cancelled government hatchery support for recovery efforts for severely stressed populations. Each of the programmes is described.

An overview of optimization

The natural world of the Atlantic salmon is highly variable, which results in fluctuating numbers of salmon surviving to spawn in home rivers each year. At various points in the species life cycle, humans are seeking opportunities to catch and/or harvest salmon. For salmon managers, this poses a number of dilemmas. Pressure is on them to 'maximize' the possible economic and social benefits obtained from the Atlantic salmon. This requires complete knowledge of salmon population status and agreement in society on costs and benefits. Neither is likely to be possible.

In Canada, the Sparrow Decision by the Supreme Court determined that the number one concern for salmon management was conservation of the species. No other right to harvest exceeded this imperative. This refocused the debate about salmon management on the conservation issue, and salmon managers were charged with arriving at an operational definition of 'conservation'. This had to be done in circumstances where information about salmon abundance in rivers was incomplete, and frequently measured with uncertainty. Scientists often do not understand how the factors that determine salmon abundance interact, and many of the environmental variables that affect salmon are not under human control.

Given the uncertainty around the status of fish populations, and the natural variability due to unpredictable environmental factors, some biologists have focused on the concept of optimization as opposed to maximization as a management paradigm (e.g. Symons, 1979; Hilborn & Walters, 1992; O'Grady et al., 2000). Salmon optimization requires recognizing the priority of the conservation imperative, natural environmental quality and variability, the knowledge available about wild fish populations and its limitations, and providing a clear definition of management goals. In a nutshell it involves identifying to the best of our ability the natural production potentials of salmon habitats especially in fresh water where we have the potential to manage things, determining whether or not present production meets these potentials, and if not, identifying why and fixing it. Subsequently, input from stakeholders and managers determines how to deal with surplus fish.

Salmon scientists recognize environmental variation and the variable quality of information available about fish populations on population status, and as part of their optimization strategies are actively developing key biological reference points for salmon populations (Potter, 2000; Prévost & Chaput, 2001). The focus is the conservation thresholds, or biological limit points, which define a minimum number of spawning adults, or eggs deposited in a stream, that should be achieved each year (Potter, 2000, 2001). These points are potentially river-specific, and practical experience has shown that highly undesirable declines in future salmon returns can result in rivers where we have let the population sink below the limit reference point (e.g. DFO, 2002). The limit points clearly focus on the conservation priority, and it is believed that in most years they will provide the necessary egg deposition required to fill natural production potential. However, in any given year or suite of years, the production potential may unpredictably increase or decrease due to natural variability. This inevitably results in bust or boom years for the fisheries. We have neither the knowledge nor the resources to track this annual production capacity variability at a fine-scale time level, and hence fall back on a robust reference point that time has shown will work in most years and which avoids doing permanent damage to the population.

For much of Atlantic Canada, the limit reference point is insuring a minimum egg deposition of 2.4 eggs/m^2 for available stream habitat. In areas where salmon may also grow up in lakes or ponds within a river system (Newfoundland and Labrador), additional reference points of 368 and 105 eggs/ha of lake or pond habitat for insular Newfoundland and Labrador and the Great Northern Peninsula, respectively (CAFSAC, 1991; O'Connell & Dempson, 1995) are needed. Quebec sets its egg deposition levels based on a combination of evaluation of habitat and relationships it has established between the number of spawners in a given year, and the number of returning adults those spawners generate (ICES, 1999). In general, these points have been evaluated for a selection of rivers and then exported to others. At present, for many if not most Atlantic salmon rivers, we do not have the knowledge to directly fix a river-specific reference point.

Once biological imperatives have been met, surplus fish are available to the fisheries. Fisheries management has a little to do with fish management, and a very great deal to do with people management. At this point, optimization becomes most difficult because different people have different priorities and values (e.g. sport versus commercial fisheries and the benefits that accrue to whom from each). Final management plans attempt to provide the greatest good to the greatest number.

Helping the cause

A small organization of limited means also has a need to optimize the use of its scarce resources in the interests of optimizing salmon management. Our view of optimality in this sense involves employing our investment to attack significant problems for salmon, helping to provide the best possible information to decision makers, to lever new and necessary resources from other interested parties and to implicate local watershed groups and other stakeholders in the efforts. Below I describe three programmes in which we are attempting to do this.

Collaborative rearing of adult fish for river-specific recovery efforts

A great deal of attention has been focused upon the conflicts that exist between wild fish interests and the salmon farming industry. While this tends to capture the headlines, there are also extremely positive examples of collaborations between salmon farmers and wild fish interests for wild salmon conservation.

One of the more positive contributions of the industry lies in its rearing of river-specific strains of salmon to the adult stage for reintroduction to their home rivers.

In the salmon rivers in northern Maine and the Bay of Fundy region, wild salmon populations have collapsed. Eight systems in northern Maine have been officially designated as endangered by the US government, as have 32 rivers in the inner Bay of Fundy region. This area happens to be home to most of the North American east coast Atlantic salmon farming industry.

Much has been written about the value of hatcheries to wild salmon populations, not all of it complimentary (e.g. Ford, 2002). However, artificial rearing, used in a focused, planned manner, can be of benefit to wild salmon populations (Independent Multidisciplinary Science Team, 2000). A principal advantage of rearing fish in captivity is the reduced mortality rates in the culture facilities, compared to those observed in the wild. For example, wild egg to smolt survivals can range from 0–40%, and wild smolt to adult survivals from 0.01–14% (e.g. Cunjak & Therrien, 1998). By contrast, salmon farmers' egg-to-smolt survival, and smolt-to-adult survival, typically exceeds 70%. When wild populations are desperately low, and broodstock is rare, anything that results in greater survival of the remaining fish is attractive.

The goal of releasing hatchery-reared wild adults into rivers is to let them spawn naturally and insure that their progeny are subjected to the full range of natural selection that normally occurs in the wild. Thus genetic adaptation to the hatchery environment is minimized, and the resulting generation of fish is subjected to the full range of survival challenges found in the wild.

A group of volunteers from the Big Salmon River Association (BSRA), led by Rita Almon, developed the pioneering programme for rearing fish of wild origin in commercial sea cages for reintroduction purposes. It started with prodigious fundraising to purchase sea cages, and other necessities. Their programme ran from 1993–95, and involved the capture of wild smolts as they were moving out to sea, and their transfer to sea cages at the salmon farming industry's Atlantic Salmon Demonstration farm in Lime Kiln Bay, New Brunswick. Here the first year-class was reared to maturity [at sizes of up to 5.4 kg (12 lb)] by the growers, and released back into the river. These fish subsequently spawned, and a measurable increase in parr numbers was noted (DFO, 1999; Kenchington, 1999). Flush with this success, the group returned to grow out a second year class, but bacterial kidney disease was detected in the fish as they were being reared, and they had to be destroyed. This closed the programme and was a deep disappointment to participants.

Two other groups have built on this initial effort. In New Brunswick, the Atlantic Salmon Federation in collaboration with the New Brunswick Salmon Growers Association has moved to develop an adult grow-out programme in the Magaguadavic River. Populations in this river have declined from runs of about 1000 fish in the 1980s to a low of 14 fish in 2000 (up to 17 in 2001).

An initial effort in 1999 to collect wild broodstock to spawn and rear the progeny to the adult stage foundered when infectious salmon anaemia (ISA), a viral disease that has devastated the salmon farming industry (Lovely et al., 1999), was detected in the adult wild fish prior to spawning. ISA was not believed to be passed on from parents to offspring through the gametes (Melville & Griffiths, 1999), so with emergency financial support from the New Brunswick Wildlife Trust Fund, we spawned the surviving fish and moved their eggs into quarantine for rearing. These juveniles subsequently tested positive for ISA using a reverse-transcriptase polymerase chain reaction (rt PCR) test that detected what was believed to be DNA fragments from the virus. Most of these fish were destroyed, but a group was quarantined to follow the evolution of the disease. To our surprise, it never evolved, even when the fish were deliberately stressed. It now appears likely that the original rt PCR tests gave false positive results for the virus. It is possible that the tests detected a related, non-virulent virus. Before the remaining fish could be removed from quarantine, a chlorine malfunction in the water supply killed them. The genetic contributions and egg depositions for almost an entire year class of fish in this river disappeared with them.

We tried again, this time with second-generation Magaguadavic fish that had initially been moved into hatcheries in 1997, been spawned once, and whose juveniles were available for grow-out. Some of these offspring were reared to the smolt stage and then moved to industry cage sites in 2001. However, in autumn 2002, an ISA control strategy was adopted by the industry that required closing of the cage site the fish were at in an effort to bring the ISA epidemic in the industry under control. The fish suddenly had to be moved, with nowhere to go. Once again, New Brunswick Wildlife Trust Fund stepped into the breach and provided emergency funding to move the fish to a land-based facility at the Huntsman Marine Science Center. They prospered there and have grown to the 4.5-kg (12-lb) size. In the summer and autumn of 2002 they were released back into the river. Some were fitted with sonic tags, and we are presently analysing data on their post-release movements. In 2003, we will be surveying the river to determine the spawning success of these fish.

The National Marine Fisheries Service (NMFS) of the National Oceanic and Atmospheric Administration (NOAA), the Maine Atlantic Salmon Commission, the St. Croix Watershed Commission and commercial grower Atlantic Salmon of Maine have also been collaborating on a release of adult fish reared from native stocks into rivers in Maine. The targets are the Dennys, Machias and St. Croix Rivers. The wild run to the St. Croix River had been exterminated, so native strains from the other sites are being used as donors for the ongoing restoration effort.

Culture of the fish began in 1998, with the first release to the rivers in 2000. Biologists have been monitoring spawning and movements of the fish and the fate of the kelts as they leave the rivers. This programme has also had its problems, including ISA scares, and escape of fish from cages due to storm damage.

The lessons from this are that collaborations are possible, the technique has potential, but there is a steep learning curve and many things can go wrong.

Equipment pools

A lack of the appropriate tools may significantly hinder salmon conservation efforts. In these circumstances, a small organization like ASF may act as a significant catalyst for productive efforts if it focuses itself on obtaining and making available equipment that can be used to address identified problems. We have developed equipment pools to meet two of these needs. The inspiration for both came from ASF Regional Director Danny Bird, whose on-the-ground reconnaissance and liaisons with local watershed groups identified the needs.

Temperature recorders

In recent times, concern has been expressed about the impacts of live-release angling upon salmon during hot-water periods. Protocols have been developed in parts of Canada to close recreational angling at times when water temperatures exceed 22°C (e.g. DFO, 2001a). Closures have enormous impact upon the recreational angling industry and the wealth that it generates; hence they are a very unattractive option.

Temperature regimes in rivers can be highly variable. Some tributaries are much cooler than others, and areas within a tributary may differ considerably from each other due to the addition of cold spring water. In many river systems, we do not have a good or any record of temperature that shows seasonal trends or the areas of rivers that are cool when other parts are warm. In the absence of such data, government managers may have to assume the worst and close all fishing in a river when temperatures are shown to be high at a single site.

With the support of the New Brunswick Wildlife Trust Fund, the ASF was able to obtain a suite of water temperature recorders manufactured by VEMCO, a Nova

Scotia company (VEMCO model TR 80). These units have an internal clock and memory, and store temperatures at preprogrammed intervals. The data are downloaded directly onto a computer.

The recorders have for the most part been placed in key salmon pools within various watersheds in New Brunswick, including the Miramichi and Restigouche Rivers. They record temperatures at half-hour intervals during the salmon fishing season. The units are provided to local watershed groups, who deploy and retrieve them, and in some cases download the data and conduct their own analyses. Otherwise ASF processes the information and provides daily minimum, maximum and average water temperatures.

Local watershed groups and government have also purchased and installed temperature recorders in other rivers or at other sites within rivers that already have recorders in them. Over a number of years, all these units will provide a comprehensive picture of water temperatures in key fishing areas, that will help managers develop strategies to balance fishing with conservation concerns at hot temperatures. The data may also help with evaluating the impacts of global warming upon Atlantic salmon populations.

Smolt wheels

Limited data have indicated that one strong contributor to recent salmon declines stems from an unexplained drop in survival after the smolts have entered the sea. When and where the losses are occurring remain unidentified at this time.

Estimates of wild smolt survival in North America were available from a limited number of rivers in Quebec, and Newfoundland. The absence of any data from New Brunswick, in the southern part of the species range, left a big hole in our understanding of survival rates for North American smolts. ASF began searching for ways to contribute to estimating smolt outputs from sites in the province. Mark–recapture estimates seemed to offer the most reasonable approach, especially in big rivers with swift currents and varying water levels. We needed a capture device that would trap reasonable numbers of smolts, and keep them alive and healthy during and after capture. The smolt wheel, also known as a rotary screw fish trap, is an effective but expensive tool that accomplishes the job.

The device floats on a catamaran platform that is typically tethered to a cable strung high above the stream. This lets the wheel float up and down with varying water levels, while maintaining its position. Unlike a counting fence, the wheel cannot be overtopped by high water, and does not block navigation on the stream or river. One end of the wheel looks like a sawn-off cement-mixer drum. This doctored drum is suspended between the two pontoons of the catamaran, with its wide end facing upstream into the current. As the current strikes an ingenious series of baffles welded inside the drum, it forces the wheel to rotate. The baffles also trap the fish in a series of water-pockets within the drum, lift them up with the rotation of the drum and dump them into a live well built on the unit's back end. Once in the well, the fish are comfortable and safe until they can be processed. Generally the part of the drum in the water is submerged to a maximum depth of 1 m, which is the depth at which smolts travel during the downstream migration.

Smolt wheels were developed on the west coast of North America, and to our knowledge Dr. John Kocik of the US National Marine Fisheries Service first brought them to an Atlantic salmon river. He used them on Maine's Narraguagus River as part of a programme investigating the precipitous declines of this river's salmon (its population was listed as endangered in 2001). John Anderson and Danny Bird of ASF and Norm Dube of the Maine Atlantic Salmon Commission and Energy went down to have a look, and returned with a glowing report. Dr. Kocik then very kindly arranged a short-term loan of a wheel to ASF's Research and Environment Department, and we were immediately convinced of the units.

Knowing the wheels were expensive (> $12 000 US each), ASF conceived of the idea of a smolt wheel pool, where we would raise the money to buy a number of them that could be loaned out to researchers and watershed groups interested in working on smolts. We approached the New Brunswick Wildlife Trust Fund with a series of grant requests, they were convinced of the idea, and we eventually created a pool of wheels for the province. Proposals to use the wheels were judged by a committee of scientists from New Brunswick's Department of Natural Resources and Energy, the University of New Brunswick and ASF. The available wheels have been allocated to local groups working in collaboration with the Department of Fisheries and Oceans on the Tobique, Nashwaak and Miramichi Rivers. The Miramichi Salmon Association and ASF have purchased additional wheels with the support of the International Paper for the Miramichi River, ASF has placed one on the Magaguadavic River, and ASF and the Listuguj First Nation have obtained wheels for the Restigouche River in time for the 2002 field season.

We also passed on John Kocik's favour by arranging a short-term loan of one of our wheels during our 'off season' to the Quebec government's Société de Parcs et de la Faune to help monitor the smolt run of the Trinité River. They too were convinced about the wheels, and bought their own.

Important scientific results are coming from the operation of these wheels. For example, on the Tobique River, the movements of salmon juveniles in spring and autumn have now been documented well enough that the results can help plan controlled spills at the river's hydroelectric dams during juvenile salmon movements. We hope to get more salmon back to the river as adults because we will bypass more smolts away from the turbines as they are leaving for the sea.

Meanwhile, the Nashwaak and Miramichi wheels are helping biologists to estimate the numbers of smolts leaving these rivers, and calculate the return rate for adults coming back. Results are showing rates (e.g. 1.79% on the Nashwaak River in 2000) that are very low. Managers are taking this into account as they set conservation regulations. The wheels are also being used to capture fish that are fitted with sonic 'pinger' tags and tracked as part of the collaborative ASF/DFO at-sea Bay of Fundy smolt tracking programme.

Originally, we planned to pass the wheels from river to river, as needed, but New Brunswick has recently run into a rash of disease problems [infectious salmon anaemia in the Bay of Fundy region in 1996 (Lovely et al., 1999); furunculosis arriving in the Miramichi River in 2000 (DFO, 2001b)]. The wheel's pontoons are not watertight. Instead, they are filled with extruded styrofoam. While this makes them unsinkable, it also creates a home for disease organisms and a refuge from the disinfectants that we spray a wheel with before it can be transferred to a new river. Until a method is found to disinfect the pontoons, we will not move the wheels from their present dispositions. There they do valuable service, providing us with information on smolt survival trends with time.

J.D. Irving, Ltd.–ASF collaborative programme

In 1996, J.D. Irving Ltd., Forest Products Division, in collaboration with ASF, initiated a programme to optimize Atlantic salmon production within their timber freeholds in the Miramichi (Clearwater Brook) and Restigouche (Little Main Restigouche) Rivers. The collaboration with this private company brought significant new resources to bear on salmon conservation in this region (e.g., Whoriskey et al., 1999; Tinker et al., 2001).

Optimization for this project was defined as identifying the natural production potentials of the river systems, determining whether or not present production meets these potentials, and, if not, identifying why. Strategies to correct any bottlenecks could then be proposed and implemented. A specific goal of this approach is to minimize costly, artificial interventions. We wished to help the system meet its natural potential.

The aim of the initial 5 years of work, from 1996–2000, was to establish methodologies and determine habitat availability, numbers of adult salmon and grilse moving into the rivers, present juvenile salmon densities and population structure, and the degree to which these vary naturally from year to year. A key interest of the ASF was to determine the survival rate of 'satellite-reared' parr stocked in the rivers.

Satellite rearing involves local stakeholders establishing streamside tanks along sections of the target river, and caring for from 5000–50 000 salmon fry from first feeding through the summer, for a release in the autumn after the mergansers had migrated south. Only limited verification of the success of the programme had been obtained (e.g. Whoriskey et al., 1998), primarily because there was no way to count returning adult salmon in most of the rivers where the fish were being stocked.

The programme generated some very interesting information. This included:

- The Clearwater watershed, which had been under intensive forestry management for over 50 years with an emphasis on sustainable forestry, has had a consistently healthy adult salmon run and consistently exceeded egg deposition conservation requirements. Satellite-reared fish were among those returning annually to the rivers, and a Master's degree programme was initiated to fit them with passive integrated transponder tags to see if they homed back to the points in the river where they were stocked. Analysis of this work is in progress.

- Large sections of suitable habitat in both Clearwater Brook and the Little Main Restigouche had low or no juvenile salmon in them. In some cases the causes were obvious as in areas where adult migration was blocked by beaver dams. When they were removed prior to spawning season, the next year fry out of these areas increased by 600%. In other cases, notably in headwater zones, there were no obvious reasons other than the possibility that adults did not reach them during spawning, and satellite-reared fry stocked into the areas did well. Hopefully fry that were planted in these areas will survive and return to spawn in them.

- On the Little Main Restigouche, long sections of river channel lacked suitable holding pools for salmon, possibly due to damage resulting from historic log drives. This may have impeded the upstream movements of adults, and resulted in the low observed numbers of juveniles. Work is underway to restore some of these pools, and hopefully move spawners into the upstream areas.

- In Clearwater Brook, out-migrant smolts from the satellite-rearing programmes that stocked fry resulting from the matings of about 12 parents in 1 year made up about 20% of the smolt run. While indicative of the success that could be realized with a well-implemented stocking programme, it did suggest that even small efforts could significantly change the genetics of a wild population. It was immediately recommended that the number of parents used in the satellite rearing be increased to mitigate for this possibility.

Results from this work, especially counts of adult fish, benefit downstream fisheries and contribute to the annual assessments conducted for salmon populations in these two rivers. Thus the work optimizes benefits for all who depend on the river's salmon. This programme is now a fixture of J.D. Irving, Ltd. annual operations.

Conclusions

Scarce resources and differing interests will continue to make salmon management difficult. Conservation is now the number one priority for salmon populations. A failure to conserve them benefits no one. Cooperative programmes such as those described above bring people together to work for salmon conservation while generating useful new knowledge. These programmes bring many difficulties to the participants, and funding agencies such as the New Brunswick Wildlife Trust Fund that can step in in an emergency are critical for success.

References

CAFSAC (1991) *Quantification of conservation for Atlantic Salmon*. CAFSAC Advisory Document 91/16, 15 pp. Canadian Atlantic Fisheries Scientific Advisory Committee (DFO – Ottawa).

Cunjak, R.A. & Therrien, J. (1998) Inter-stage survival of wild juvenile Atlantic salmon, *Salmo salar* L. *Fisheries Management and Ecology*, **5**, 209–23.

DFO (1999) *Atlantic salmon maritime provinces overview for 1998*. DFO Science Stock Status Report D3-14. Department of Fisheries and Oceans, Ottawa.

DFO (2001a) *Newfoundland and Labrador Atlantic salmon stock status for 2000*. DFO Science Stock Status Report D2-01. Department of Fisheries and Oceans, Ottawa.

DFO (2001b) *Atlantic salmon maritime provinces overview for 2000*. DFO Science Stock Status Report D3-14 (2001) (revised). Department of Fisheries and Oceans, Ottawa.

DFO (2002) *Atlantic salmon maritime provinces overview for 2001*. DFO Science Stock Status Report D3-14. Department of Fisheries and Oceans, Ottawa.

Ford, M.J. (2002) Selection in captivity during supportive breeding may reduce fitness in the wild. *Conservation Biology*, **16**, 815–25.

Hilborn, R. & Walters, C.J. (1992) *Quantitative Fisheries Stock Assessment*. Chapman and Hall, New York, 570 pp.

ICES (1999) *Report of the Working Group on North Atlantic Salmon*. ICES CM 1999/ACFM: 14, 288 pp.

Independent Multidisciplinary Science Team (2000) *Conservation hatcheries and supplementation strategies for recovery of wild stocks of salmonids: report of a workshop*. Technical Report 2000-1 to the Oregon Plan for Salmon and Watersheds. Oregon Watershed Enhancement Board, Salem, Oregon.

Kenchington, E. (1999) *Proceedings of the Inner Bay of Fundy Salmon Working Group, Regional Advisory Process, Maritime Region*. Canadian Stock Assessment Proceedings Series 99/29, 39 pp.

Lovely, J.E., Dannevig, B.H., Falk, K., et al. (1999) First identification of infectious salmon anaemia virus in North America with haemorrhagic kidney syndrome. *Diseases of Aquatic Organisms*, **35**, 145–8.

Melville, K.J. & Griffiths, S.G. (1999) Absence of vertical transmission of infectious salmon anaemia virus (ISAV) from individually infected Atlantic salmon *Salmo salar*. *Diseases of Aquatic Organisms*, **38**, 231–4.

O'Connell, M.F. & Dempson, B.A. (1995) Target spawning requirements for Atlantic salmon, *Salmo salar* L., in Newfoundland rivers. *Fisheries Management and Ecology*, **2**, 161–70.

O'Grady, M.F., O'Neill, J. & Molloy, S. (2000) Optimizing natural production. In: *Managing Wild Atlantic Salmon: New Challenges – New Techniques* (eds F.G. Whoriskey & K.E. Whelan). Atlantic Salmon Federation, St. Andrews, New Brunswick, pp. 139–51.

Potter, T. (2000) New challenges and new techniques for the management of interceptory salmon fisheries. In: *Managing Wild Atlantic Salmon: New Challenges – New Techniques* (eds F.G. Whoriskey & K.E. Whelan). Atlantic Salmon Federation, St. Andrews, New Brunswick, pp. 74–97.

Potter, T. (2001) Past and present use of reference points for Atlantic salmon. In: *Stock Recruitment and Reference Points: Assessment and Management of Atlantic Salmon* (ed. É. Prévost & G. Chaput). INRA Editions, Hydrologie et Aquaculture, INRA, Paris, pp. 195–223.

Prévost, É. & Chaput, G. (eds) (2001) *Stock Recruitment and Reference Points: Assessment and Management of Atlantic Salmon*. INRA Editions, Hydrologie et Aquaculture, INRA, Paris, 223 pp.

Symons, P.E.K. (1979) Estimated escapement of Atlantic salmon (*Salmo salar*) for maximum smolt production in rivers of different productivity. *Journal of the Fisheries Research Board of Canada*, **36**, 132–40.

Tinker, S., Whoriskey, F., Connell, C. & Hackett, S. (2001) Atlantic Salmon Federation/J.O. Irving, Limited collaborative research program, Little Maine Restigouche River and Clearwater Brook (Miramichi River system). *Report on 2000 field work*. Atlantic Salmon Federation, St. Andrews, New Brunswick.

Whoriskey, F.G., Bangsgaard & Brown, K. (1998) Integration of satellite reared Atlantic salmon (*Salmo salar*) parr to the Upsalquitch River, New Brunswick. ICES CM 1996/T:3.

Whoriskey, F.G., Connell, C. & Perley, L. (1999) Atlantic Salmon Federation/J.D. Irving, Limited collaborative research program, Little Main Restigouche River and Clearwater Brook (Miramichi River system). *Report on 1998 field work*. Atlantic Salmon Federation, St. Andrews, New Brunswick.

Pointers for the Future

Chapter 19
Stream Restoration for Anadromous Salmonids by the Addition of Habitat and Nutrients

B.R. Ward[1], D.J.F. McCubbing[2] and P.A. Slaney[3]

[1] *Ministry of Water, Land and Air Protection, Fisheries Research and Development, 2204 Main Mall, University of British Columbia, Vancouver, British Columbia, V6T 1Z4, Canada*
[2] *InStream Fisheries Consultants, 223–2906 West Broadway, Vancouver, British Columbia, V6K 2G8, Canada*
[3] *Ministry of Water, Land and Air Protection, Watershed Restoration Program, 2204 Main Mall, University of British Columbia, Vancouver, British Columbia V6T 1Z4, Canada*

Abstract: In an evaluation of the salmonid response to watershed rehabilitation treatments at the Keogh River, we document the positive trends in juvenile density, growth, survival and smolt yield of steelhead trout and coho salmon observed in comparison to the untreated neighbouring Waukwaas River, on northern Vancouver Island, British Columbia, Canada. Juvenile fish abundance in the Keogh River indicated positive effects of the increased watershed restoration, particularly that from the addition of habitat structures and nutrients. Steelhead parr densities in the Keogh River were significantly higher compared to untreated (both rivers) and pre-treatment values, and highest in reaches treated with both restoration techniques. Despite reductions in adult escapement, the abundance of coho fry in the Keogh River exceeded that in the Waukwaas River; densities in preferred habitat exceeded those of past surveys. Inorganic nutrient addition led to significant increases in salmonid fry and smolt weights. Increase in length and weight of steelhead parr improved survival over winter, culminating in increased smolt yield and a shift to predominantly 2-year-old smolts in 1999, 2000 and 2001. Smolt yield reflected significant improvements in juvenile production and survival in the freshwater phase in the Keogh River despite low brood year strength, and proved the better response variable; juvenile density was highly variable. Steelhead smolt yield in 2001 was > 2000 smolts. Coho smolt yield increased in 2001 from the Keogh River, but less so than in 2000, over the historically poor yield observed in 1998. Steelhead smolts produced per spawner in the Keogh River have risen from historic lows of < 3 smolts per spawner (i.e. below replacement) from the 1996 brood to > 50 smolts per spawner from the 1998 brood year, the highest production per spawner of the 27-year record, offering hope for recovery despite low smolt-to-adult survivals. Further evaluation of effects to salmonid smolts will require a continued analysis of smolts-per-spawner recruitment to at least 2004, to more fully describe the benefits of the watershed ecosystem approach to restoration.

Introduction

Salmonids of the northern hemisphere, and particularly within their southern distribution in North America, have declined dramatically over the last decade. Walters & Ward (1998) presented evidence that salmon and trout of both the east and west coasts of Canada experienced sudden, dramatic and persistent declines after 1990. Atlantic salmon of the Bay of Fundy on Canada's east coast and coho salmon of the Thompson River on the west coast were listed at risk this year by the Committee on the Status of Endangered Wildlife in Canada (COSEWIC, 2002). The map of the USA denoting listed salmonids under the Endangered Species Act is also painted red with threatened or endangered status for Pacific and Atlantic salmon (Anderson, 1999; Colligan et al., 1999). In Europe, declines in survival in wild Atlantic salmon stocks (Friedland et al., 2000) parallel those observed on the west coast of Canada.

Reasons for these declines remain as elusive as effective management responses, although theories abound [e.g. > 60 hypotheses were tabled to explain the decline in Atlantic salmon of eastern Canada; O'Neil et al. (2000)]. In the Pacific Ocean, the changes appear coincident to a series of El Niño events, the Pacific Decadal Oscillation and global climate warming (Ward, 2000; Tyedmers & Ward, 2001), the interactions among which are poorly understood; likewise for the interaction of these climate measures with salmonid ecology. The ability of a management agency to develop an effective and timely response to these rapid changes is a challenge (Welch, 1999). The altered recruitment patterns and subsequent reductions in harvestable surplus may not be detected with sufficient rapidity, as suggested for Thompson River coho salmon (Bradford & Irvine, 2000). Anthropogenic impacts will increase, and society remains undaunted in its growth and development, factors that will continue to result in declines in salmonid abundance (Hartman et al., 1999; Lackey, 1999). While attempting to comprehend the reasons for the declines, applied strategies of freshwater habitat rehabilitation may assist in forestalling or perhaps reversing these trends.

Reductions in survival rate at one salmonid life stage may be compensated by improvements in another, but this has rarely been demonstrated. Recovery plans include methods to improve productivity and capacity in fresh water. Programmes of watershed restoration have emerged in British Columbia, Washington, Oregon, California and elsewhere (e.g. Slaney & Martin, 1997). However, few of these programmes present sufficient monitoring and evaluation to assess effectiveness in appropriate response variables. On the Keogh River, monitoring of life stage survivals of steelhead and comparisons of juvenile abundance, growth and yields of steelhead and coho smolts in treatment and control streams have provided preliminary indications of the utility of attempts to stall or reverse extinction trajectories in anadromous salmonids (Ward et al., 2003).

A programme of watershed rehabilitation at the Keogh River was assessed in comparison with the untreated and neighbouring Waukwaas River. Ward et al. (2003) demonstrated that the initial responses in steelhead trout and coho salmon juvenile abundance and size, smolt yield and smolts per spawner were positive, but that at least an additional year of monitoring of the highly variable juvenile response was required to statistically validate the positive trends, followed by several more years of smolt enumeration and comparison as the key response variables. Here, we report on the final year of evaluation of the juvenile response and an additional year of the response by smolt recruits.

Methods

The Keogh and Waukwaas Rivers, two oligotrophic fourth-order streams, are situated at the northern end of Vancouver Island, Britich Columbia, Canada (Fig. 19.1). The logging history, river size and fish species present have been described previously [Ward et al. (2003), and references therein], along with the annual sampling methodology, which is briefly described.

The experimental design considered two treatment types: habitat structures and slow-release nutrient addition. The plan of restoration within the Keogh watershed was adapted to the staircase-like experimental design (Walters et al., 1988), with structure placement commencing in the upper reaches, nutrient addition beginning in the lower reaches and both treatments expanding annually until the whole watershed was treated. Density, growth and smolt yield of salmonid juveniles were analysed by ANOVA.

The entire watershed has undergone habitat restoration treatment from 1997–2001 to address logging impacts (riparian trees had been removed within > 50% of the basin) and to compensate for a reduction in marine-derived nutrients (mainly pink salmon carcasses, now much reduced). Inorganic nutrients were added in 2001 as slow-release briquettes (905 kg). A total length of 36.5 km of the Keogh River mainstem, from the river mouth to 4 km upstream of Keogh Lake, was treated, as well as 11 km of key tributaries entering the mainstem from Muir Lake to the river mouth. This was the third year of complete watershed treatment with nutrient briquettes. Over 450 habitat structures [a variety of boulder and large woody debris designs (Slaney & Martin, 1997)] have been added to the mainstem. Final structure placement was undertaken in the lower river in 2001, with 16 structures constructed over a 0.25-km length of reach X (Fig. 19.1).

To evaluate the effectiveness of watershed rehabilitation techniques three primary methods were utilized. First, salmonid juvenile density in fresh water was assessed by mark–recapture using electroshocking and seine-netting techniques (McCubbing & Ward, 1997). In addition, representative structures were sampled (electroshocked and minnow trapped) to determine salmonid use during winter. Second, smolt yield was assessed through operation of a full-river counting fence near the mouth of the Keogh River, and by mark–recapture estimates using rotary screw traps on the Waukwaas River (Ward et al., 2003). Finally, we analysed the smolt yield per spawner as a function of the number of spawners in the Keogh River, in comparison with the historic record.

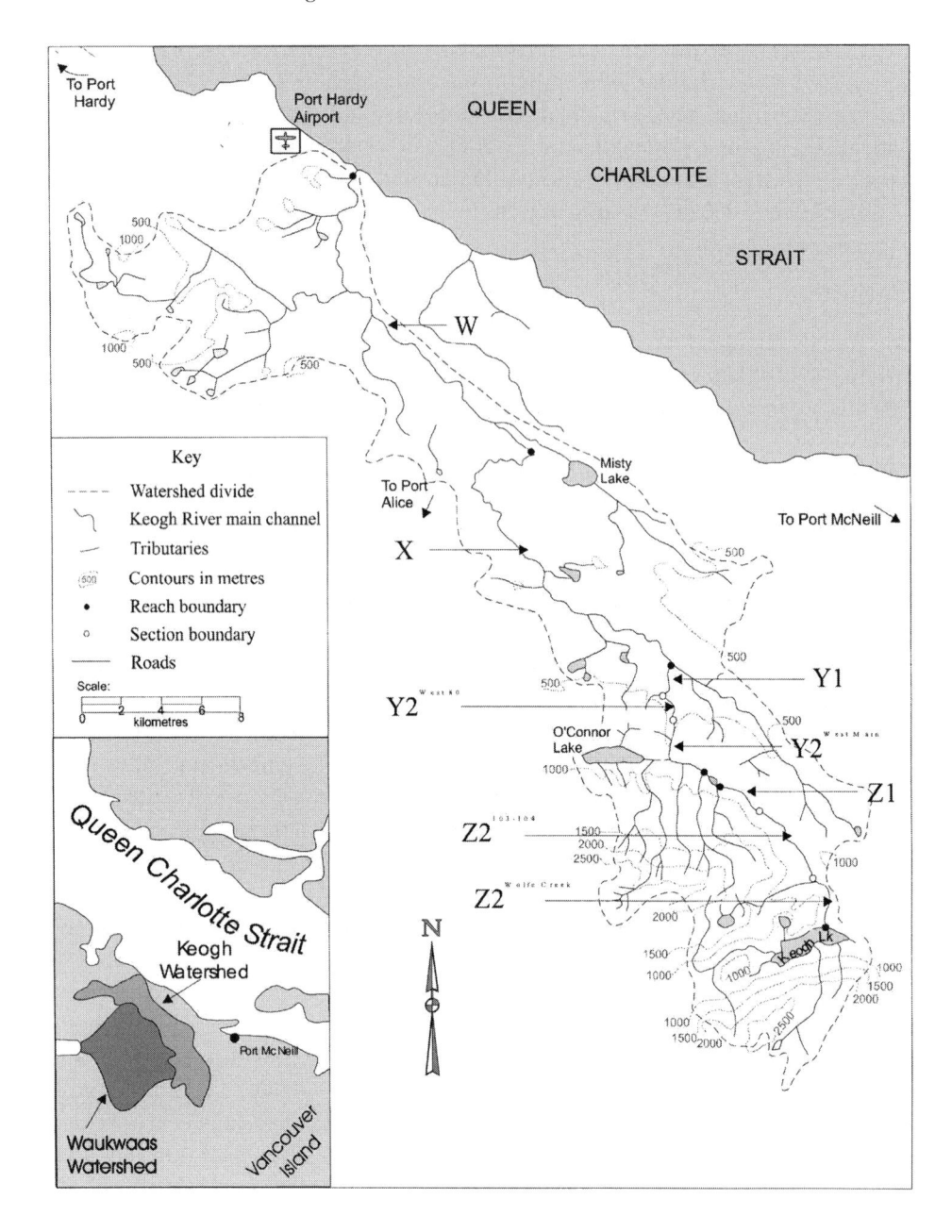

Fig. 19.1 The Keogh River watershed, indicating reach boundaries and section boundaries used in this study, elevations, the watershed divide and location on the British Columbia coast (inset), as well as relative position of the neighbouring Waukwaas River, British Columbia.

Representative lengths (250 m) within main reaches of the river were assessed for habitat frequency. Habitat classes (pool, flat, riffle and run) were summed to give total lengths of each type for all reaches. The relative frequency of habitat types was calculated from the total length of habitat in each class and the total length of each reach (Ward et al., 2003). Fish sampling within each reach (100 m except reach W on the Keogh where 200 m was sampled) was undertaken in proportion to the frequency of occurrence of habitat.

Forty six (12%, of 380 sites) of the total habitat structure sites on the Keogh River were sampled in 2001. These log and boulder structures were selected according to type and location, and were representative of the treatments in each reach. They also represented the most commonly introduced structure types. Some were resampled in 2001 for comparison with results from 1997–2000. Where possible, structures were sampled individually but included proximal river area near the structure.

Population estimates were calculated using the mark–recapture equation as described in Ward et al. (2003). Coho and steelhead juveniles were split into age classes of fry ($0+$) and parr ($>0+$) based on length–frequency distributions. Steelhead >80 mm were estimated to be parr, and coho juveniles >70 mm were classed as parr rather than fry. Ages from scale samples later confirmed these estimates of age designation.

Growth of juvenile steelhead and coho was examined among the treated reaches of the Keogh River (W, X, Y and Z) and among the untreated reaches (1–4) on the Waukwaas River, in a 'Before–After Control–Impact' analysis. Reach means were calculated; both summer and autumn sampling represented a sub-sample of the reach population. Statistical analysis of the length of steelhead and coho fry by late autumn was undertaken with a t-test on populations with equal variance. Two population groupings were used, Keogh fertilized (reaches X, W, Y and Z), and Waukwaas unfertilized (reaches 2 and 4).

Smolt yield from the Keogh River was measured as total counts. Smolt yield from the Waukwaas River from 1996–2001 was estimated based on mark–recapture estimates derived through operation of two rotary screw traps. Methods were as described in Ward et al. (2003).

Methods of estimating adult steelhead abundance in the Keogh River were unchanged from previous years. Briefly, the total numbers of adult males and females were estimated separately by mark and recapture using the adjusted Petersen estimate by marking upstream migrant adults and by capturing and examining kelts during their downstream migration. In addition, steelhead trout were counted electronically from 1997 through 2001, and coho and pink salmon from 1998 through 2001, using the Logie 2100C resistivity fish counter [Aquantic Ltd.; McCubbing et al. 1999)].

Keogh steelhead smolt recruitment was recorded using age, based on scale analyses, and tabulation into brood year of origin. These data were used to estimate the smolts per spawner in comparison to the number of brood spawners, as explained in Ward et al. (2003). After transformation to linearize the relationship, a comparison of slopes and intercepts using covariance analysis was used to test differences in

Results

Adult steelhead declined significantly during the 1990s, but have increased moderately over the last few years (Fig. 19.2). Coho salmon escapement declined steadily through the period 1998–2000, one full cycle of returns, but increased slightly in 2001 (Fig. 19.2). However, steelhead escapement was skewed heavily to male fish in 2001, resulting in a slight reduction in female spawners compared to 2000 (McCubbing, 2001).

production regimes and from values obtained from the years of watershed rehabilitation.

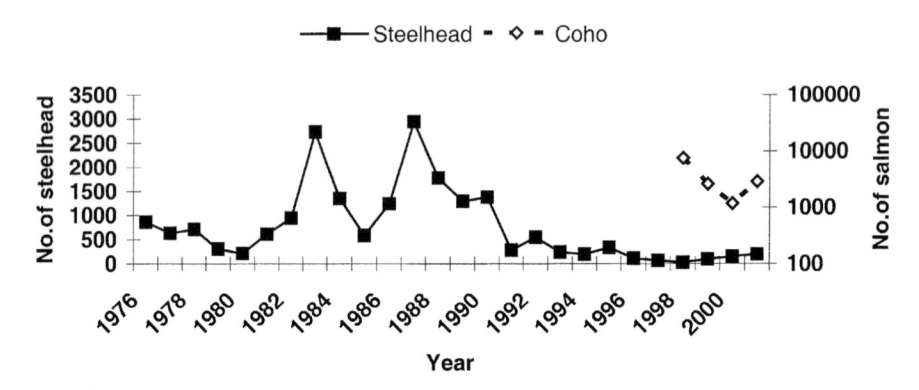

Fig. 19.2 Adult escapement estimates of steelhead trout (solid line, squares) and coho salmon (dashed line, open diamonds) to the Keogh River from 1976–2001.

Steelhead fry were observed in greatest densities in the reach X of the Keogh River, but in reach 3 in the Waukwaas River (Fig. 19.3a). This differed from 1997 and 1998 on the Keogh River when steelhead fry had been found in greatest abundance in the lower Keogh (reach W), yet was similar to results obtained in 1999 and 2000. The middle reaches of the Waukwaas (reaches 2 and 3) have consistently produced greatest steelhead fry densities in this watershed throughout the 5 years of investigation. Steelhead fry abundance declined slightly in 2001 compared to 1999 and 2000 in the Keogh River on reaches W, X and Y, while reach Z showed a slight increase in fry abundance (17 fry/100m; Fig. 19.3a). These variations were significant among reaches ($P=0.03$) but not among years ($P=0.21$). Two of the four sampled reaches on the Waukwaas River had higher fry abundance than those observed in 2000, whilst two had lower densities. Differences among years were not significant ($P=0.61$), but differences among reaches were ($P=0.002$). Although there was no significant difference in steelhead fry abundance in 2001 in the Keogh River reaches compared to the Waukwaas River ($P=0.26$), the Waukwaas River has indicated significantly higher steelhead fry production over the whole experimental term ($P=0.04$).

Steelhead parr were more abundant in reach W on the Keogh River in 2001 than in other reaches. Steelhead parr abundance among reaches of the Keogh River fell slightly, to an average of 35 parr/100 m. Parr were distributed unevenly among reaches of the Waukwaas River in 2001, with upper reaches exhibiting greatest parr densities, as observed in 1997, 1999 and 2000 (Fig. 19.3b). Two of the four sample reaches on the Waukwaas River indicated further increases in parr abundance in 2001 following improved fry densities in 1999. However, mean parr densities were only slightly higher than in the Keogh, at 39 parr/100 m. Changes in parr abundance were not significant among Waukwaas reaches ($P=0.06$) or years ($P=0.11$). Differences on the Keogh were also not significant ($P=0.10$) among reaches or years ($P=0.14$). Overall, there was no significant difference in steelhead parr abundance in 2001 between rivers ($P=0.63$).

Coho fry abundance in the Keogh River in 2001 was similar to that observed in 1998, with reach Z highest. On the Waukwaas River, reach 2 was the area of most abundant coho fry in 2001, as in 2000. Differences among years were not significant (Fig 19.3c). Densities were similar to those observed in 1997 and 1998 (pre-restoration conditions); differences on the Keogh River were significant among reaches ($P=0.01$, upper reaches most productive) but not among years ($P=0.14$) and were, on average, > 24% higher than densities observed on the Waukwaas River in 2001. Coho fry abundance on the Waukwaas River was among the highest recorded throughout the five survey years. Two of the four sampled reaches on the Waukwaas River also had higher fry abundance than those observed in 1999 (by 126 and 358 fry/100 m), whilst two had lower densities (by 46 and 69 fry/100 m; Fig. 19.3a), i.e. differences were significant ($P=0.007$) among reaches, but not years ($P=0.37$). Overall, there was no significant difference in coho fry abundance in 2001 in the Keogh River reaches compared to the Waukwaas River ($P=0.63$), in contrast to the result when all sample years were combined ($P=0.002$).

Standardized density [(count $-$ mean)/SD] to zero mean at the watershed level indicated a decrease in the standing crop of steelhead fry in the Keogh and Waukwaas Rivers in 2001. Steelhead fry standing crop in the Keogh watershed was similar to that recorded in 1998 and 1999 but higher than in pre-restoration (1997) conditions, yet below the overall mean. Steelhead fry standing crop in the Waukwaas River was reduced slightly over 2000 data but was similar to previous years (1997 and 1998) and much lower than the high values observed in 1999. No significant difference was observed between watersheds among all survey years ($P=0.11$). Data from the Waukwaas River indicated below-average standing crop in the last two survey years, with a significant reduction in parr standing crop. Again, differences between watersheds were not significant ($P=0.49$). Standardized trends of coho fry standing crop in the Waukwaas River indicated that 2001 results were similar to the mean of the five sample years; lower than seen in 1998 and 1999, but above that observed in 2000. A significant reduction (> 1 SD) in coho fry standing crop was observed in the Keogh River to the lowest value in the five sample years. Overall, Keogh River standing crop was significantly greater among sample years ($P=0.01$) than that of the Waukwaas River coho.

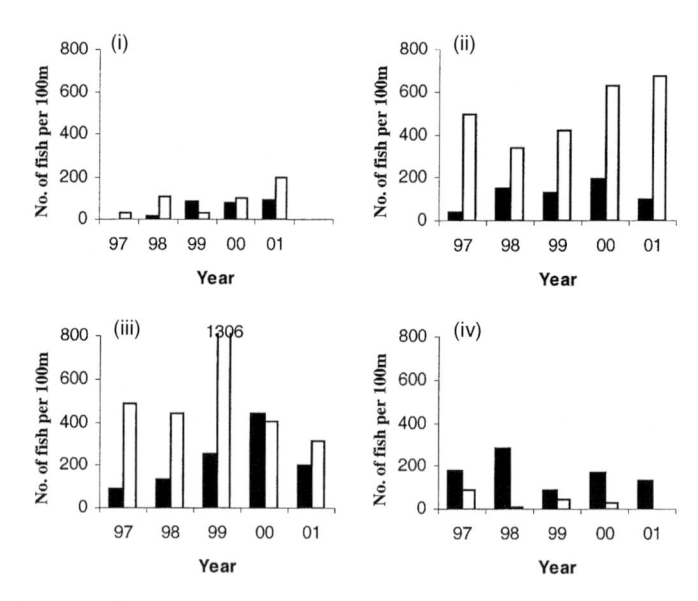

Fig. 19.3a Steelhead fry abundance (no. fry/100 m) on the Keogh River (black bars) and Waukwaas River (open bars) on reaches (i) W and 1, (ii) X and 2, (iii) Y and 3 and (iv) Z and 4, from 1997–2001.

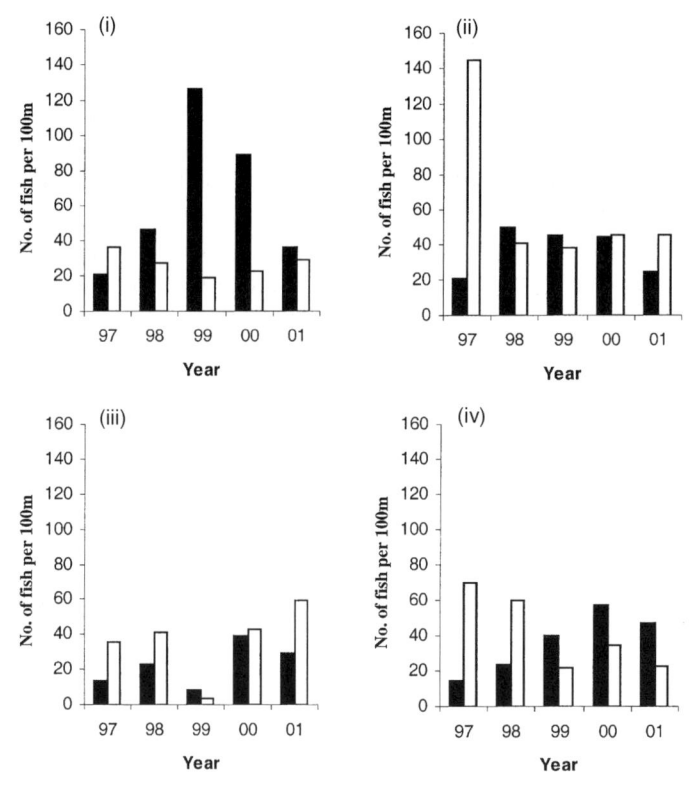

Fig. 19.3b Steelhead parr abundance (no. fry/100 m) on the Keogh River (black bars) and Waukwaas River (open bars) on reaches (i) W and 1, (ii) X and 2, (iii) Y and 3 and (iv) Z and 4, from 1997–2001.

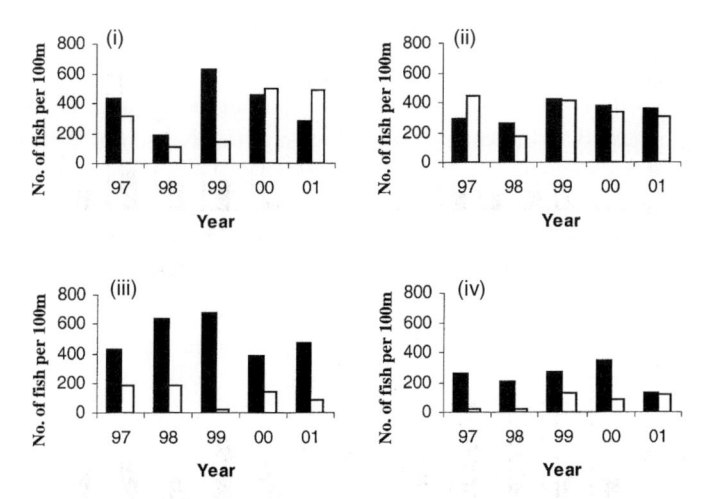

Fig. 19.3c Coho fry abundance (no. fry/100 m) on the Keogh River (black bars) and Waukwaas River (open bars) on reaches (i) W and 1, (ii) X and 2, (iii) Y and 3 and (iv) Z and 4, from 1997–2001.

Mean densities of fish (no./100 m^2) among and within the seven main structure types were not significantly different. Range in mean density was broad among common structure types in 2001 as in other sample years (coho fry, 15 to 231 fry/100 m^2; steelhead parr, 0.7 to 4 parr/100 m^2), similar to the range in values in previous years, but lower than the high values recorded in 2000. Greatest steelhead parr densities in 2001 were recorded in lateral debris jams (LDJ) and A-log (AL) structures. For the period 1997–2001, steelhead fry, coho fry and steelhead parr densities in LDJ ($n=9$), boulder clusters (BC, $n=13$), single log deflectors (SLD, $n=5$), riffle reconstructions (RR, $n=4$) and AL ($n=6$) were similar ($P=0.67$, 0.60 and 0.31, respectively). Throughout the five survey years, steelhead fry were most abundant in BC and AL structures, whilst steelhead parr favoured both BC and LDJ. Coho fry were most abundant in double deflector logs, LDJ and root wads, although there was high variance in these data as well.

Significant differences in steelhead fry weights ($P=0.001$) between the Keogh and Waukwaas Rivers were apparent from sample data collected during summer 2001. Keogh fish were larger by an average of 14 mm. Winter sampling effort was over a 1-week period, at temperatures when growth was limited (4–6°C). At that time, steelhead fry sampled in all reaches of the Keogh River (fertilizer added) exhibited weights and lengths significantly greater than (unfertilized) reaches of the Waukwaas River ($P<0.01$). Average weight of steelhead fry in the Keogh by autumn exceeded that in the Waukwaas River by over 50%, as observed in 1999 and 2000. Coho fry were also larger on the Keogh River, by 40% in areas of nutrient addition. Average condition factors during winter were also significantly higher on the Keogh, for both steelhead fry (1.08) and coho fry (1.10), than from the Waukwaas River (1.01, both species).

Growth differences in steelhead parr were difficult to interpret because several age classes may have been included in the population within any reach. The age-class structure may also vary by reach on the Keogh River, given the progressive treatment of fertilizer in the watershed, and the number of years in which resident parr were exposed to conditions of higher nutrient levels. Length–frequency plots over time (1997–2001) suggested an increase in parr size-at-age (University of British Columbia, data on file).

Steelhead scale-age data taken during smolt sampling was used to interpret the main response to nutrient addition. Results from the Keogh River in the spring of 1997–2001 indicated an increase in smolt length-at-age in years where whole river fertilizer addition was undertaken, and an increase in the proportion of 2-year-old smolts in the sample. The proportion of younger smolts was altered in the Keogh River in 2001 with age 1 and 2 smolts representing over 67% of the smolt yield compared to an average age 2 composition of 32% in years where no nutrient addition occurred. The calculated proportion of age 2 smolts in the Waukwaas River remained similar through the 1997–2001 sample period, and averaged 47% (Fig. 19.4a, b). In years where partial or full river fertilizer addition was undertaken (1998–2001) on the Keogh River, smolt length at age was increased by 5% to 10%; age 2 smolts averaged 173 mm and age 3 smolts 204 mm. Length at age of Waukwaas smolts remained stable through the 4 years; age 2 smolts averaged 165 mm and age 3 smolts 187 mm (Fig. 19.5), i.e. no different than Keogh smolts prior to nutrient addition.

Coho fry mean weights were, on average, 10% greater in the Keogh River in 2001 than in the untreated Waukwaas River in summer and autumn samples, but not significantly different between rivers for either length or weight ($P = 0.45$, ANOVA). Variance was high among mean weights of fry from different reaches on both rivers (Table 19.1).

Table 19.1 Mean length and weight of steelhead (SHF) and coho (COF) fry in treated and untreated sections of the Keogh River and within the untreated Waukwaas River during 2001. (f) = Fertilizer added; ND = no sample data.

Reach	SHF Summer weight (g)	SHF Summer length (mm)	SHF Autumn weight (g)	SHF Autumn length (mm)	COF Summer weight (g)	COF Summer length (mm)	COF Autumn weight (g)	COF Autumn length (mm)
Keogh								
W (f)	1.7	53.4	4.3	70.2	2.7	57.7	5.5	78.3
X (f)	1.7	50.3	ND	ND	2.4	54.8	ND	ND
Y (f)	2.2	56	5	77.4	2.4	56.6	4.9	75.8
Z (f)	2.3	56.9	ND	ND	2.3	55.9	ND	ND
Waukwaas								
1	ND	ND	ND	ND	3	52	ND	ND
2	0.7	40.7	2.02	59.3	1.6	49.1	2	70.7
3	0.8	37	ND	ND	1.6	49.9	ND	ND
4	1.4	40.4	3.08	66.9	1.7	58.4	3.1	73

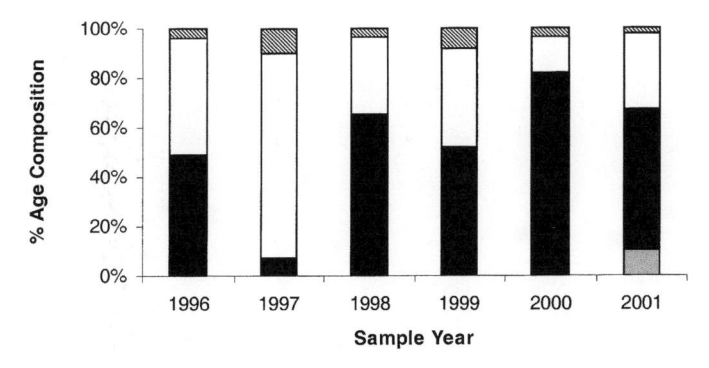

Fig. 19.4a Change in relative abundance of smolts of three age-classes before and during nutrient enhancement studies in the sampling period 1996–2001 on the Keogh River; grey bars, age 1; black bars, age 2; open bars, age 3; hatch bars, age 4.

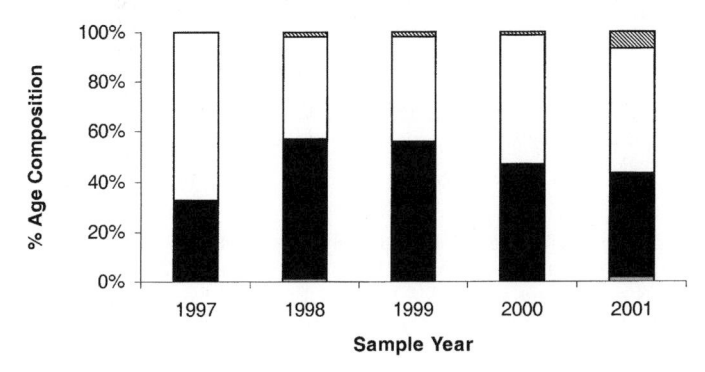

Fig. 19.4b Change in relative abundance of smolts of four age-classes in the sampling period 1996–2001 on the Waukwaas River; grey bars, age 1; black bars, age 2; open bars, age 3; hatch bars, age 4.

Data from the combination of treatment methods on the Keogh River (i.e. no treatment, fertilizer addition only, structure placement only, fertilizer and structure placement) and the untreated sections on the Waukwaas River were used to evaluate the average response to treatments. Steelhead fry densities were, on average, over 10-fold greater in areas treated with fertilizer and structures on the Keogh River than in control sections of the same watershed. Fertilized areas (without structures) also, on average, exhibited greater abundance of fry than areas with structure placement only. Average levels of fry abundance in areas treated with both fertilizer and structures in the Keogh River still fell short of average reach densities in the Waukwaas River, but less so than pre-treatment (Table 19.2). Steelhead parr densities, in comparison, were, on average, nearly twice as abundant in areas treated with both fertilizer and structures on the Keogh River than in sections treated with only fertilizer or structures. However, both treatments individually resulted in greater abundance of parr than in control (untreated) sections on the Keogh River. Average levels of parr abundance in areas treated with both fertilizer and structures in the Keogh River now exceed average reach densities in the Waukwaas River, by over 15%.

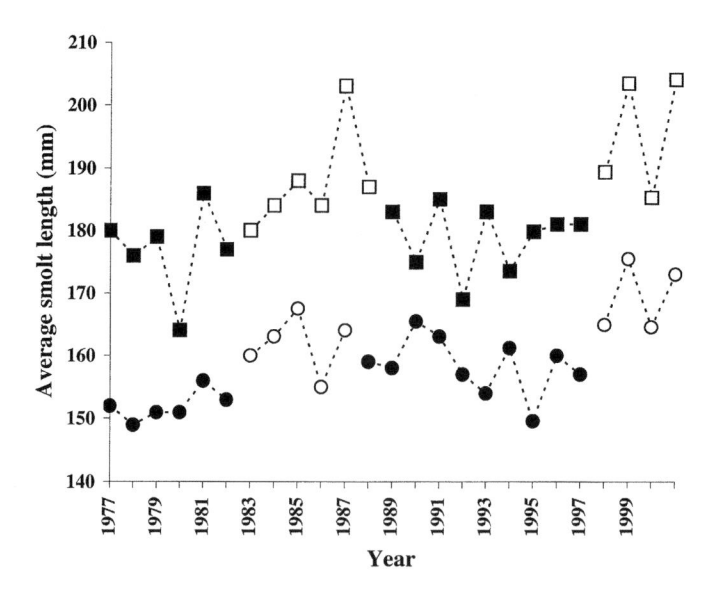

Fig. 19.5 Mean length of age 2 (circles) and age 3 (squares) smolts from the Keogh River, 1977–2001. Year classes present during fertilizer addition are indicated with open markers, non-fertilized year classes with solid markers.

Table 19.2 Mean abundance of salmonid juveniles [steelhead fry (SHF), coho fry (COF)] (no./100 m) in treated and untreated reaches of the Keogh and Waukwaas Rivers from 1997–2001. S = Structures added; F = fertilized; SF = structures and fertilization; SD = standard deviation. Numbers in parentheses are number of samples.

		Waukwaas No treatment (20)	Keogh No treatment (3)	Keogh S (3)	Keogh F (5)	Keogh SF (12)
SHF	Average	298	14	14	145	168
	SD	311	20	8	94	105
SHP	Average	42	15	23	30	49
	SD	29	6	24	18	32
COF	Average	208	281	633	271	390
	SD	152	149	380	91	158

Coho fry abundance was less responsive to nutrient replacement in the Keogh River (5% lower abundance than Keogh River control reaches), whilst the addition of 'structures only' appeared as an effective restoration technique, as well as where structures were combined with nutrient replacement. Reaches treated with structures (and in some cases fertilizer) had coho fry abundance that was, on average, as much as three-fold higher than untreated sections, on the Keogh or Waukwaas Rivers. Untreated and fertilizer-only reaches on the Keogh River had higher (30–40%) coho fry abundance than observed in the untreated Waukwaas River.

The response to rehabilitation treatments in the watershed was evident in the smolt yield which integrated effects through the fishes' freshwater life stage (Fig. 19.6a, b). Coho smolt numbers rose steadily for both watersheds through the first 3 years of investigation, but fell sharply in the Keogh River in 1998. An increase in coho smolt yield was observed in 1999 (more than two times 1998 yield), to > 53 000 fish. This trend continued in 2000, when 74 459 smolts were enumerated, the highest yield since 1987. Coho smolt yield in 2001 was reduced compared to the high value observed in 2000, to 59 993, which was 95% of the mean (1977–2001). In the Waukwaas River, reductions in coho smolt yield in 1999 and 2000 were followed in 2001 by a decline to 21 574 smolts (44% decrease), or 40% of the 7 year average. Steelhead smolt yield decreased slightly on the Keogh River to 2010 fish in 2001. This followed increases in yield in 1999 and 2000, versus a pre-treatment record low yield of < 1000 in 1998. In

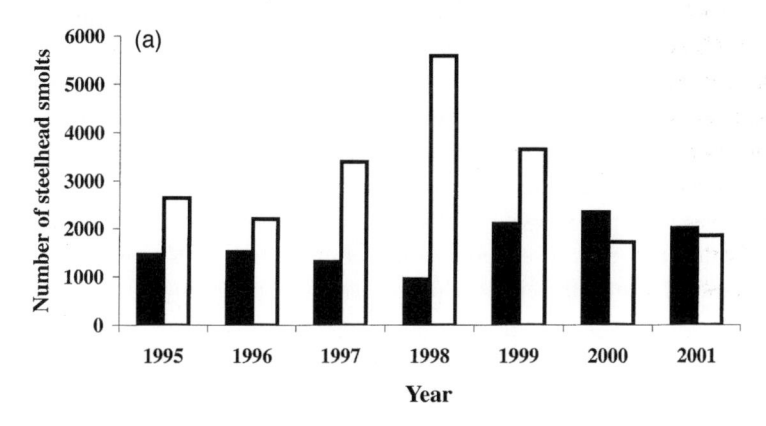

Fig. 19.6a Steelhead smolt yield from the Keogh and Waukwaas Rivers (1995–2001). Open bars represent Waukwaas River smolts and solid bars Keogh River smolts.

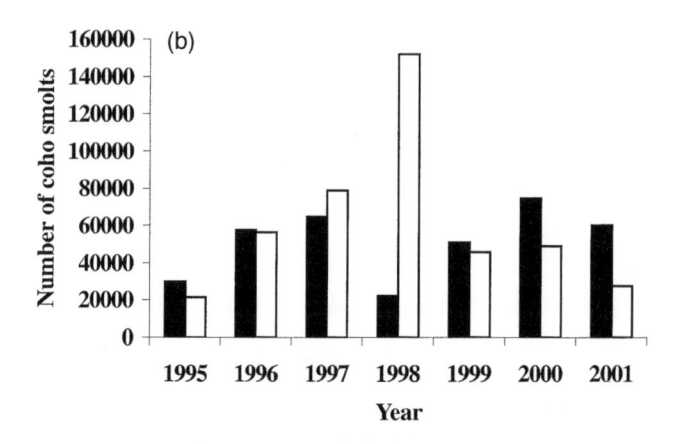

Fig. 19.6b Coho smolt yield from the Keogh and Waukwaas Rivers (1995–2001). Open bars represent Waukwaas River smolts and solid bars Keogh River smolts.

comparison, estimated yield on the Waukwaas River remained low at 1859 fish, the second lowest in the six sample years.

Differences in smolt recruitment were a key result. Information on smolt yield per spawner as a function of the number of spawners in the Keogh River is incomplete for year classes where steelhead trout juveniles were produced under partial or totally restored conditions. Significant differences in slopes and intercepts of the transformed relationship precluded further analysis. Also, results from the Watershed Restoration Program (WRP) treatment were outside the range of previous data (very low numbers of spawners). Pre-treatment data from the 1990s indicated very low yield, with < 2 smolts per spawner. This was quickly exceeded under even partial treatment, when > 14 smolts per spawner were produced from the 1996 brood year (Fig. 19.7). Further increases to 42 smolts per spawner from the 1997 brood year indicated the highest smolt recruitment in 25 years of study. More recent yields are incomplete with an unknown number of older smolts in 2002 (4-year-old smolts from the 1998 brood year and 3-year-old smolts from the 1999 brood year). However, significant increases in freshwater production are evident, with a minimum yield of 53 smolts per spawner from the 1998 brood year, and 15 smolts per spawner from the 1999 brood year. Similar data are not available from the Waukwaas River, because no spawners are enumerated. Data on the recruitment of coho smolts per spawner are only available for the 1998 and 1999 brood year as yet, the first years of accurate adult enumeration at Keogh. Yield was estimated at 9 and 21.5 smolts per spawner respectively, assuming 90% of smolts were 1 year in age.

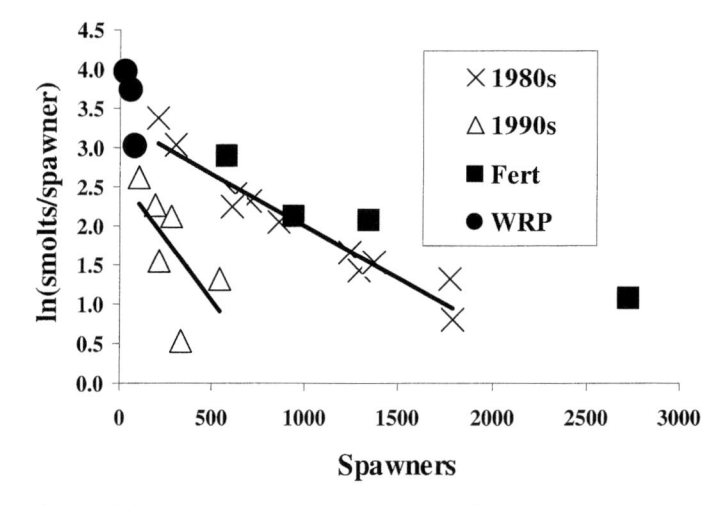

Fig. 19.7 Relationship between the natural logarithm of steelhead smolts per spawner and number of spawners in Keogh River during the production regimes of the 1980s (crosses) and the 1990s (triangles), during nutrient experiments (squares) in the mid-1980s, and preliminary results from the period of the Watershed Restoration Program (WRP) (circles).

Discussion

Five years of data (1997–2001) from the sampling of juvenile salmonids during the summer and 7 years of smolt yield data (1995–2001) were used to determine the effects of restoration techniques on salmonid densities, growth and yield per spawner. Trends of in-stream juvenile abundance and smolt yield from the Waukwaas River, draining to the west coast of Vancouver Island, were likely related to climatic conditions and adult spawning escapements; the latter were relatively high. Trends in fish abundance in the Keogh River indicated positive effects of the incrementally increasing amount of watershed restoration work, although tempered with fry density changes related to low and varied adult escapement to this east coast Vancouver Island stream (Smith & Ward, 2000). All main stem reaches in the river have now undergone habitat improvement treatments (i.e. where sufficient large woody debris was lacking). Additionally, the whole main stem has now had nutrient replacement treatment (for at least 3 years); thus almost all steelhead parr in the river have resided under enriched conditions.

In contrast to the suggestion of Ward et al. (2003), an additional year of juvenile sampling did little to statistically assist the interpretation of benefits during that life stage. Average steelhead fry abundance in the Keogh River decreased over 2000 results in three of the four sample reaches (the fourth, reach Z, indicated a slight increase in production). There was a 50% decline in fry abundance in the middle Keogh River (reaches X and Y) compared to 2000 data, although densities were higher than in pre-treatment conditions. Reasons for these among-year variations are complex, but nutrient addition was a key positive factor. Decreased numbers of mature steelhead in 2001 were also responsible for these changes, because escapement decreased in 2001 to an estimated 70 wild female spawners, from 88 in 2000. Reaches on the Keogh River treated with nutrient enrichment produced, on average, 167 fry/100 m, compared to 14 fry/100 m in untreated areas. Despite these increases, abundance was still lower than that observed in the untreated Waukwaas River, where, on average, 297 fry/100 m were observed over the study period. Keogh River fry densities (9–28 fry/100 m^2, dependent on habitat type) were in the lower range of historical data [12–90 fry/100 m^2; Ward & Slaney (1993)] and lower than in 2001. Increases (two- to three-fold) in adult steelhead escapement, i.e. greater than the recent escapement of 148 adults, are required to effectively seed the full river length to produce the full benefits of the habitat rehabilitation.

Steelhead parr densities in the Keogh River were higher when compared to 1997 (pre-treatment data) but lower than observed in 1998, 1999 and 2000. Average abundance was 35 parr/100 m, a two-fold increase over pre-treatment data (McCubbing & Ward, 1997). This was similar to the average densities recorded on the untreated Waukwaas River, 39 parr/100 m. Densities of parr (3–5 parr/100 m^2) by habitat type on the Keogh River were lower than historic densities recorded in preferred habitat (boulder clusters and riffles) both where naturally occurring or experimentally introduced (Ward & Slaney, 1993). These annual variations in steelhead parr abundance may have been due to climate, treatment and escapement effects. The abundance of steelhead parr in Keogh reaches treated with both restoration techniques were highest and exceeded those observed in the untreated Waukwaas River. Restoration efforts appear to have dramatically and persistently improved the fry-to-parr survival rates (perhaps by as much as three-fold) in steelhead; thus the yield of parr and smolts has improved above that expected based on smolt recruitment in the decade of the 1990s (Ward, 2000).

Differences in spawner escapement may explain some of the variations in coho fry abundance among reaches and years. Despite a reduction in adult escapement, abundance of coho fry in the Keogh River exceeded that on the Waukwaas River (adult escapements unknown). Densities in preferred habitat exceeded those in historic surveys [0.6 fry/m^2 compared to 0.4 fry/m^2; Irvine & Ward (1989)]. A significant increase in the frequency of pool and run habitat was associated with the additions of large woody debris; thus fish densities increased. Fry abundance in reaches treated with structures supported, on average, 300–500 fry/100 m compared to 250–280 fry/100 m in untreated areas (both watersheds, regardless of nutrient replacement). Similar results have been observed elsewhere for individual structure placements (Ward & Slaney, 1993).

Significant differences in salmonid juvenile densities were not observed among structure types, regardless of structure complexity, within the 5 years of study (pooled data). Reasons for this include variation in adult escapement and subsequent egg deposition among years and locations (McCubbing et al., 1999; Ward, 2000), the effects of nutrient addition on some structures sites in some years but not in others, and variation in use of structures among fish species during periods of extremely low stream flow during summer. In general, structures associated with higher velocity water (riffles and runs in summer months) exhibited increased steelhead parr abundance, as previously observed by Ward & Slaney (1993). Complex debris jams and double deflector logs associated with low velocity (pool and flat habitat) were associated with a greater abundance of coho fry. This information, as in previous years, indicated that a variety of structure types and placements within a range of habitat types provided a diversity of niches for stream-dwelling salmonids, and that individual structure types may be less important than restoration plans that provide a diversity of habitats.

Inorganic nutrient replacement led to significant increases in fry and smolt weights compared to data from the untreated Waukwaas River. Increasing fry densities since 1998 do not appear to have increased competition for resources to a level that has restricted growth. There are insufficient data at low fry density under conditions of natural nutrient levels prior to these studies to conclusively separate the density effect from the nutrient addition effect. Nevertheless, fry densities in areas that had received nutrient addition were similar to values obtained under much higher adult escapement (Ward & Slaney, 1993), yet weight and condition factor of fry in these areas was dramatically elevated over pre-treatment and Waukwaas results. This suggests there may be a growth and a numerical response to the nutrient addition, as observed and reported in earlier Keogh studies (Johnston et al., 1990; Slaney & Ward, 1993). The effects on coho fry weight, length and condition factor, while less pronounced than that in steelhead fry, were positive. Higher densities of larger fry in the Keogh River than the Waukwaas River resulted in a significant difference in the assessed biomass for this species between rivers. Even modest increases in fry length and weight can have significant effects on overwinter survival rates (Scrivener & Brown, 1993); thus coho smolt yield on the Keogh River will likely remain high.

Length-at-age of steelhead smolts from the Keogh River had increased annually during nutrient addition until 2000 when a slight downturn in size at age was observed. This change was attributed to the lack of additional natural nutrient input to the lower Keogh River in 1999 from a very poor pink salmon escapement, an association identified previously (Ward & Slaney, 1988). Smolt length-at-age increased again in 2001 for both 2-year and 3-year smolts, to among the highest recorded in > 27 years of study (Slaney & Ward, 1993; Ward, 2000) on the Keogh River and in other coastal streams where nutrient addition has been monitored (Slaney et al., 2003). The increase in length and weight of steelhead parr documented here also appeared to be related to improved overwinter survival, culminating in sustained increases in smolt yield, particularly 2-year-old smolts in 1999, 2000 and 2001. Historically, these larger smolts have survived at higher rates in the ocean, producing a greater escapement of adult steelhead upon return than smaller smolts (Ward & Slaney, 1988; Slaney & Ward, 1993).

Coho smolt yield increased again on the Keogh River in 2000 over the historically poor yield observed in 1998, to the highest since 1987. This was likely related to the large escapement of adults recorded in 1998; the majority of smolts migrate at age 1 from the Keogh River (Irvine & Ward, 1989; McCubbing & Ward, 1999). Coho smolts were similar to steelhead in their response to nutrient addition, and were, on average, larger for their age than in previous unfertilized years or in the untreated Waukwaas River. Coho fry in the Keogh River may have thus experienced better-than-average overwinter survival rates due to their larger size, a consequence of treatments. Improved diversity of habitat, both in-stream and off-channel, may also be responsible for the observed improvements in fry-to-smolt survival.

Limited data on smolts produced per spawner in the rehabilitated stream are now available from the 1997, 1998 and 1999 brood years and the 1999, 2000 and 2001 smolts. Further smolt data (to 2004) are required for a more complete analysis. However, initial results for both steelhead trout and coho salmon are positive to date. Steelhead smolts per spawner in the Keogh River have risen from the dangerously low < 3 smolts per spawner from the 1996 brood to > 53 smolts per spawner from the 1998 brood year, the highest production per spawner on record (Ward, 2000). Current data indicate continued high levels of production from low adult escapement, although data from 2002 smolt production is required to provide 1999 brood-year results. Low numbers of spawners and reduced competition for resources by adult spawners, fry and parr may in part be responsible for these observations, but if future spawner numbers increase correspondingly, these confounding factors should be reduced, and restoration efforts may become increasingly evident as responsible for maintaining higher levels of smolt yield, as observed thus far.

Coho smolt-per-spawner data are as yet restricted to two brood years. Approx. 90% of coho smolts migrate at age 1 (University of British Columbia, data on file); thus the 1998 brood year produced approx. 9 smolts per spawner. This increased significantly for the 1999 brood year, with an estimate of 19 smolts per spawner. In comparison, coho smolt recruitment averaged 83 smolts per female spawner at Black Creek, near Courtney, on the east coast of Vancouver Island, a highly productive stream with abundant coho habitat (K. Simpson, DFO, Nanaimo, pers. comm.).

Positive effects of watershed restoration treatments, particularly nutrient addition and stream habitat restoration, were again apparent from studies of juvenile salmonids in the Keogh River in 2001 when compared to the untreated Waukwaas River, although lower escapement of adult steelhead and coho salmon confounded results. Annual variations in adult salmon and trout escapements and in freshwater survival were evident in both watersheds, likely related to varied climatic conditions. Thus, trends in juvenile densities and smolt yield might be expected to likewise vary. However, the results presented here suggest that negative environmental factors may be buffered by stream habitat restoration and nutrient replacement. This, combined with occasional years of improved ocean survival, may sustain salmonid populations in coastal streams against poor ocean conditions. Protection of the spawning stock, habitat protection and rehabilitation, and a better understanding of the limits to production in the marine environment remain of paramount importance.

Acknowledgements

Research on salmonids in the Keogh River has been supported through British Columbia's Habitat Conservation Trust Fund. The Watershed Restoration Program funded the rehabilitation work and its in-stream evaluations. We are most grateful to the field staff and, in particular, to Lloyd Burroughs.

References

Anderson, J.J. (1999) Decadal climate cycles and declining Columbia River salmon. In: *Sustainable Fisheries Management: Pacific Salmon* (eds E.E. Knudsen, C.R. Steward, D.D. MacDonald, J.E. Williams & D.W. Reiser). Lewis Publishers, Florida, pp. 467–84.

Bradford, M.J. & Irvine, J.R. (2000) Land use, fishing, climate change, and the decline of Thompson River, British Columbia, coho salmon. *Canadian Journal of Fisheries and Aquatic Sciences*, **57**, 13–16.

Colligan, M., Nickerson, P. & Kimball, D. (1999) Endangered and threatened species; proposed endangered status for a distinct population segment of anadromous Atlantic salmon (*Salmo salar*) in the Gulf of Maine. *Federal Register*, **64** (221), 62627–40.

COSEWIC (2002) *Canadian species at risk, May 2002*. Committee on the Status of Endangered Wildlife in Canada, Canadian Wildlife Service, Environment Canada, Ottawa, 34 pp.

Friedland, K.D., Hansen, L.P., Dunkley, D.A. & MacLean, J. (2000) Linkage between ocean climate, post-smolt growth, and survival of Atlantic salmon (*Salmo salar* L.) in the North Sea area. *ICES Journal of Marine Science*, **57** (2), 419–29.

Hartman, G.F., Groot, C. & Northcote, T.G. (1999) Science and management in sustainable salmonid fisheries: the ball is not in our court. *Sustainable Fisheries Management: Pacific Salmon* (eds E.E. Knudsen, C.R. Steward, D.D. MacDonald, J.E. Williams & D.W. Reiser). Lewis Publishers, Florida, pp. 31–74.

Irvine, J.R. & Ward, B.R. (1989) *Patterns of timing and size of wild coho smolts (*Oncorhynchus kisutch*) migrating from the Keogh River watershed on northern Vancouver Island.* Canadian Journal of Fisheries and Aquatic Sciences, **46**, 1086–94.

Johnston, N.T., Perrin, C.J., Slaney, P.A. & Ward, B.R. (1990) Increased juvenile salmonid growth by whole river fertilization. *Canadian Journal of Fisheries and Aquatic Sciences*, **47**, 862–72.

Lackey, R.T. (1999) Salmon policy: science, society, restoration, and reality. *Renewable Resources Journal*, **17** (2), 6–16.

McCubbing, D.J.E. (2001) *Adult steelhead trout and salmonid smolt migration at the Keogh River during spring 2001.* Province of British Columbia Ministry of Environment, Lands and Parks, Fisheries Research and Development Section Contract Report by the Northern Vancouver Island Salmonid Enhancement Association, 25 pp.

McCubbing, D.J.F. & Ward, B.R. (1997) *The Keogh and Waukwaas Rivers paired watershed study for B.C.'s Watershed Restoration Program: juvenile salmonid enumeration and growth 1997.* Province of British Columbia, Ministry of Environment, Lands and Parks, and Ministry of Forests Watershed Restoration Project Report no. 6, 33 pp.

McCubbing, D.J.F., Ward, B.R. & Burroughs, L. (1999) *Salmonid escapement enumeration on the Keogh River: a demonstration of a resistivity counter in British Columbia.* Province of British Columbia Fisheries Technical Circular no. 104, 25 pp.

O'Neil, S., Ritter, J. & Robichaud-LeBlanc, K. (eds) (2000) *Proceedings of a Workshop on Research Strategies into the Causes of Declining Atlantic Salmon Returns to North American Rivers.* Fisheries and Ocean Canada Canadian Stock Assessment Secretariat Proceedings Series 2000/18, 80 pp.

Scrivener, J.C. & Brown, T.G. (1993) Impact and complexity from forest practices on streams and their salmonid fishes in British Columbia. In: *Le Développment du Saumon Atlantique au Québec: Connaître les Règles du Jeu pour Réussir* (eds G. Shooner & S. Asselin). Colloque International de la Fédération Québécoise pour le Saumon Atlantique, Quebec, December 1992. Collection *Salmo salar* no. 1, pp. 41–9.

Slaney, P.A. & Martin, D.A. (1997) Planning fish habitat rehabilitation: linking to habitat protection. In: *Fish Habitat Rehabilitation Procedures* (eds P.A. Slaney & D. Zaldokas). Province of British Columbia, Ministry of Environment, Lands and Parks, and Ministry of Forests, Watershed Restoration Technical Circular no. 9.

Slaney, P.A. & Ward, B.R. (1993) Experimental fertilization of nutrient deficient streams in British Columbia. In: *Le Développment du Saumon Atlantique au Québec: Connaître les Règles du Jeu pour Réussir* (eds G. Shooner & S. Asselin). Colloque International de la Fédération Québécoise pour le Saumon Atlantique, Quebec, December 1992. Collection *Salmo salar* no. 1, pp. 128–41.

Slaney, P.A., Ward, B.R. & Wightman, J.C. (2003) Experimental nutrient addition to the Keogh River and application to the salmon river in coastal British Columbia. In: *Nutrients in Salmonid Ecosystems: Sustaining Production and Biodiversity* (ed. J.G. Stockner). American Fisheries Society, Symposium 34, Bethesda, Maryland, pp. 111–26.

Smith, B.D. & Ward, B.R. (2000) Trends in adult wild steelhead (*Oncorhynchus mykiss*) abundance for coastal regions of British Columbia support the variable marine survival hypothesis. *Canadian Journal of Fisheries and Aquatic Sciences*, **57**, 271–84.

Tyedmers, P. & Ward, B.R. (2001) A review of the impacts of climate change on B.C.'s freshwater fish resources and possible management responses. *University of British Columbia Fisheries Centre Research Reports*, **9** (7), 12.

Walters, C.J. & Ward, B.R. (1998) Is solar UV radiation responsible for declines in marine survival rates of anadromous salmonids that rear in small streams? *Canadian Journal of Fisheries and Aquatic Sciences*, **55**, 2533–8.

Walters, C.J., Collie, J.S. & Webb, T. (1988) Experimental designs for estimating transient responses to management disturbances. *Canadian Journal of Fisheries and Aquatic Sciences*, **45**, 530–38.

Ward, B.R. (2000) Declivity in steelhead trout recruitment at the Keogh River over the past decade. *Canadian Journal of Fisheries and Aquatic Sciences*, **57**, 298–306.

Ward, B.R. & Slaney, P.A. (1988) Life history and smolt-to-adult survival of Keogh River steelhead trout (*Salmo gairdneri*) and the relationship to smolt size. *Canadian Journal of Fisheries and Aquatic Sciences*, **45**, 1110–22.

Ward, B.R. & Slaney, P.A. (1993) Habitat manipulations for the rearing of fish in British Columbia. In: *Le Développement du Saumon Atlantique au Québec: Connaître les Règles du Jeu pour Réussir* (eds G. Shooner & S. Asselin). Colloque International de la Fédération Québécoise pour le Saumon Atlantique, Québec, Décembre 1992, Collection *Salmo salar* no. 1, pp. 142–8.

Ward, B.R., McCubbing, D.J.F. & Slaney, P.A. (2003) Evaluation of the addition of inorganic nutrients and stream habitat structures in the Keogh River watershed for steelhead trout and coho salmon. In: *Nutrients in Salmonid Ecosystems: Sustaining Production and Biodiversity* (ed. J.D. Stockner). American Fisheries Society Symposium 34, Bethesda, Maryland, 24–26 April 2001, pp. 127–47.

Welch, D. (1999) Global warming and the prospects for fisheries management. *Newsletter of the Canadian Aquatic Resources Section of the American Fisheries Society*, **8** (2), 8.

Chapter 20
Prospects for Improved Oceanic Conditions

S. Hughes and W.R. Turrell

Fisheries Research Services Marine Laboratory, PO Box 101, Victoria Road, Aberdeen, AB11 9DB, Scotland, UK

Abstract: Observed and modelled sea-surface temperature data from the north Atlantic have been collected to identify trends and investigate the spatial variability of these trends. Data obtained from the Reynolds sea-surface temperature data set have been used to examine temperature trends over the last 20–40 years. Predicted sea-surface temperature data, obtained from the climate model HADCM3, are used to examine trends over the next 20–50 years. Understanding of the cause of past trends is the key to predicting future changes. The modelled data are presented as our best estimate so far of the changes that will take place in the marine habitat of Atlantic salmon. It is clear that the last decade has been a period of warming and it is likely that this trend will continue in the future. The rate of warming in the future may be lower than we have seen over the past decade and will vary across the different regions, with some areas experiencing cooling.

Introduction

Over the past decade the ocean habitat of Atlantic salmon has experienced sustained change, characterized by warming within regions at the southern boundary of their ocean habitat, and cooling within more northern areas. These changes have been accompanied by changes in oceanic plankton populations and production, and both the environmental and ecosystem changes have been linked to large-scale climatic forcing by processes such as the North Atlantic Oscillation (NAO) (Dickson & Turrell, 1999). With the present view of the ocean migration of Atlantic salmon, the changes over the past decade suggest a less favourable ocean habitat.

The last decade may, however, not be characteristic of the past century, or the coming decade. It is extremely problematic to extrapolate present trends into the future, but we must make a start by trying to understand the trends we have seen in the past. There appear to be three ways in which future prospects for 'improvements' in ocean climatic conditions might be predicted: by extrapolating the present global changes into the future using assumptions about the world's climate system and greenhouse gas emissions; by assuming past changes have been the result of cyclical phenomena, and using these cycles to predict future change; and by examining possible mechanisms that might affect the north Atlantic over the coming decades, using records of past change or studies of present-day active mechanisms.

In this paper we examine the present changes occurring throughout the ocean habitat of north Atlantic salmon using the best available data set describing monthly variation of sea surface temperature. We go on to use the internationally accepted predictions of future trends based on various economic models of global growth and its effect on the climate, published by the Intergovernmental Panel on Climate Change (IPCC), to examine probable future trends. Finally, we mention some results of recent research which suggest we are entering a period of increasingly uncertain climatic conditions.

Third report of the Intergovernmental Panel on Climate Change (IPCC)

In 2001, the IPCC published its third assessment report on global climate change (IPCC, 2001). Compared with the second assessment report (IPCC, 1995), the IPCC has been able to make more definite statements about global warming as there is now sufficient data to describe the trends with statistical confidence. Using temperature data from instrumental records that extend back to 1861, and historical records prior to that date, the IPCC report demonstrates quite clearly that overall the global trend in temperature is one of warming. In fact, the global average surface temperature (the average of near-surface air temperature over land and sea surface temperature over the oceans) increased by around 0.6°C during the 20th century and the warming trend in the northern hemisphere was even more significant, with the decade of the 1990s being the warmest decade of the 20th century and probably the warmest decade of the last millennium (IPPC, 2001).

Despite the overall global warming trend, the IPCC notes that there is a good deal of temporal and spatial variability. Regional-scale processes such as the NAO have introduced decadal-scale variability as well as spatial differences in the temperature trend throughout the north Atlantic area. Globally, there have been two distinct periods of warming, 1910–45 and 1976–2000. Although there has been a general warming trend over the north Atlantic between 1976 and 2000, some regions did experience cooling (most notably in the Labrador Sea, south-west of Greenland). These regional patterns will be analysed in more detail in the first part of this paper.

Future trends

The recent warming is almost certainly due to the increased concentrations of greenhouse gases in the atmosphere, gases that were emitted as a result of human activities and which have increased rapidly since the beginning of the 1900s and the dawn of the industrial revolution. Hence the key to understanding future trends in the global climate is to make accurate predictions of the concentration of greenhouse gases in the atmosphere. The IPCC has published a set of scenarios based on a variety of socio-economic models of the future, including different rates of global economic growth and resource utilization. For example, some of the scenarios assume that in the future, economies will still be dependent upon fossil fuels, whereas others assume a switch to newer and greener energy.

These scenarios have been used to predict the future emissions of greenhouse gases and the resulting buildup of concentrations of these gases in the atmosphere. The IPCC report presents the results from 40 different economic scenarios used to drive various different climate models. The results of these analyses show that, despite the wide range of scenarios, all of the outcomes result in the continued increase in greenhouse gas concentrations over the next 50 years (IPCC, 2001).

Climate models

In order to interpret the predicted changes of greenhouse gas concentration in terms of its effect on the global heat budget and hence temperature, the IPCC uses complex computer models of the global climate system. Over the last 30 years there has been a huge amount of effort put into building such climate models to test our knowledge of the present climate system and help make predictions for the future. Any numerical model is a mathematical representation of the real world, and the complexity of climate models is limited by scientific knowledge of the interactions in the ocean–atmosphere system and the computing power available to run the model. Climate models are constantly developing as knowledge and computing power improve by the inclusion of more levels of detail and better spatial resolution. In the 1970s the early climate models only included atmospheric processes, but by the 1990s, models included the effect of oceanic and terrestrial surfaces, detailed ocean and sea ice processes, as well as information about the carbon cycle and atmospheric aerosols. In the second part of this paper, we use the results of one of the climate models utilized by the IPCC in order to describe the probable future development of north Atlantic sea surface temperatures.

Data sources and methods

Reynolds sea surface temperature data sets

Sea surface temperature (SST) data for the north Atlantic region between 1950 and the present day were obtained from the United States National Oceanic and Atmospheric Administration (NOAA) National Centre for Environment Prediction (NCEP). These data have been obtained by merging estimates obtained from satellite observations and data measured *in situ* from buoys or ships. The merged data have then been optimally interpolated to provide monthly average values on an even grid spacing (Reynolds & Smith, 1994).

Two data sets of different spatial resolution have been merged to produce time series of data at a 2° grid spacing, from January 1950 until May 2002. Before 1981, data from the Reynolds Historical Reconstructed (HR) SST data set were used. The northward extent of this data set is limited to 69°N on the eastern side and 59°N on the western side of the Atlantic. From January 1981 until May 2002, the Reynolds Optimally Interpolated (OI) SST data set was used.

HADCM3 model data

Modelled SST data for the north Atlantic region from 1960–2050 were obtained from the HADCM3 global circulation model developed at the Hadley Centre for Climate Prediction and Research, part of the United Kingdom Meteorological Office. The HADCM3 model is a second-generation ocean circulation model that is able to give fairly accurate representation of the ocean currents in the north Atlantic. The oceanic component of the HADCM3 model has 20 levels and has a resolution of $1.25° \times 1.25°$.

The HADCM3 model is able to represent the north Atlantic reasonably well, although there are some localized errors as the model does not always accurately predict the location of frontal regions. The model also appears to underestimate the northward turn of the north Atlantic current around the Grand Banks (Gordon et al., 2000). Despite this, the benefits of the HADCM3 model are its ability to represent the thermohaline circulation in reasonable agreement with oceanographic observations and give a reasonable representation of circulation in the Nordic Seas.

The HADCM3 predictions of future ocean conditions used in this study assume two of the greenhouse gas emission scenarios as developed by the IPCC. These were the medium-high (scenario A2) and medium-low (scenario B2) emissions scenarios. Despite the uncertainties involved, the information from these scenarios should provide a reasonable envelope of the expected developments during the next few decades.

The HADCM3 temperature data were provided on a grid of 96 longitudes (3.75°) and 73 latitudes (2.5°). Data are skin surface temperature, equivalent to sea surface temperature and near-surface temperature over land. In order to compare the data with that of the Reynolds data set, the model data were interpolated onto a $2° \times 2°$ grid. Land surface temperatures were first removed and replaced by representative sea temperatures. Once the data were sub-sampled at a 2° resolution, data over land were removed by applying a land–sea mask.

Statistical analysis

Simple statistical analysis of the data sets described above has been undertaken for the figures presented below. Mean SSTs have been calculated by either using all the data for a fixed period (annual means) or selecting only data for a particular season (seasonal means). For these analyses, each season covers three calendar months, starting with winter, which has been defined as December, January and February. Short-term trends have been calculated using a 1-year running mean. Longer-term trends in the data sets are expressed in terms of degrees Centigrade per decade and have been calculated from the slope of a linear regression through the data over a selected period. Care has been taken to ensure that the periods cover a whole number of years to avoid biasing a trend by including an extra value from a particular month.

Results

North Atlantic circulation

When considering the future climate change in our region in terms of SST, as well as when interpreting past change, we need to understand the general features of the surface circulation of the north Atlantic (Fig. 20.1). The north Atlantic current transports warm surface water from the western North Atlantic eastwards towards northern Europe. A portion of this current turns northwards, entering the Nordic Seas near Scotland and Faeroe, and joined by waters flowing northwards along the edge of the European continental shelf in the shelf edge current. The warm surface waters enter the Nordic Seas and flow further north, towards the Arctic Ocean, cooling and freshening along the route. The region where cold waters coming from the Arctic directly meet the warmer northwards-flowing Atlantic waters is termed the Arctic or sub-Arctic front. Surface waters in the Arctic and sub-Arctic undergo a variety of transformation processes, and the resulting cold surface outflow occurs southwards down the eastern coast of Greenland in the east Greenland current. This cold water circulates around Greenland and the Labrador Sea to eventually flow southwards and join the northern boundary of the north Atlantic current as it leaves the Grand Banks region.

Atlantic salmon

For the purposes of this paper, we wish to interpret past, present and future ocean changes that are most relevant to the ocean life of Atlantic salmon. Hence we need to understand the paths of salmon ocean migration, in relation to the circulation pattern described above. There is, however, still very little knowledge about the life of Atlantic salmon in their marine phase. On the eastern side of the Atlantic, post-smolt salmon enter coastal waters from their rivers and it is probable that they make use of the north-eastward-flowing shelf edge current for their northerly migrations (Turrell & Shelton, 1993). Northward of 64°N the distribution of post-smolt catches is less clear. Salmon from eastern Atlantic stocks have been found feeding in the Norwegian Sea, north of the Faeroe Islands and west of Greenland and it is possible that they travel along the frontal regions separating the warmer Atlantic waters from the cooler Arctic waters during their open ocean migrations.

Observed average conditions 1961–1990

The Reynolds SST data set provides us with a good high-resolution estimate of the typical temperature distribution over the north Atlantic region. Plate 20.1 shows the

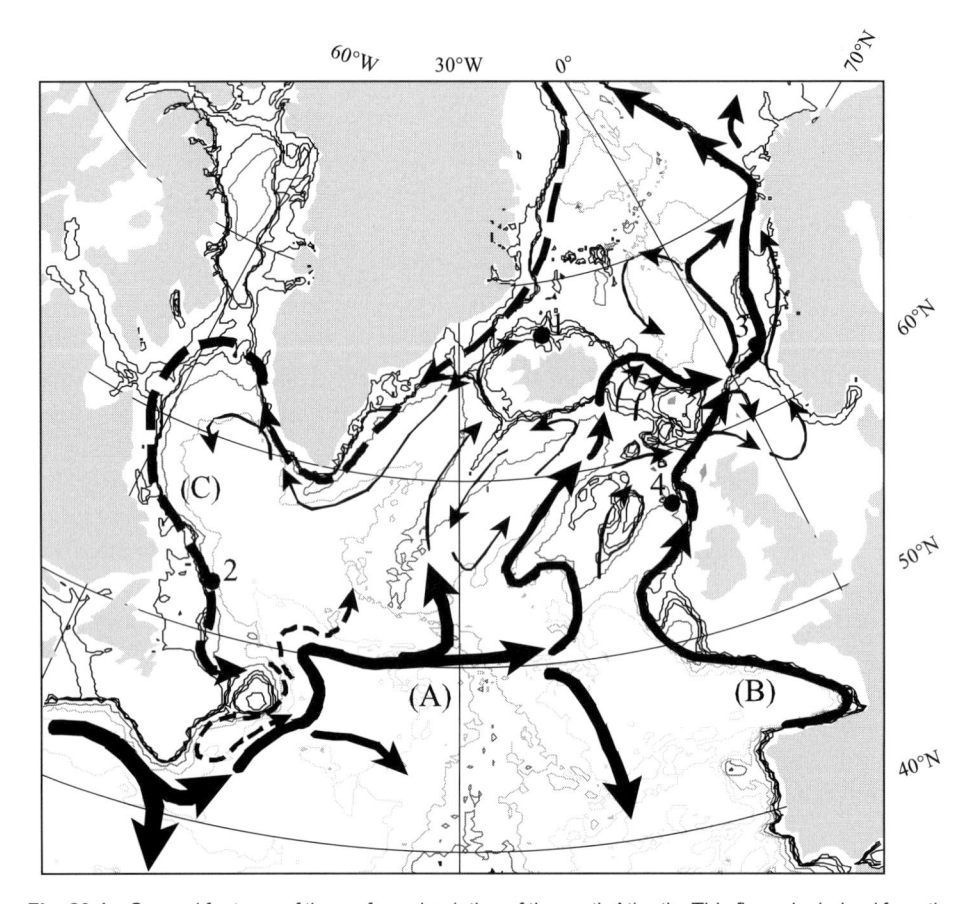

Fig. 20.1 General features of the surface circulation of the north Atlantic. This figure is derived from the classical picture of north Atlantic circulation (Dietrich et al., 1975) with detail in the more northerly region adapted from Hansen & Østerhus (2000). Thickness of each line provides an indication of relative strength of each part of the current flow. The solid lines indicate path of northward-flowing warm water; the dashed lines indicate cooler waters returning from the Arctic. The three main currents are indicated as follows: (A) North Atlantic current, (B) shelf edge current, (C) East Greenland current. Four locations, selected as indicators of key climactic conditions encountered by Atlantic salmon, are indicated by black dots. Location 1, to the north of Iceland, indicates position of front between cooler Arctic waters and warmer Atlantic waters. Location 2, in the Labrador Sea, indicates extent of cooler waters returning from the Arctic. Location 3 is within the Norwegian coastal current and shows how close this current flows to the Norwegian coast. Location 4 is along the UK continental shelf edge and is indicative of conditions in the shelf-edge current around Scotland.

annual mean (Plate 20.1a) and seasonal mean (Plate 20.1b) SST between 1961 and 1990; this represents the average climactic conditions in the region prior to the most recent period of warming. The area with average temperatures between 4°C and 13°C has been highlighted in the Plates to indicate the areas in which there is a 'favourable' thermal habitat for salmon, although this does not exclusively define the areas in which salmon can be found.

The pattern of the annual mean SST across the north Atlantic clearly reflects the

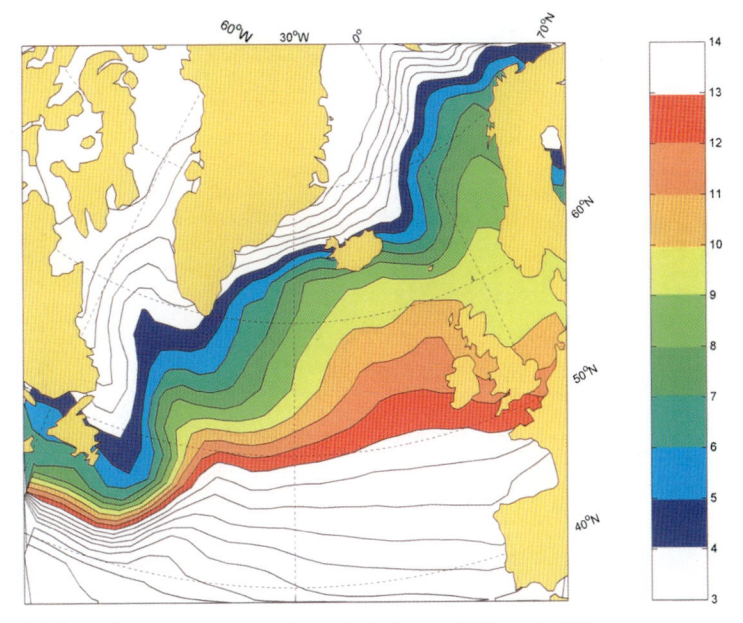

(a) Annual mean: average of all data between 1961 and 1990.

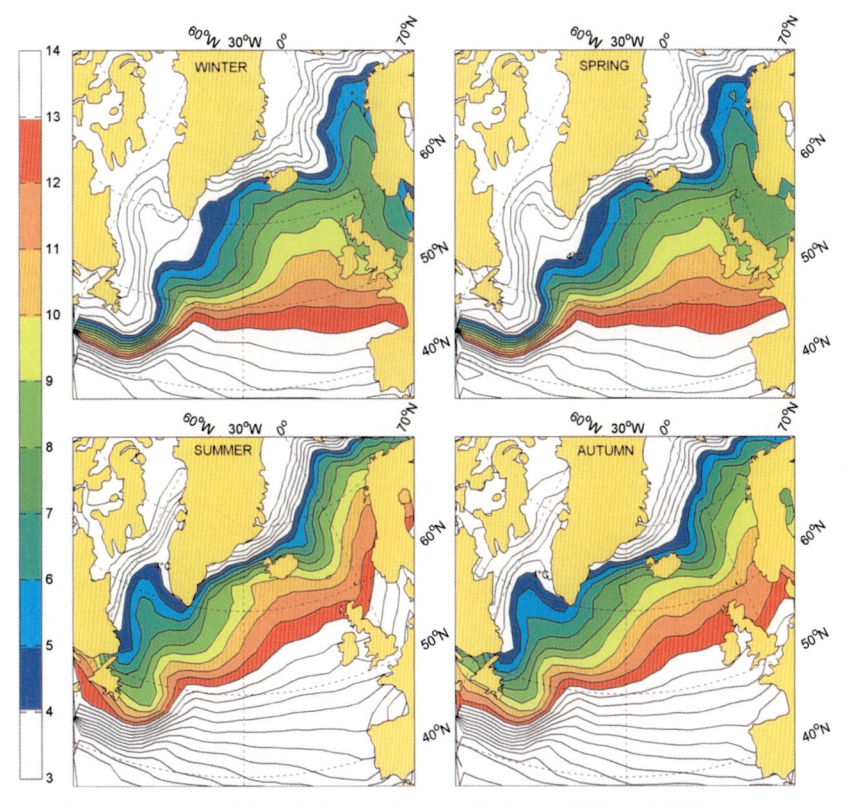

(b) Seasonal mean: average of data during each season between 1961 and 1990.

Plate 20.1 Average sea surface temperature (SST) patterns derived from the combined Reynolds data sets for the period 1961–1990. The temperature contours are at 1°C intervals. The area between the 4°C and 13°C contours has been highlighted in colour. Winter: Dec, Jan, Feb. Spring: Mar, Apr, May. Summer: Jun, Jul, Aug. Autumn: Sept, Oct, Nov.

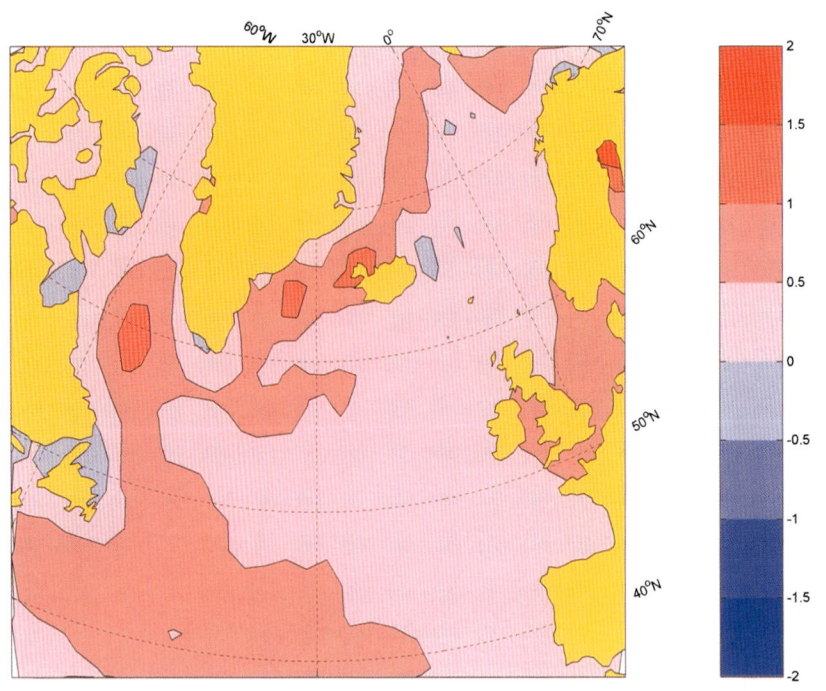

Plate 20.2 Map of temperature trend derived from the combined Reynolds data sets for the period 1981–2000. Red values indicate warming (a positive trend in temperature) and blue values indicate cooling (a negative trend in temperature). Trend values are °C per decade.

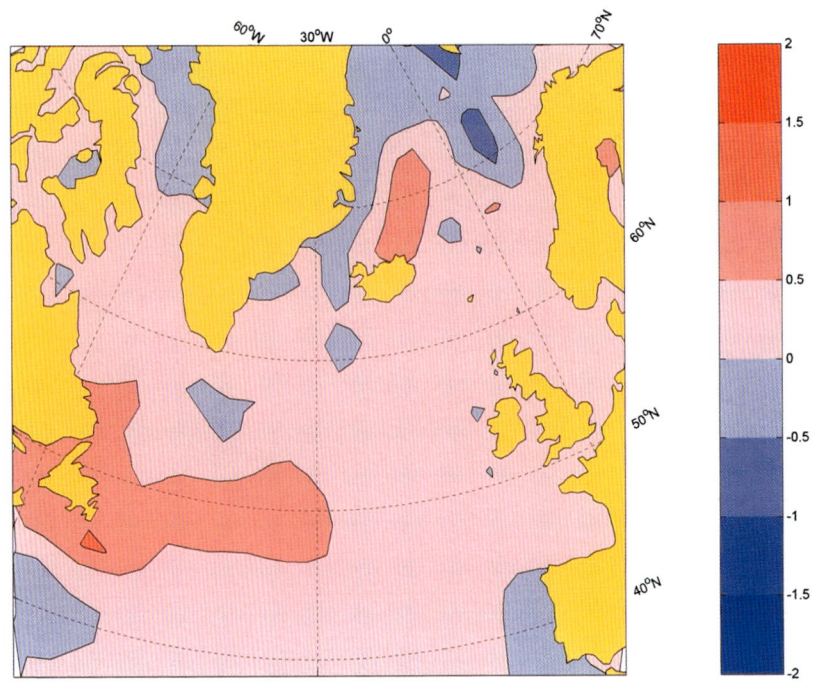

Plate 20.3 Map of temperature trend derived from the HADCM3 modelled dataset, using IPCC scenario A2, for the period 2001–2020. Red values indicate warming (a positive trend in temperature) and blue values indicate cooling (a negative trend in temperature). Trend values are °C per decade.

path of the north Atlantic current as it flows from west to east (Plate 20.1a). To the south of Newfoundland, where the temperature contours are tightly packed together, there is a strong front that marks the start of the north Atlantic current after it has split from the Gulf Stream. The temperature contours curve with the path of the current, northwards around the Grand Banks and then across to the east Atlantic. The warm water travels northwards from Scotland along the Norwegian coast, shown by the northward bulge of the temperature contours in this area.

The position of the 13°C contour line on the eastern side of the north Atlantic changes quite markedly between seasons (Plate 20.1b). During the winter, the 13°C contour sits along the northern coast of Spain, but by the end of the summer, the waters have warmed enough to move the 13°C contour as far as the southern coast of Norway. In contrast, on the eastern side of the Atlantic, the 4°C contour lies within the sub-arctic front which remains fairly stable to the north of Iceland. Across on the western side of the Atlantic the 4°C contour varies in position from south of Newfoundland in the winter, moving northward between Labrador and Greenland in the summer.

Observed temperature trends 1961–2000

Time series of SST over the period 1961–2000 have been extracted from the Reynolds data set at the four locations shown in Fig. 20.1 (northward of 69°N, the time series start in 1981). These four locations have been selected in order to represent key climate conditions encountered by salmon during their ocean migrations.

The seasonal cycles of SST can clearly be seen in the time series plots (Fig. 20.2a), the maximum temperatures occurring in the summer and minimum temperatures in the winter. The time series at location 1 (north of Iceland) shows a high level of inter-annual variability, due to its location close to the sub-arctic front. Small changes in the position of the front from month to month will result in marked changes in the SST.

The 1-year mean time series clearly shows the year to year (inter-annual) differences and the decadal variability at each location. At location 1 (north of Iceland), the inter-annual variability in the annual mean temperature is quite large, in the order of 6°C. At locations 2, 3 and 4 the variability is less (1–2°C), but there are still clearly marked changes from year to year and on a decadal scale.

Since 1981, the long-term trend at locations 1 and 2 (north of Iceland and the Labrador Sea) has been one of warming at around 0.95°C per decade, whereas on the eastern side of the Atlantic (Norwegian coast and Scottish shelf edge) the trend has also been of warming but at a lower rate of around 0.20°C per decade.

Spatial patterns of observed temperature trends 1981–2000

The annual and seasonal trends in SST at all sites calculated over the period 1981–2000 have been mapped in Plate 20.2. Again, it is clear from this figure that the overall trend over the past 20 years has been one of warming.

a) OBSERVED DATA   b) PREDICTED DATA

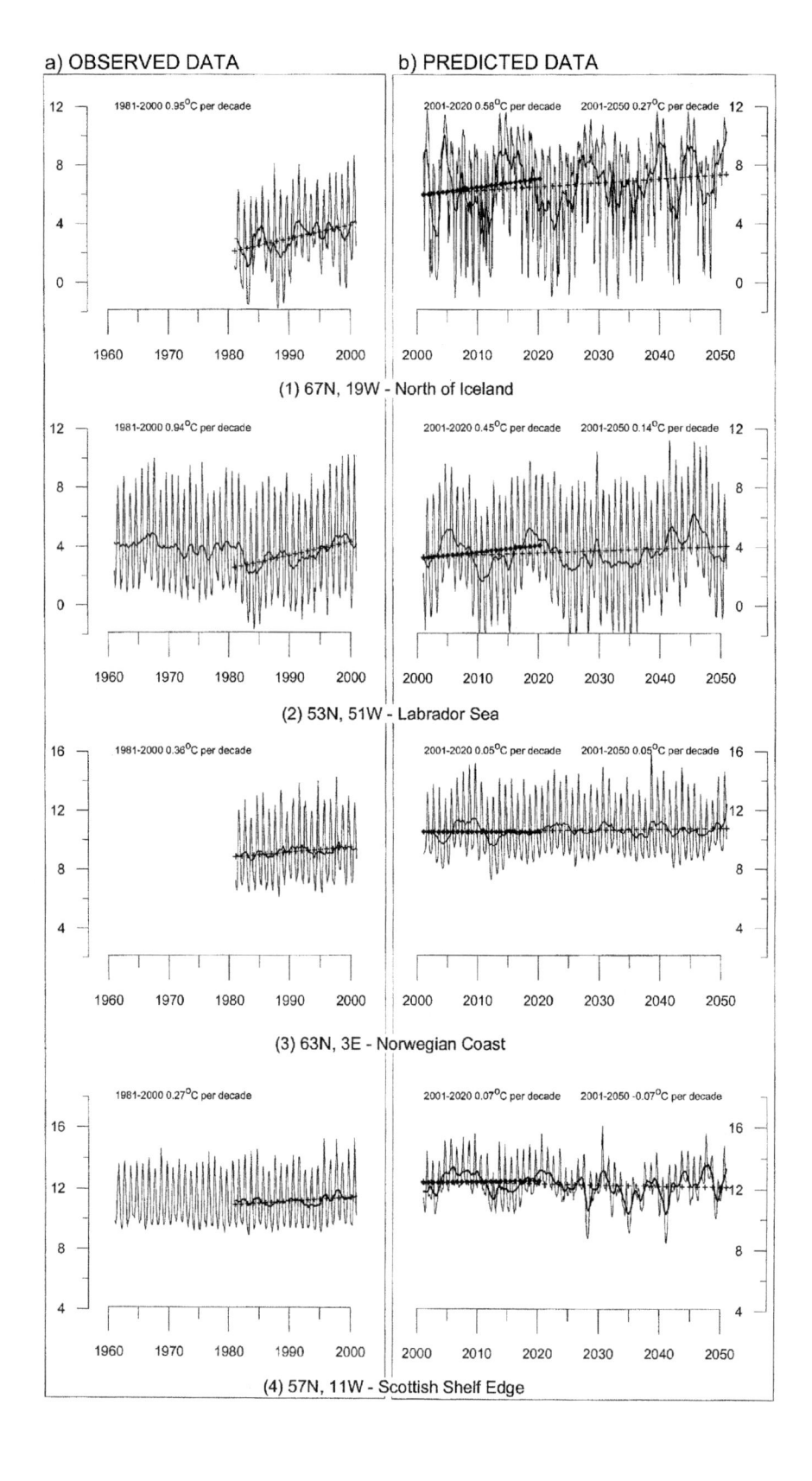

1981-2000 0.95°C per decade   2001-2020 0.58°C per decade   2001-2050 0.27°C per decade

(1) 67N, 19W - North of Iceland

1981-2000 0.94°C per decade   2001-2020 0.45°C per decade   2001-2050 0.14°C per decade

(2) 53N, 51W - Labrador Sea

1981-2000 0.36°C per decade   2001-2020 0.05°C per decade   2001-2050 0.05°C per decade

(3) 63N, 3E - Norwegian Coast

1981-2000 0.27°C per decade   2001-2020 0.07°C per decade   2001-2050 -0.07°C per decade

(4) 57N, 11W - Scottish Shelf Edge

Over most of the north Atlantic, the warming trend was between 0 and 0.5°C per decade. On the western side of the Atlantic there was a band where the warming rate was higher, between 0.5 and 1°C per decade, running from the northern Norwegian Sea, down between Iceland and Greenland and into the Labrador Sea. At some points in this band, the warming trend over the last 20 years was greater than 1°C per decade (1–1.5°C per decade). There were some regions, such as the area to the east of Iceland and the coastal region around Newfoundland, in which the long-term trend in temperature was one of cooling (the trend was between 0 and −0.5°C per decade).

In the waters around the UK, it can be seen that the warming trend has been more pronounced in the southern North Sea and the Irish Sea (between 0.5° and 1.0°C per decade) than elsewhere (between 0° and 0.5° per decade).

HADCM3 model data

Overall, it is clear from Fig. 20.2 and Plate 20.2, that the trend during the last 20 years has been one of warming, consistent with the data presented in the IPCC report (IPCC, 2001). Data from the HADCM3 model have been used to investigate future trends in SST in the north Atlantic region.

Analysis of the two scenarios has shown that over the next 20–50 years, the medium–low emissions scenario (B2) results in the strongest warming trend in SST over the north Atlantic region. To simplify the results, time series and maps from only one scenario have been presented. We have chosen to present the results from scenario A2, which gives the most conservative estimate of future warming trends. The trends from both scenarios A2 and B2 over each time period are presented in Table 20.1.

Temperature trends over the period 1981–2000 are calculated from both the Reynolds data set and the HADCM3 data set for comparison (Table 20.1). At these four locations, for this comparison period, the trends for the HADCM3 data do not match the Reynolds data very well. However, for the A2 scenario, the scale of the trends is well represented, with the highest temperature trends occurring on the western side of the north Atlantic in both the HADCM3 and Reynolds results.

The HADCM3 data reproduces the range and average values of the seasonal cycle at each of the four key locations fairly well (Fig. 20.2). Some differences would be expected due to the resolution of the model and the difficulties inherent in modelling

---

Fig. 20.2 (*Left*) Time series of sea-surface temperature (SST) at four key locations. The positions of each location are marked on Fig. 20.1. The sharply varying line shows the monthly SSTs (y-axis is °C, x-axis is year). The thicker black undulating line is a one-year running mean. The straight lines with symbols indicate the trends in temperature over selected data periods; the actual value of the trend (°C per decade) is marked on each figure. (a) Observed SSTs from the combined Reynolds datasets for the period 1961–2000. (Northwards of 69°N, data are only available from 1981 onwards.) Trend lines are shown for the period 1981–2000 (+). (b) Predicted SSTs from the HADCM3 model output using IPCC scenario A2 for the period 2001–2050. Trend lines are shown for the periods 2000–2020 (♦) and 2000–2050 (+).

264   Salmon at the Edge

the complex system of the north Atlantic. For example, at location 1, which is close to the sub-arctic front, larger differences would be expected between the range and variability of modelled data and observed data than at location 3, on the Norwegian coast. The 1-year mean trend in the predicted time series from HADCM3 has a pattern of inter-annual and decadal variability similar in scale to that seen in observed data between 1961 and 2000.

Predicted temperature trends 2001–2050

The temperature trends over the next 50 years as predicted by the HADCM3 model using emission scenario A2 are plotted in Fig. 20.2b. The general trend at the four key sites is one of warming, with the greatest increase in temperature occurring in the western regions of the north Atlantic.

The predicted long-term trends in temperature (20 and 50 years) are higher on the western side of the north Atlantic than on the eastern side (Table 20.1, Fig. 20.2b). At locations 1 and 2, the predicted long-term trend in SST is around 0.2°C per decade over the next 50 years. In contrast, at location 3 the predicted trend in temperature is an order of magnitude less, closer to 0.05°C per decade. At location 4, the long-term trend in temperature for scenario A2 is for cooling, a trend of −0.07°C per decade.

Over the next 20 years (2001–2020), the trend at all four locations is for warming. At locations 1 and 2, the warming trend is predicted to be around 0.5°C per decade, and at locations 3 and 4 the trend will be around 0.05°C per decade.

Spatial patterns of predicted temperature trends 2001–2020

The trends in SST at the four key locations are representative of the pattern across the rest of the north Atlantic. The spatial pattern of temperature trends is similar for both the A2 and B2 scenarios, although only the A2 scenario has been presented here. In

Table 20.1 Trends (°C per decade) in sea-surface temperature at four key locations in the north Atlantic, over selected time periods. Reynolds data are observations derived from a combined Reynolds data set (see text). HADCM3 data are model output from two economic scenarios A2 (medium–high emissions) and B2 (medium–low emissions).

Location	Position	Reynolds 1981–2000	HADCM3 1981–2000		HADCM3 2001–2020		HADCM3 2001–2050	
			A2	B2	A2	B2	A2	B2
North of Iceland	67°N, 19°W	0.95	0.34	−0.66	0.58	1.99	0.27	0.63
Labrador Sea	53°N, 51°W	0.94	0.20	1.09	0.45	0.80	0.14	0.48
Norwegian coast	63°N, 03°E	0.36	−0.14	−0.14	0.05	0.14	0.05	0.09
Scottish shelf edge	57°N, 11°W	0.27	−0.20	0.10	0.07	0.49	−0.07	0.09

general, over the next 20 years the trend in temperature over the whole north Atlantic is one of continued warming at a rate of between 0 and 0.5°C per decade (Plate 20.3). The areas in which there is predicted to be the highest trends in SST are those on either side of strong frontal boundaries such as the sub-arctic front and along the path of the north Atlantic current.

The predicted warming trend is greatest on the western side of the Atlantic. Off the coast of Newfoundland, where temperatures have decreased during the last decade, the temperatures are predicted to increase by between 0.5 and 1.0°C per decade over the next 20 years. The model predictions also show regions of cooling, in the Norwegian Sea and on each side of the Atlantic south of 45°N.

Other possible future outcomes

The data presented so far have been from the surface seawater temperature data set, which is to some extent justified, as salmon spend most of their time in surface waters. However, to fully understand possible future changes in ocean circulation and hence ocean climate, it is necessary to consider the pattern of changes over the whole depth of the ocean.

Figure 20.1 shows the warm surface waters flowing across the north Atlantic in a north-easterly direction. The surface currents have the effect of transporting heat northward to keep the climate of northern Europe between 5 and 10°C warmer than other sites around the world at equivalent latitudes (Plate 20.1). The annual mean SST off the western coast of Scotland is around 10–11°C, whereas at similar latitudes off the east coast of Canada, the annual mean SSTs are much lower at between 2 and 3°C.

Warm surface waters flow into the Norwegian Sea through the Faeroe–Shetland channel. In the winter these waters cool and become dense, sinking down and then flowing at depth, back out of the Norwegian Sea into the north Atlantic. This sinking occurs at two sites in the north Atlantic, the Norwegian Sea and the Labrador Sea. The sinking of these dense waters effectively acts as a pump, driving the thermohaline circulation of the world oceans.

Fisheries Research Services has been monitoring the temperature and salinity at many depths and positions across the Faeroe–Shetland channel for the last 100 years. These observations have indicated that the deep outflow of water from the Norwegian Sea into the Atlantic has been warming and freshening (Turrell et al., 1999). Other observations have shown a decrease in the measured outflow over the last 5 years and using modelled data have implied a decrease in the last 50 years (Hansen et al., 2001).

A decrease in outflow from the Arctic would have to be balanced by a decrease in the inflow. If this decrease happens it could lead to cooling of the waters around Scotland. In the past, cooler periods have been linked to decreased Arctic inflow. Data from ice cores have shown that (before the last Ice Age) these changes happened very quickly, a mechanism known as rapid climate change.

Discussion

The changes in SST of the north Atlantic during the past 20 years have been described using one of the best available data sets. For the next 20–50 years, predictions of the SST in the north Atlantic from the HADCM3 model have also been presented.

It is possible to identify long-term trends in these data which indicate that there has been warming of a scale that is not explained by natural variations in our climate and is most likely to have been caused by emissions of greenhouse gases into the atmosphere. There will always be some uncertainty in predicting future levels of greenhouse gas emissions; however, over the next 20 years, every economic scenario presented by the IPCC will result in warming. It appears that despite our best efforts to limit emissions of greenhouse gases in the atmosphere, it is already too late to affect a change in the warming trend over that period.

In addition, there has been decadal variability in SST driven by changes in the atmosphere–ocean system of the north Atlantic. The North Atlantic Oscillation (NAO) has been used to explain this decadal variability and presented as a possible indicator for predicting future changes. The decadal variability has, in the past, masked longer-term trends and made it more difficult to investigate trends in the climate data. It is likely that this decadal variability will continue to be a feature of the climate of the north Atlantic in the future.

Furthermore, long-term monitoring of temperature and salinity in key regions such as the Faeroe–Shetland channel has shown that there may be other mechanisms that are important for determining the pattern of SST in the north Atlantic. Ocean circulation models such as HADCM3 are still limited; better resolution will allow models to include more detail such as the Faeroe–Shetland channel and allow us to predict changes with more confidence.

Atlantic salmon: prospects for the future

Hence we see the prospects of a continuation of the warming that salmon have encountered during their ocean migration over the past three decades, with the possibility that the warming rate will accelerate, but with strong regional differences caused by more local circulation changes. SST data have been presented here, but in fact this is only an indicator of the wider changes that are predicted. Whilst it is still difficult to make accurate prediction on the scale of change, it is certain that there are changes happening and these are unlikely to return the climate system to conditions similar to those in the past.

However, relating these changes to changes in salmon populations is difficult and may even be impossible. In the past many studies have demonstrated simple correlation between environmental parameters and various aspects of salmon biology. Generally these correlations hold for a certain period, and then break down, presumably either because they arose through chance, or because the causative mechanisms changed. It is only through a better understanding of both the processes controlling ocean climate change and the mechanisms in the sea by which ocean climate change impact Atlantic salmon can we hope to assess the future prospects for an improved ocean life for this species.

References

Dickson, R.R. & Turrell, W.R. (1999) The NAO: the dominant atmospheric process affecting variability in home, middle and distant waters of European Atlantic salmon. In: *The Ocean Life of Atlantic Salmon* (ed. D. Mills). Fishing News Books, Oxford, pp. 92–115.

Dietrich, G., Kalle, K., Drauss, W. & Siedler, G. (1975) *General Oceanography*. Wiley, New York, 626 pp.

Gordon, C., Cooper, C., Senior, C.A., et al. (2000) The simulation of SST, sea-ice extents and ocean heat transport in a version of the Hadley Centre coupled model without flux adjustments. *Climate Dynamics*, **16**, 147–68.

Hansen, B. & Østerhus, S. (2000) North Atlantic–Nordic Seas exchanges. *Progress in Oceanography*, **45**, 109–208.

Hansen, B., Turrell, W.R. & Østerhus, S. (2001) Decreasing overflow from the Nordic Seas into the Atlantic Ocean through the Faroe Bank channel since 1950. *Nature*, **411**, 927–30.

IPPC (1995) *IPPC second assessment report: climate change 1995. Synthesis of scientific–technical information relevant to interpreting Article 2 of the UNFCCC*. IPPC, Geneva, Switzerland, 64 pp.

IPPC (2001) IPPC third assessment report: climate change 2001. Synthesis report (eds R.T. Watson and the Core Writing Team). IPPC, Geneva, Switzerland, 184 pp.

Reynolds, R.W. & Smith, T.M. (1994) Improved global sea surface temperature analyses using optimum interpolation. *Journal of Climate*, **7**, 929–48.

Turrell, W.R. & Shelton, R.G.J. (1993) Climactic changes in the north-eastern Atlantic and its impact on salmon stocks. In: *Salmon in the Sea and New Enhancement Strategies* (ed. D. Mills). Fishing News Books, Oxford, pp. 40–78.

Turrell, W.R., Slesser, G., Adams, R.D., Payne, R. & Gillibrand, P.A. (1999) Decadal variability in the composition of Faroe–Shetland channel bottom water. *Deep Sea Research*, **46**, 1–25.

PUTTING IT TOGETHER

Catchment management

Chapter 21
The European Water Framework Directive and its Implications for Catchment Management

J. Solbé

Environmental, Dol Hyfryd, The Roe, St. Asaph, Denbighshire, LL17 0HY, UK

Abstract: If salmon are 'at the edge', in the sense of 'under threat', some of the causes of this situation must be sought in the quality of the various waters in which they spawn, spend their early lives, migrate through *en route* to the open sea and on their return to their natal streams. Such waters – defined in the Water Framework Directive (WFD) as 'rivers', 'lakes', 'transitional waters' (typically estuaries) and 'coastal waters' – their status and plans for their improvement and protection are the subject of the Directive. It is no longer sufficient (and never has been) to make isolated studies of potential stresses on wild populations: the broadest possible view is required or resources may be used up on tackling the consequences and not the root cause of ecological damage. The clean-up of European estuaries seems to be making good progress, but migratory salmonid populations are still a cause for concern.

This paper presents the objectives of the WFD, timetables for its implementation and key points in its *modus operandi,* such as the concept of river basin management and improvement targets based on optimal ecological quality for each area, concluding with a discussion of the question: 'What does this all mean for the Atlantic salmon?'

Introduction

The regulation of surface waters and groundwaters in the European Union (EU) has developed in a largely *ad hoc* and unintegrated manner over more than 40 years, responding to particular needs and political pressures. For example, one of the Dangerous Substances Directives (76/464) (European Council, 1976) which concerned discharges into the aquatic environment was largely the result of the Seveso accident and recognized the need to control extremely dangerous chemicals. In a different process, the Freshwater Fisheries Directive (European Council, 1978) was drafted over a number of years to meet a long-term strategic need to protect freshwater fisheries. Now a major change is under way in the protection of the aquatic environment in Europe. Existing laws will be replaced by a single holistic body of legislation. Directive 2000/60/EC of the European Parliament and of the Council of 23 October 2000 has established a 'framework for community action in the field of water policy': the Water Framework Directive (WFD) (European Council, 2000).

An early clause in the Directive states that '... Water is not a commercial product like any other but, rather, a heritage which must be protected, defended and treated as such.' The Atlantic salmon, its habitats and migration routes inland and in coastal waters are part of this 'heritage'.

Origins of the Water Framework Directive

The development of environmental concerns and policies may be traced through the various treaties of the EU. A few points that are significant to the evolution of a Water Framework Directive follow. The original EU Treaty of Rome in 1957 made no specific mention of environmental policy. Europe was too much occupied in recovering from World War 2 and in establishing the Community to worry about such matters. Through the 1960s and 1970s European legislation developed and in 1988 the European Council asked the Commission to submit proposals to improve ecological quality in Community surface waters. In 1991 the need for action was recognized to avoid long-term deterioration of freshwater quality and quantity. A programme of actions was called for, to be implemented by the year 2000, aiming at sustainable management and protection of freshwater resources.

Gradually environmental awareness and the consequent political agenda gathered pace and at the Maastricht Treaty of 1992 a number of important agreements were reached:

- Adoption of sustainable development as Community policy.
- Environmental policy is a Community matter within limits of subsidiarity.
- As distinct from total agreement, qualified majority should be required for decisions concerning most environmental issues.
- Environmental policy must be integrated into the other Community policies.
- A high level of protection was required in all environmental measures.

Three principles had emerged over the years and were re-stated at Maastricht: the *precautionary principle*, the principle of *prevention of pollution at source* and the *'polluter pays'* principle.

In 1995 the Council of Ministers and the Environment Committee requested an action programme for groundwater as part of an overall coherent policy on freshwater protection, recognized the increasing pressures from growth in demand for sufficient quantities of good-quality water for all purposes and invited the Commission to come forward with a proposal for a new 'framework Directive' establishing the basic principles of sustainable water policy in the EU. The Amsterdam Treaty of 1997 strengthened the integration of environmental policy and allowed member states to apply more stringent national legislation than the European norm.

A great deal of work at the Commission and in the member states ensued and eventually, in December 2000, the WFD was published (European Council, 2000).

Objectives and scope of the Water Framework Directive

The principal objectives, as they relate to the Atlantic salmon, are to:

- Achieve/maintain *'good' ecological status*, within a time frame.
- Base freshwater management on *river basins*, over-riding other administrative structures or boundaries.
- Combine the approaches of *Emission Limit Values* and *Environmental Quality Standards*.

In addition, the WFD requires regulators to involve all interested parties in the protection of water, to streamline the legislation (repealing some existing Directives, including the Freshwater Fish Directive), expand the scope of protection to groundwater as well as surface waters and find a system of pricing and charging for all water-use services which reflects the true economic cost. For the Atlantic salmon, it is the achievement of 'good ecological status' or even 'high ecological status' that is the key marker for success of the WFD.

The scope of this emerging legislation is very broad: rivers and lakes, transitional waters, coastal water and groundwaters. For each and all of these Water Quality Objectives (WQO) must be provided, together with a programme of measures to achieve the WQO.

An important recognition in the WFD is the unit of management of the *river basin* for inland waters. The river basins are natural entities and do not have to conform to political boundaries. For each river basin there must be a River Basin Management Plan.

Timetable of implementation

Such a broad measure is going to take many years to implement fully, but valuable steps, with consequent environmental benefits, should become visible along the way. In brief, the steps are:

- WFD in force   22 December 2000
- Define the basins   2003
- Appoint the competent authorities   2003
- Review human impact   2004
- Start monitoring   2006
- Define the issues and objectives   2007
- Repeal of first tranche of Directives   2007
  - Drinking water
  - Methods/frequency of sampling fresh waters
- Derive the 'measures'   2008
- Consult on the plan   2008

- Implement the plan   2009/2012
- Commission publishes first progress report   2012
- Review the plan   2013/2015
- Repeal of second tranche of Directives   2013
  – Groundwater pollution
  – Mercury and mercury discharges
  – HCH (hexachlorocyclohexane) discharges
  – Dangerous substances discharges
  – Freshwater fish
  – Quality required of shellfish waters
- Balance abstraction and recharge   2015
- Aim for good status for groundwaters   2015
- Aim for good status for surface waters   2015
- Commission publishes second progress report   2018
- Zero emissions of priority list substances (see below)   2020
- Eliminate priority list substances   20??
- Near-background levels for natural substances in sea   20??

There is pressure to speed up this process but it is complex and needs to be done correctly first time.

Ecoregions and river basins

Europe may not have the same land area as other continents, but it has a huge range of climatic and geomorphic conditions. It is necessary to recognize these if appropriate target conditions are to be defined that constitute good ecological status. For rivers and lakes Europe is divided into 25 'ecoregions', at least ten of which are relevant to the Atlantic salmon. For example region 17 comprises Ireland and Northern Ireland, region 18 Great Britain, region 19 Iceland, and region 20 the Borealic uplands of Norway. In the same way, there are six marine ecoregions: the Atlantic Ocean, Norwegian Sea, Barents Sea, North Sea, Baltic Sea and Mediterranean Sea. For each of these, separate consideration will be given by the member states.

River basins define their own areas, but it is necessary to group smaller basins together for management purposes, as we have always done in England and Wales, for example in west Wales or north-west England where a series of very small (by European standards) basins discharge to the sea. On the other hand, the Rhine and the Danube are major, international river basins which stand alone, and, indeed, for a number of years have been the subject of International Commissions. Member states, individually or together, establish the river basins in their regions. Advantage can be taken here to separate out smaller but very important basins, as has been done for the Welsh Dee, a small part of which is in England.

Types of surface water body

Each river basin may be divided up into individual 'types'. The Directive requires the water body to be classified on a geomorphological basis but allows a choice of two methods. In the first – system A – the ecoregion is given as in the Directive and the type is defined on the basis of three criteria, as follows:

(1) Altitude: below 200 m, 200–800 m, above 800 m
(2) Area of catchment: 10–100 km^2, > 100–1000 km^2, > 1000–10 000 km^2, > 10 000 km^2
(3) Geology: calcareous, siliceous, organic

In the second – system B – as well as specifying the ecoregion and using a similar classification to system A, the type of water body can be described using a series of optional factors such as distance from the sea, energy of flow, water depth/width/slope, buffering capacity, etc. To fisheries scientists and managers this will all be familiar territory: it is the kind of description that is made during any freshwater fishery investigation.

In the case of other water bodies – lakes, transitional waters (typically estuaries) and coastal waters, systems A and B have their own criteria. For lakes, shape, residence time, mixing characteristics, nutrient status and water-level fluctuation may be given if system B is chosen, and for transitional waters and the sea the salinity and (for the former) tidal range are included.

There is a further classification, and that is the recognition that although the bulk of waters in Europe may be basically natural, there exist quite large bodies that are 'heavily modified' or 'artificial'. Straightened rivers, canals and reservoirs would come into these classes and are distinctly treated in the WFD.

Surface water elements considered

Given the ecoregion and the type of water body, it is now appropriate to consider what should be measured in preparation for planning to improve the quality of the ecological status to at least 'good', or to maintain it if it is already 'high'. Table 21.1 summarizes these elements. The Directive sets out recommended frequencies of sampling to monitor the characteristics in Table 21.1.

Surface water objectives

Three objectives are stated:

(1) To achieve good surface water quality status for all natural waters.
(2) To achieve good ecological potential with good chemical status for all heavily modified waters.
(3) The cessation of all discharges and emissions of priority hazardous substances.

Table 21.1 Summary of surface water elements considered in the Water Framework Directive (2000/60/EC).

	Rivers	Lakes	Transitional waters	Coastal waters
Phytoplankton	✓	✓	✓	✓
Macrophytes and phytobenthos	✓	✓	Macro-algae and angiosperms	
Benthic invertebrates	✓	✓	✓	✓
Fish	✓	✓	✓	
Hydrological regime	✓	✓	Tidal regime	✓
River continuity	✓			
Morphological condition	✓	✓	✓	✓
General physico-chemical conditions	✓	✓	✓	✓
Specific synthetic pollutants	✓	✓	✓	✓
Specific non-synthetic pollutants	✓	✓	✓	✓

The priority substances referred to in 'Timetable of implementation' above are identified by the current reviews of new and existing chemicals, by the Dangerous Substances Directive and future legislation. No doubt the provisions of the White Paper on the Strategy for a Future Chemicals Policy for Europe will introduce procedures by which further priority substances are identified, following risk assessment.

But what constitutes 'high status'? This is explained in the Directive as follows:

'There are no, or only very minor, anthropogenic alterations to the values of the physico-chemical and hydro-morphological quality elements for the surface water body type from those normally associated with that type under undisturbed conditions. The values of the biological quality elements for the surface water body reflect those normally associated with that type under undisturbed conditions, and show no, or only very minor, evidence of distortion. These are the *type-specific conditions and communities*.'

For 'good status' the wording demonstrates some difference from the type-specific conditions:

'The values of the biological quality elements for the surface water body type show low levels of distortion resulting from human activity, but deviate only slightly from those normally associated with the surface water body type under undisturbed conditions.'

For 'moderate status' a further distance from the ideal conditions is described:

'The values of the biological quality elements for the surface water body type deviate moderately from those normally associated with the surface water body type under undisturbed conditions. The values show moderate signs of distortion resulting from human activity and are significantly more disturbed than under conditions of good status.'

Clearly, at face value these are highly subjective judgements, but work is under way to establish benchmarks to aid interpretation. In the British Isles we have the advantage of an existing system available to aid one of the processes – the definition of high quality invertebrate communities. This is the River Invertebrate Prediction and Classification System (RIVPACS) of the UK, created in the Freshwater Biological Association/ Institute of Freshwater Ecology (Wright et al., 2000). It utilizes information from over 600 reference sites to allow prediction of the expected species list at any running water site. This then gives a target and can be compared with the actual species present at a monitoring site to compare the monitoring site with the ideal, i.e. 'high status'.

Definitions of status for fish fauna

Considering the Atlantic salmon, the definitions for fish fauna in the Directive are of interest. They cover:

- Species composition
- Abundance
- Presence of sensitive species
- Age structure.

For *high status* these read: Species composition and abundance correspond totally or nearly totally to undisturbed conditions. All the type-specific disturbance-sensitive species are present. The age structures of the fish communities show little sign of anthropogenic disturbance and are not indicative of a failure in the reproduction or development of any particular species.

For *good status*: There are slight changes in species composition and abundance from the type-specific communities attributable to anthropogenic impacts on physico-chemical and hydro-morphological quality elements. The age structures of the fish communities show signs of disturbance attributable to anthropogenic impacts on physico-chemical or hydro-morphological quality elements, and, in a few instances, are indicative of a failure in the reproduction or development of a particular species, to the extent that some age-classes may be missing.

For *moderate status*: The composition and abundance of fish species differ moderately from the type-specific communities attributable to anthropogenic impacts on physico-chemical or hydro-morphological quality elements. The age structure of the fish communities shows major signs of anthropogenic disturbance, to the extent that a moderate proportion of the type-specific species are absent or of very low abundance.

We are all familiar with such signs of stress in a population or community. Expert judgement may be needed in each case to determine just where on the scale *moderate—good—high* a given fishery is placed.

Significance of the Water Framework Directive for the Atlantic salmon

If salmon are indeed 'at the edge', in the sense of 'under severe threat', some of the causes of this situation must be sought in the quality of the various waters in which they spawn, spend their early lives, migrate through *en route* to the open sea and on their return to their natal streams. The sixth International Atlantic Salmon Symposium is a suitable occasion to consider all the impacts on the Atlantic salmon and whether the WFD will bring about a benefit to stocks in the long term. The answer surely must be that benefit will accrue, because all nations with an interest in the Atlantic salmon are either legally obliged to conform to or at least interested in this regulatory initiative of the EU. On the other hand, there may be issues of great importance to Atlantic salmon survival that are not covered, even by such a broad framework as the WFD.

It is to be applauded that Environmental Quality Objectives (EQOs) (i.e. the targets of good ecological status) remain a cornerstone of this legislation. Fixed Emission Levels have limited applicability for the most toxic and persistent chemicals only but do not take into account the nature of the receiving environment, its different sensitivities, assimilation capacities and ability to create permanent or temporary 'sinks'. EQOs make no predictions about what is causing the problem and the Environmental Quality Standards that flow from them should be seen not just as numerical values limiting named chemicals in water but also as the successful application of codes of good practice for all uses of land that may affect fresh waters, groundwater and the sea.

In summary, among major potential influences on the species today may be mentioned:

- Overexploitation at sea or in transitional and fresh waters.
- Water pollution by strong biodegradable organic wastes (depleting dissolved oxygen).
- Soil erosion causing silting and smothering of habitats.
- Destruction of spawning sites by fording vehicles, gravel extraction, etc.
- Diseases including parasites, exacerbated by human activities.
- Introduction of alien species.
- Eutrophication in fresh waters from phosphorous compounds and saline waters from nitrogen compounds.
- Physical barriers to movement, reducing access to spawning areas.
- Water pollution by toxic substances.
- Endocrine disruption(?)
- Climate change.

These are not listed in any agreed priority order and there are of course natural hazards to salmon populations to add to this list, such as:

- Predation by other wild animals in fresh waters and the sea.
- Competition in fresh waters with other fish.

Some of the issues are the result of long-term management or socio-economic problems; others are typically short-term catastrophic events.

The *existence* of a problem can be identified by the systems put in place by the WFD. The challenge will be to assign correctly the cause or causes to the effect.

There are already in place most of the management tools needed to overcome the problems: Forestry and Water Guidelines, Codes of Good Agricultural Practice, waste-water treatment technology, etc. The need may often be not for more guidelines but for education of those who inadvertently cause damage to the salmon's life cycle by lack of attention to the existing guidelines and codes of good practice. Some problems, however, remain intransigent and some may just cost too much to rectify at present.

It is no longer sufficient (and never has been) to make isolated studies of potential stresses on wild populations: the broadest possible view is required or resources may be consumed on tackling the consequences and not the root causes of ecological damage. The clean-up of the diverse problems of European estuaries seems to be making good progress, but migratory salmonid populations are still a cause for concern, and the causes may be far upstream or far out to sea.

The Water Framework Directive will give us a series of management tools, soundly based on the right principles (e.g. of using the river basin as the optimal unit, of using the ecological status as the best indicator of environmental stress). Three things are now needed: the political will to keep to the timetable, sufficient experts to interpret the observed ecological status, and the resources to eliminate the causes of environmental damage in a cost-effective manner. Then the Atlantic salmon will have a chance to draw back from the edge.

References

European Council (1976) Directive 76/464/EC of 4 May 1976 on pollution caused by certain dangerous substances discharged into the aquatic environment of the Community. *Official Journal of the European Communities*, **L129**, 23–9.

European Council (1978) Directive 78/659/EEC of 18 July 1978 on the quality of fresh waters needing protection or improvement in order to support fish life. *Official Journal of the European Communities*, **L222**, 1–10.

European Council (2000) Directive 2000/60/EC of the European Parliament and of the Council of 23 October 2000 establishing a framework for Community action in the field of water policy. *Official Journal of the European Communities*, **L327**, 1–72.

Wright, J.F., Sutcliffe, D.W. & Furse, M.T. (eds) (2000) *Assessing the Biological Quality of Fresh Waters: RIVPACS and Other Techniques*. The Freshwater Biological Association, Ambleside, 400 pp.

Reports and Recommendations

The Outlook for Post-Smolts

D. Mills

Atlantic Salmon Trust, Moulin, Pitlochry, Perthshire, PH16 5JQ, UK

The outlook for post-smolts starts when they enter estuarial waters and then move out into a truly marine environment. At an Atlantic Salmon Trust Workshop on *The Ocean Life of Atlantic Salmon*, Moore, Lacroix and Sturlaugsson listed a number of questions and specific hypotheses concerning post-smolt behaviour and survival in the marine environment, some of which could be tested by using tracking technology and others by capture using nets. These included:

(1) Does the timing of smolt migration from fresh water affect the initial survival of post-smolts in coastal areas?
(2) What is the extent of post-smolt mortality in coastal waters and what are the potential causes, and how long do they remain in coastal waters?
(3) Do origin or age/size of smolts influence the timing and route of post-smolt migration?
(4) Do post-smolts use tidal streams and shelf edge currents to migrate to feeding grounds?
(5) Do post-smolt migration routes follow preferred temperature and salinity fronts?
(6) Do extremes in the range of environmental conditions interrupt migration?
(7) Is the distribution of post-smolts correlated with prey distribution and abundance?
(8) What effect has climate?

How far have we come towards answering some of these questions?

(1) Low river temperatures may delay smolt migration. For example in northern Canadian and Icelandic rivers smolts only enter the sea in August when sea surface temperatures (SSTs) finally reach 2–4°C. Smolts may remain in the estuary even longer until the differential between river and sea temperatures disappears. These fish will have reached lengths of 30–35 cm and this may increase their chances of sea survival. Delay in river migration by dams may affect time of sea entry and survival. This may be critical where seasonally increasing estuarine and coastal water temperatures affect water quality and oxygen levels due to pollution and where, consequently, there is only a narrow

time window for smolts to reach coastal waters. Water flow at time of migration is also important.

(2) A study of the foraging of ranched post-smolts in Icelandic coastal waters as observed through the proportion of feeding fish and their forage ratio reflected the strong urge of post-smolts to start seawards migration immediately. This was shown repeatedly by immediate outward migration of the majority of fish after their release. A study undertaken in Norwegian waters on released ranched smolts indicated that migratory motivation was higher for the larger size-groups. The detailed study of post-smolt movements in the Bay of Fundy has provided invaluable data for a rather unique environment and has highlighted the importance of water temperature with respect to the effect of warm water from the Gulf of Maine 'pushing' the smolts back into the Bay of Fundy.

It was concluded that the observed low forage efficiency ratio among post-smolts during their migration in Icelandic coastal waters is unlikely to have any great effect on their survival owing to the short time spent in this environment. However, a number of factors probably affect their survival at this stage. For example, in acidified rivers, accumulation of aluminium during the freshwater phase may lead to an impaired ability to osmoregulate and, therefore, adapt to marine conditions. Recent studies in New Brunswick also suggest that 4-nonylphenol may adversely affect the later stages of smolting and lead to delayed mortality at sea. There has also been some concern over the occurrence of cataracts in post-smolts which it was thought might be linked to water quality problems.

As was raised by Sir Frederick Holliday, conditions prevailing in coastal waters are, I fear, likely to become worse.

Reddin suggested that annual SSTs inshore around the coast of Newfoundland may have a relationship with post-smolt survival. High mortalities during early sea life may also be caused by localized predation and disease and parasite infection (aspects covered by Anderson and Windsor, respectively).

(3) Gilles Lacroix demonstrated that the origin of smolts entering the inner and outer Bay of Fundy influenced the timing and rate of post-smolt migration. Post-smolts caught in the Faeroe–Shetland Channel and western sector of the Norwegian Sea had lower river ages than post-smolts caught in the eastern Norwegian Sea. On this evidence, fish from southern Europe (principally the British Isles, France and the Iberian peninsula) dominate the western sector while a larger proportion of post-smolts from northern Europe (principally Norway) are present in the eastern sector. In the north-west Atlantic it was found that post-smolts in the Labrador Sea originated in rivers from Maine to Labrador.

(4) The observed distribution of post-smolts considered in relation to the prevailing hydrographic regime suggests a close correlation between strong northerly or

north-easterly surface currents and post-smolt migrations in the north-east Atlantic. From the available catches and current knowledge of the slope current running north-east in the areas west of Scotland, the general pattern appears to be that most of the post-smolts are caught close to the core of the current which may enable them to reach their northern feeding areas with the least expenditure of energy in migratory swimming. Gilles Lacroix also showed quite clearly the use of tidal streams by post-smolts.

(5) Marianne Holm has shown a post-smolt distribution related to temperature and salinity where the optimum salinity is at least 35 ppm and temperature 11°C. Vertical distribution for maximum abundance of smolts was between 0 and 1 m.

(6) Extremes in the range of environmental conditions can interrupt migration, particularly water temperature.

(7) Marianne Holm found that post-smolts move alongside the 'food-stream' (moving in a food basket). Such behaviour minimizes energy expenditure. Post-smolts have been shown to modify their diet in response to restrictions imposed by thermal structure or to the opportunities created by the concentration of food along oceanographic fronts. It is suggested that the salmon is an opportunistic feeder, and therefore is not dependent on one particular prey, and so food is expected to be a limiting factor only in extreme cases.

An external and man-made effect on food availability is the potential effect of the industrial fisheries mainly operating in coastal areas and on continental shelves, which obviously involves the removal of potential prey species. A change in age and size structure of the prey populations indirectly affects the salmon by changing the predator–prey relationship. However, the likely effect might be less than expected owing to the presumed opportunistic forage behaviour of salmon.

However, in the north Pacific several scientists have indicated density-dependent survival due to food limitation for Pacific salmon. In the north-east Atlantic no such relationships have been found. There are, however, differences between the Pacific and Atlantic ecosystems, particularly concerning salmonids.

Other open seas pelagic fisheries, such as pair-trawling for mackerel, have other implications for post-smolts in that they become part of the by-catch, and some estimates of the quantity taken are extremely alarming.

(8) Sarah Hughes, in her consideration of the prospects for improved oceanic conditions, has stressed that changes in salmon populations through the effect of climate change may be impossible to predict. However, in a study financed by the Atlantic Salmon Trust and undertaken by Reid and Beaugrand of the Sir Alister Hardy Foundation for Ocean Science, there was evidence that the recent decline in salmon may result from the impact of climate warming on ocean dynamics, although mechanisms behind the relationship are not clear. The authors suggest that the salmon stock is expected to continue its decline,

especially in the southern boundary of its spatial distribution, if the warming and the associated changes in plankton composition continue.

The following Recommendations for further study are suggested:

- Further study of factors affecting survival of post-smolts in coastal waters including water quality. Attempt to refine estimates of early marine mortality to be conducted in an organized, progressive approach from time of entry to coastal waters to moving into the open sea.
- Extend work on tracking and capture in coastal waters and experiment with telemetry using new techniques. Obtain more information on age and origin to include genetic studies.
- Increase range of tracking and capture in open seas and instigate a large international micro-tagging programme to facilitate knowledge and origin of post-smolts caught in trawl surveys.
- Studies to relate post-smolt distribution to food abundance incorporating data from plankton surveys.
- Obtain more precise information on presence of post-smolts in the mackerel and other pelagic trawl fisheries and, if necessary, charter a commercial pelagic trawl vessel with scientists in attendance.
- Co-operation with Pacific salmon investigations. This has already to some extent been set up through establishment by NASCO of an International Co-operative Research Board which will identify research gaps and priorities.

Overcoming Estuarial and Coastal Hazards

J. Anderson

Atlantic Salmon Federation, PO Box 5200, St. Andrews, New Brunswick, E5B 3S8, Canada

The underlying purpose of this session was to explore the possibility to which predation could be a contributing factor to the crisis in Atlantic salmon abundance.

Dr. Bill Montevecchi's paper was an important contribution to the literature because it dealt with avian predation well out to sea – by the gannet, which has a foraging range of up to 180 km offshore. Most of the data came from Funk Island, one of six gannet colonies in Atlantic Canada. During the early 1990s, a marked increase in consumption by the birds of post-smolt salmon occurred. The question was, did this lead to a decrease in adult salmon abundance? Analysis of the data suggested not, that concurrent with an increase in gannet numbers was an increase in smolt numbers, which led naturally to an increased consumption of post-smolts. No smoking gun here.

As Dr. Russell Brown pointed out in his paper, predation is likely the ultimate source of natural mortality of salmon in the estuarine and marine environment, but determining a cause-and-effect relationship between post-smolt abundance and fish predation is statistically difficult because so few salmon are eaten per predator. Salmon is not the main prey of any marine fish. He did indicate, however, that there were three ways in which fish predation on salmon could be affected, two caused by in-stream barriers, such as dams, the other caused by fluctuations in the abundance of other prey species which normally serve as predation buffers for salmon. In none of these examples, however, was there any evidence for a negative correlation between fish predation and subsequent adult salmon abundance. Again, no smoking gun.

Dr. John Armstrong showed in his paper that, as with estuarine and marine fish predation, marine mammals are opportunistic feeders on post-smolt salmon, the natural abundance of which, compared to other prey species, is so low as to present no significant part of any marine mammal's diet. Of the six species considered, only harp seals, because of their presence in the millions, could offer any practical threat to salmon abundance; but they do not seem to. Seal predation, particularly that of the harbour seal, on returning adults is another matter. Here, significant, highly localized predation can take place by what Armstrong referred to as 'in-river specialists'. Thus, some evidence for a gun having been fired does exist, but it is not strong enough to account for the global decline in salmon stocks.

Without question, a marine predator whose efforts *are* negatively correlated with salmon abundance does exist in the form of commercial salmon fishermen. Andrew Whitehead dealt at length with the long-standing efforts to close the English drift-net fishery of the north-east coast of England. Some progress has been made. The reduction in the number of netsmen from 124 to 70 is certainly a solid step in the right direction. For reasons not made entirely clear, however, despite this reduction in fishing effort, the total catch has remained constant, apparently a basis for the argument by the fishermen that there is no need to close the rest of the fishery. On the other side of the debate, however, is the fact that annually about 29 000 adult salmon are prevented from spawning (mostly in Scotland). Some success has been made in buying out commercial fisheries in other parts of the UK. The most significant drift-net fishery still in operation is that off the west coast of Ireland, where, we were told from the audience, something like 237 000 pre-spawners were harvested last year. Dead fish do not spawn. There is certainly a smoking gun here.

Stephen Chase explained that commercial fishing for Atlantic salmon in New England ceased, uncontroversially, in 1948. In Atlantic Canada, it took 26 years (1966–92), and much controversy and money [$72 million (Canadian)], for the federal government in co-operation with provincial governments, to accept the conservation imperative – in which ASF played a leading role – and close the commercial fisheries in each of the five provinces. A small aboriginal fishery has re-emerged in Labrador. Moreover, France, from its islands off southern Newfoundland in the Gulf of St. Lawrence – St. Pierre and Miquelon – conducts a small fishery. While the principle of the undesirability of interceptory salmon fisheries anywhere remains, these two fisheries cannot account for the decline in North American salmon stocks.

Recommendations

- Predation at sea has to be playing an important role in post-smolt marine survival. It is recommended that research continue in this area so we can obtain a better understanding of the role that predation plays in controlling wild salmon abundance.
- Interceptory fisheries for Atlantic salmon, and the management of Atlantic salmon based on sound science, are mutually exclusive actions. It is recommended that efforts continue to eliminate interceptory Atlantic salmon fisheries in the home waters of salmon-producing countries.

Resolving Conflicts Between Aquaculture and Wild Salmon

M. Windsor

North Atlantic Salmon Conservation Organization, 11 Rutland Square, Edinburgh, EH1 2AS, Scotland, UK

Salmon aquaculture is certainly not the only threat to wild salmon populations. However, wild stocks are in a very weakened state and there are particular concerns about the disease and parasite impacts of salmon aquaculture and the genetic impacts of escapees on wild stocks. Salmon populations in the vicinity of salmon farms (< 20 km) appear to have particular problems which have resulted in local extinctions, threats of further extinctions and serious economic losses. The rapid growth of the salmon farming industry has meant that it has outpaced the development of regulations.

Safeguarding the wild stocks is in the aquaculture industry's own interest since the wild stocks form the seed core on which the industry is based. This will also protect industry from public criticism. If the industry is seen as damaging wild stocks it will do nothing for its future prosperity and its sales.

Encouraging progress has been made in controlling sea lice numbers on some fish farms, particularly in the spring. While the numbers per fish are lower than previously, they still represent a threat to wild salmonids. Salmon lice emanating from fish farms can cause high levels of mortality in wild salmonid populations. The level of such infestation is variable but remains high in densely farmed areas. The rapid growth of the industry means that lice control will need to be increasingly effective, with consideration of alternative treatment strategies and special protection for particularly threatened/valuable stocks (e.g. zones free of aquaculture). There is a need to consider improvements in monitoring lice levels in farms and for transparency of the information obtained. Such transparency can produce trust and confidence.

New experimental findings delivered at this Symposium confirm that escapees from salmon farms pose a serious genetic threat to the fitness and viability of wild salmon populations and that the hybrid vigour or 'new blood' argument is not justified. It is too great a risk to accidentally tamper with genetic diversity which has taken at least 10 000 years to evolve.

Repeated intrusions of farmed salmon may lead to extinction of locally adapted populations. Progress has been made in improving containment, but escapees may exceed the number of wild salmon. The continuing growth of the industry means that containment must significantly improve and consideration should be given to use of sterile salmon and, in the longer term, the possibility of using land-based units, although these are presently not economically viable.

Stocking with non-native fish for 'enhancement' purposes may be equally damaging to wild salmonid populations and should be actively discouraged.

After some years of denial by the industry of an impact on wild salmonids, there is now improved collaboration between wild and farmed fish interests, e.g. local area/bay management agreements, international containment codes. Such progress and action must be maintained and urgently enhanced so as to ensure that aquaculture is conducted in a sustainable manner that in future does not pose a threat to the wild stocks, while permitting development of the industry and safeguarding the economic benefits it can bring.

Given the competitive nature of the industry, international agreements may help to ensure a level playing field, and should be backed up by national and local regulations and agreements that are enforceable, with management regimes that are independently audited, and based on best practice.

Can We See a Brighter Future?

D. Clarke

Environment Agency, Rivers House, Aztec West, Almondsbury, Bristol, BS12 4UD, UK

Introduction

I originally had some doubts about whether I could see any future, let alone a brighter one! However, I offer the following, in the form of the good news, the bad news and, I hope, a balanced conclusion.

The good news

Firstly the good news – and there was plenty of it in the Symposium sessions.

Water quality

Water quality in the UK has improved and continues, in general, to do so. This is a direct result of regulatory action; better legislation and tighter regulation combined with massive investment. In England and Wales alone, some £8 billion is being invested by the water industry over the current 5-year period. This follows substantial prior investment since 1989, and more is in the pipeline.

Fish stocks

Fish stocks are also responding to these changes. Ross Doughty described the recovery of the Clyde and the redevelopment of the fishery. One of the posters described the recovery of some other Scottish fisheries.

For England and Wales, Guy Mawle described the return of salmon to cleaner rivers. In south Wales, rivers such as the Tawe, Neath, Ogmore, Ebbw and Taff have seen recovering runs and fisheries. In the north east, the Tyne and Tees have made major recoveries. The Tyne is now the most productive rod fishery in England and Wales.

We are also seeing the first returns for many years to major industrial systems such as the Trent, Mersey and Yorkshire Ouse. There are now more salmon rivers in the UK than there have been in living memory. In fact I predict that in 30 years, half the England and Wales catch will be delivered from recovering rivers (though the cynics might note that the outcome of this prediction post-dates my likely retirement).

Recovery has not just been limited to UK rivers. Conrad Mullins showed how construction of fish passes on impassable falls and obstructions in Newfoundland has created new salmon runs and new stock capacity.

Survival

All of these results lead us to some very important, underlying information. There is no real doubt that sea survival is currently poor. We see this both in index river data and in catches and returns from traditional, longstanding, fisheries. However despite that, if we remove limiting barriers, whether physical obstructions or water quality, then stocks recover, and recover strongly.

So at least with sea survival at current levels, stocks will recover, and, by implication, the vast majority of stocks are sustainable, *if* we get the freshwater environment right. At a stock level, therefore, salmon populations are still robust even though established populations may vary around a lower equilibrium than in the recent past.

Habitat improvement

We also saw presentations on habitat improvement, showing that we can improve stocks and increase production of stocks that are already present. Bruce Ward presented successes and showed how combining nutrient management and improvements to habitat structure had increased stocks. At a recent European Inland Fisheries Advisory Commission (EIFAC) conference, Martin O'Grady expressed the view that smolt production in Ireland could be doubled with the right levels of investment. These studies show not only that we can restore past damage, but also that we can create production levels much better than nature – for example, by removing factors that limit one particular stage of the life cycle in the river. The key in my view is achieving the right overall balance and distribution of habitats within individual systems.

The bad news

Scale

Eighty-four percent of salmon rivers in the USA no longer contain salmon, according to Andrew Goode, and more are at risk. Russell Brown gave us a graphic illustration of why, with figures illustrating the growth in the number of dams in the Connecticut River system over time. Given the sheer scale – both numbers and size of structures – it is difficult to see even medium-term resolution of some of these problems.

Aquaculture

Aquaculture was thoroughly covered by Malcolm Windsor, and I am limiting myself to a few comments here. There has clearly been massive growth, and in the Symposium we have seen evidence of stock level impacts in western Scotland, together with data from Jens Christian Holst describing mechanisms for impact.

But Jens and others also gave us some hope. If we take a longer-term view, introducing better regulation, better standards and separation where practicable, we may minimize the damage caused by an industry that has economic value and is undoubtedly here to stay.

We should also be aware that aquaculture has brought positive benefits for many wild salmon stocks. Simple economics have reduced salmon wholesale prices to levels that are at or below those of the early 1980s, without taking account of inflation. This, combined with reduced catches, has rendered both illegal fishing and many commercial fisheries uneconomic. As a direct consequence, exploitation rates in many wild stocks have been substantially reduced, with clear benefits.

By-catches

By-catches have also been a concern at the Symposium, particularly the trawl fishery for mackerel in the north Norwegian Sea. Some may be surprised that I would like us to find that smolt by-catches in the mackerel fishery (or indeed other fisheries) are very large indeed – in fact the cause of increased sea mortality. Why? Because fisheries are factors we can do something about! Which brings me to the big issue.

Climate change

We had an excellent, very interesting presentation from Sarah Hughes, with three clear messages:

(1) Climate change is here. All the major model scenarios predict increasing temperatures. That should be no surprise to any of us – we are already seeing the impacts. The growing season around the north Atlantic has been extended by 4 or more weeks over the last 30 years or so, and this must have massive ecological impacts. In the salmonid world we are seeing these effects in the form of changing run times, smolt age shifts, changed juvenile growth rates and other parameters.
(2) The models do not fit the observed data unless we include the effects of increased atmospheric emissions. So we are probably looking at a man-made impact, and unknown territory in relation to factors such as rate or extent of change.
(3) We do not really understand the implications of further change for salmon populations. I believe it greatly increases risk to future populations, but it is also possible we might see a repeat of the gadoid outburst of the late 1960s and high sea survival – who knows?

So can we see a brighter future?

I would like to present three alternatives:

(1) *The worst case.* In this, climate change is dramatic and sea survival collapses completely. Going back to Fred Holiday's introduction, he referred to his 'Swiss cheese' model, where slices of Swiss cheese represent barriers that salmon have to pass at different life stages. The holes in a set of Swiss cheese slices need to line up if individual salmon in a population are to be successful. In this scenario, the analogy is that the slice representing the sea becomes processed cheese – no holes – and salmon populations (excluding land-locked varieties and cage-reared fish) die out.

If this happens, I doubt that salmon will be the main focus of our concern.

(2) *The middle scenario.* Here sea survival stays as it is. The outcomes are down to us – to our actions in fresh water, and in dealing with issues like by-catches and aquaculture.

(3) *The best case.* Sea phase survival recovers *and* we make improvements to the things we control. This is undoubtedly a brighter future.

Final thoughts

Where salmon are concerned, in many respects we make our own luck. There are increasing risks, not least further reductions in sea survival associated with ongoing climate change. However, we can minimize these risks by looking to increase the size of the holes in Fred Holliday's cheese slices. That means improving water quality, improving habitat (production *better than nature*?) removing barriers, sorting out the problems of aquaculture and dealing with the myriad of other challenges facing both the salmon and our fisheries.

We can only achieve this if we work together, with the conservation of salmon stocks as a common goal, whether we are in government, the NGOs or acting as individuals. We all have a part to play. If we do this – working in partnership – then we can build on the 'beacons of hope' identified in this Symposium and develop a brighter future – for us, and for the salmon. We must.

THE WAY AHEAD

Conclusions, Discussion and Resolutions

R. Shelton

Atlantic Salmon Trust, Moulin, Pitlochry, PH16 5JQ, UK

In adopting the title suggested by the Atlantic Salmon Federation, *Salmon at the Edge*, the organisers of the Sixth International Salmon Symposium achieved two things. They drew public attention to the increasing concern about the current well-being of wild salmon and sea trout resources and, at the same time, focused the attention of the meeting on the problems faced by wild Atlantic salmon and sea trout during their outward migration through the coastal zone and, in the case of salmon, during their early weeks in the open ocean. It has long been known that the conditions young salmon and sea trout encounter at this critical time have a lasting effect. This is because losses that take place during the marine phase are not mitigated by the compensatory reductions in natural mortality which help to buffer the effects of predation early in freshwater life. What this means in practical terms is that the loss of a given fraction of outward migrants 'at the edge' will reduce the number of adults returning to the river in roughly the same proportion.

One of the facts that struck me when I first became involved with salmon fishery assessment was just how stable recruitment to their stocks is compared with demersal marine species like haddock in which 10- or even a 100-fold variation in year-class size is commonplace. In salmon and sea trout the population of young fish which recruits to the sea each year as smolts consists of more than one-year class, and this mixing of year-classes helps to iron out variation. Furthermore, provided the parental population has laid enough eggs in the first place, the food and space available for them in the river sets an upper limit on smolt production. Recruitment to the fisheries themselves is then subject to the growth and survival opportunities available to the young fish in the sea, and there is increasing evidence to suggest that the early months at sea are of critical importance. That is why understanding what happens 'at the edge' matters so much.

It is only relatively recently that the likely evolutionary history of the 'at the edge' life styles of migratory salmonids has been uncovered. Although the family has an ancient lineage, its fossil record is sparse. Burial in fine silt offers the best opportunity for later fossilization, but salmonid fishes prefer to live in gravelly places. New insights have come from the science of 'cytogenetics', the specialized branch of micro-anatomy devoted to the study of chromosomes. When cytogeneticists began to study salmonid fishes, they were surprised to discover just how many chromosomes were present in each cell. They found that the largest numbers of chromosomes (96, and

Man has only 46) were found in the most primitive members of the family. They concluded that, tens of millions of years ago, the common ancestor of the salmon family must have had a double set of chromosomes. In other words, it must have been a 'tetraploid' with two sets from each parent instead of one from each. Such a doubling up of chromosomes is common in some flowering plants, especially after hybrid crossing, and is often associated with growth to a larger size. When opportunities for rapid growth are good, such an accidental change can be advantageous and will be favoured by natural selection. It seems that something like that must have happened to give rise to the first salmonid fishes which many now think may have lived a smelt-like life as 'children of the tides'.

The advantage of living 'at the edge' is the amount of food available where water masses mix. Over the millennia, the migratory salmonid fishes have improved on the life cycle of their ancestor by extending their migrations in two directions, so confining their reproduction and early growth to fresh waters, where there are relatively few predators, and growing to adulthood in the rich feeding grounds of the sea. This separation of life history stages is seen in both salmon and sea trout, but there are important differences in their life histories which affect the relative vulnerability of these species when running the gauntlet between their two worlds.

The more flexible life cycle of the sea trout is thought to be the more primitive. Despite their separate status in Scots' law, which treats them as 'salmon', sea trout are merely the migratory members of more broadly based trout populations which include some fish, especially males, that complete the whole of their lives in fresh water. The migrations of the sea-going members are coastal and often involve incursions into estuaries. In essence, sea trout are 'at the edge' for the whole of their lives at sea.

The last 30 years or so have seen substantial reductions in the survival of salmon and sea trout at sea. The challenge for fishery scientists is both to record the resultant changes in the abundance, structure and distribution of the stocks and to identify and measure the environmental changes that lie behind them. The challenge for fishery managers is to track down and deal with those environmental problems that are man-made and to conserve and enhance the depleted stocks in such a way that the worst effects of low return rates on fishing opportunities and spawning stocks are safely mitigated. The programme of the Sixth International Atlantic Salmon Symposium included:

- New information on the migration of post-smolts.
- Predation and mixed-stock interceptory fishing.
- The problems of aquaculture and the development of solutions.
- Experience in attaining higher smolt production through conventional measures such as improving water quality and riparian habitat and increasing access, as well as the use of new techniques such as the addition of measured quantities of nutrients.
- The implications of the Water Framework Directive for catchment management.

Thus, the Symposium covered the needs of both scientists and managers, but the emphasis of our discussions, at a period of unprecedented concern for the well-being of the resource, was firmly on the need for decisive management action. It will be apparent, both from the presentations themselves and from the admirable summaries by Session Chairmen, that effective management action depends upon distinguishing between those problems that are tractable and those that are not. Changes in marine climate, even those thought to be partly anthropogenic, are events that managers and their wild salmonid charges must live with. By contrast, there is no way that they should be required to live with the avoidable effects of aquaculture, near-surface pelagic trawling, interceptory netting or poorly-co-ordinated and underfunded marine research programmes. All of these topics receive due attention in the four resolutions that represent the distilled outcome of the proceedings and ensuing discussion. In presenting its resolutions, the Sixth International Atlantic Salmon Symposium expressed its grave concern at the current crisis in the abundance of wild Atlantic salmon and sea trout. The need for purposeful action in four identified areas was strongly urged:

1 Regulation of aquaculture

The Symposium recognized that the scale, location and practice of aquaculture should be controlled in such a way that the well-being of wild salmonids, and of the other fishery and biological resources that share their environment, should not be compromised. This requires a comprehensive, consistent and enforceable regulatory regime, flexible enough to take account of changes in production and developments in knowledge and technology.

It was resolved that these requirements should be met by national planning and regulatory regimes that require all operations throughout the aquaculture industry to be conducted at all times in accordance with a formal set of Codes of Environmental Best Practice, backed up by monitored targets. Mandatory and audited implementation of such codes, as a condition of continued operation, would place as much emphasis on measures that are pro-active and preventative as on those that are reactive.

2 Near-surface pelagic trawling

The Symposium viewed with concern the evidence of the potential threat to migrating post-smolt salmon from the development of commercial near-surface pelagic fisheries for other species. While accepting the need for research to refine the scale of this threat, the symposium confirmed the need to take precautionary action, without waiting for this research to be completed, to identify the means of avoiding possible major damage to stocks.

It was resolved that international government action should be taken as a matter of urgency to prepare for the implementation of technical and administrative measures to

reduce the by-catch of Atlantic salmon by near-surface pelagic trawling, while the conduct of research into the problem should be expedited.

3 International collaborative research

The Symposium endorsed the recent establishment of an International Co-operative Salmon Research Board, under the aegis of the North Atlantic Salmon Conservation Organization (NASCO), in response to the challenge of understanding the causes of wild Atlantic salmon mortality and the actions that can be taken to improve the species' survival.

Convinced that collaborative action involving both governments and the private sector is essential if salmon abundance is to be achieved, let alone increased, and while recognizing the primary responsibility of governments to identify and fund priority research initiatives, the Symposium affirmed that non-government organizations (NGOs) are willing and able to participate fully in identifying potential research projects and to seek private-sector financial support to complement primary funding by governments.

It was resolved that NASCO should be urged to invite immediate representation of NGOs as members of the International Co-operative Salmon Research Board, in order to ensure that organizations and individuals in the Atlantic salmon conservation constituency are given the opportunity to participate fully, as partners with governments, in a shared endeavour to halt the critical decline in wild Atlantic salmon abundance.

4 Mixed-stock interceptory fisheries

The Symposium reiterated the need to continue to press for the ending, with fair compensation, of remaining coastal mixed-stock interceptory fisheries.

It was resolved that governments should be strongly urged to do their utmost to achieve the early closure of fisheries in home waters which take Atlantic salmon from mixed river stocks.

Index

acidification, 183, 186, 195, 284
acoustic hydrophones, 9
acoustic signal relay buoys, 9
acoustic telemetry, 24
acoustic transmitters, 9
Advisory Committee on Rivers Pollution Prevention, 179
Åkra trawl, 10
aluminium, 27, 284
American eel, *Anguilla rostrata*, 204
Ammodytes americanus, 30
Ammodytes sp., 63
amphipods, *Themisto* spp., 30–31
Anderson, John, 228
Anticosti Island, 66
Area Management Agreements, 110, 159, 167–9
Area Management Groups, 159, 167, 169
artificial spawning channels, 206, 219
Association of Salmon Fishery Boards, 167
Association of West Coast Fishery Trusts, 167
Atlantic herring, *Clupea harengus*, 19, 33, 63, 69
Atlantic salmon, *Salmo salar*
　development, 1, 2, 46
　factors influencing survival, 36
　feeding at sea, 25
　marine diet, 30–32
　marine predators, 34, 43–56
　migrations, 26
　mortality, 8
　predation by dolphins, 51–3
　predation by seals, 47–53, 55–6
　post-smolt, 7–21
　recolonization of rivers, 180–81, 193
　straying, 34, 181, 193, 210–12, 214, 219
　survival rates, 33
Atlantic Salmon Association, 87
Atlantic Salmon Federation (ASF), 84, 87–91, 147–8, 151–2, 156, 225, 227–8
Atlantic Salmon of Maine, 152, 156
Atlantic Salmon Trust (AST), 78–9, 90, 167, 285
Atlantic saury, *Scomberesox saurus*, 63, 69
Arctic charr, *Salvelinus alpinus*, 204

Baccalieu Island, 64, 66
Baltic sea, 50, 274
Barents sea, 10, 274
barracudinas, 31

Bay of Fundy, 24, 69, 145–6, 156, 224, 228–9, 236, 284
beluga whale, *Delphinopterus leucas*, 48
Benguela ecosystem, 53
Big Salmon River Association, 225
Bird, Danny, 226, 228
Bird Rocks, 64, 66
Blantyre Weir, 178, 180, 182
blue whiting, *Micromesistius poutassou*, 31
Bonaventure Island, 66
bottlenose dolphin, *Tursiops truncatus*, 48, 50–51, 56
Brankin, Rhoda, MSP, 162
brook trout, *Salvelinus fontinalis*, 204, 220
brown trout, *Salmo trutta*, 46, 204, 218
Burrishoole fishery, 123, 141
buy-outs, 78, 80, 87, 90
by-catches, 19, 20, 34–6, 293, 299, 300

Canadian Species at Risk Act, 24
Cape fur seal, *Arctocephalus pusillus pusillus*, 53–4
capelin, *Mallotus villosus*, 63, 69, 71
Cape St. Mary, 64, 66
Carlin tags, 10, 16
catch-and-release, 209
chum salmon, *Oncorhynchus keta*, 49
Clew Bay, 121–2, 124, 129–30
Clyde Fisheries Management Trust, 182–3
Clyde River Foundation, 182–3, 185
Clyde River Purification Board, 181–2
cod, *Gadus morhua*, 34, 53, 69, 85
Codes of Good Agricultural Practice, 279
Codes of Practice, 164
coho salmon, *Oncorhynchus kisutch*, 235, 239–40, 243–7, 251
commercial salmon fishery moratorium, 213
Committee on the Atlantic Salmon Emergency (CASE), 89
Committee on the Status of Endangered Wildlife in Canada, 236
common murre, *Uria aalge*, 71
Conservation Law Foundation, 152
containment management system 152–3
Control of Pollution Act, 1974, 187
coordinated local aquaculture management system (CLAMS), 154–5
cormorant, *Phalacrocorax carbo*, 4, 26, 96

critical control points, 152
Cromarty Firth, 10, 15
Crown Estate, 166, 182
CTD profiles, 10

data storage tags (DSTs), 62–3, 74
deep water hake, *Merluccius paradoxis*, 53
Delphi fishery, 120
Department of the Environment, Food and Rural Affairs (DEFRA), 80–83
Department of Fisheries and Oceans, 88, 147, 156
depth sensitive transmitters, 18
district salmon fishery boards, 83
DNA, 56, 62, 138, 140, 148, 225
dolphins, 44
Dooley Current, 16
Dundee, 179

ecoregions, 274
Edinburgh, 178
egg deposition, 223
egg incubation boxes, 206, 219
El Niño, 236
Emission Limit Values, 273
Enard Bay, 112–13,
Endangered Species Act, 236
England, 32, 186–98
Environment Agency, 79–81, 188, 191, 196
Environmental Management Systems, 159, 170
Environmental Quality Objectives, 278
Environmental Quality Standards, 273
estuarine barrages, 194
euphausiids, 31
European Inland Fisheries Advisory Commission, 292
European Treaty of Rome 1957, 272
European Water Framework Directive, 162, 184, 271–9
evolutionary significant units, 110–11
Exclusive Economic Zone (EEZ), 11, 20–21

Faeroes Fisheries Research Institute, 10
Faeroe Islands salmon fishery, 89, 91
Faeroes–Shetland Channel, 15, 19, 265–6, 284
Falls of Clyde, 182
FAO Code of Conduct for Responsible Fisheries, 150
Federal Aquaculture Development Strategy, 150
Firth of Clyde, 30, 111–13, 177
Firth of Forth, 177, 180, 184
Firth of Tay, 179
Fish Gap, 159, 161
fish passes, 182, 184, 193–4, 200, 202, 206–9, 213, 218

fish traps, 102, 108, 123–4, 193
Fisheries Research Services (FRS), 97, 166, 265
Fisheries Research Services, Freshwater Laboratory, 181–2
Food and Agriculture Organization (FAO) 148–9
Food Standards Agency, 166
Forest and Water Guidelines, 279
food webs, 54
Forth Fisheries Federation, 183
fry planting, 180-81, 190, 193, 201, 205–6, 212, 214, 219
Funk Island, 63–71, 73, 287
furunculosis, 167–8

Galashiels, 178
gannet, *Morus bassanus*, 61–74, 287
genetic dilution, 147
genetic studies, 286
Glasgow, 177–80
globalization, 2–3
global climate change,162
global democracy, 3
global warming, 2–3, 85, 183, 256
goosander, *Mergus merganser*, 4
Grand Banks, 258–9, 261
Great Bird Rock, 64
Greenland, 8, 26, 50, 89, 259, 261, 263
Greenland salmon fishery, 84, 89–91, 151
grey seal, *Halichoerus grypus*, 43, 47–8, 50, 96
Gruinard Bay, 111–13
Gulf of Maine, 24, 284
Gulf of St. Lawrence, 7, 28, 30, 50, 66
Gulf Stream, 261
Gyrodactylus salaris, 35, 166, 168

haddock, *Melanogrammus aeglefinus*, 32, 148
Hadley Centre for Climate Prediction and Research, 258
halibut, *Hippoglossus hippoglossus*, 148
harbour porpoise, *Phocoena phocoena*, 48, 50
harbour seal, *Phoca vitulina*, 45, 48–9, 54, 56, 96
harpseal, *Phoca groenlandica*, 43, 48–50
Hebrides, 15, 31
Heritage Salmon, 152
Highlands and Islands Enterprise, 160, 166
Holy Island, 83
Humber Estuary, 187, 191
Huntsman Marine Science Center, 225
'hybrid' Atlantic salmon, 138–43
hybridization, 138–43
hydro-electric development, 204, 228

Iceland, 17, 20, 30, 47, 71, 156, 263
index of fry and parr abundance, 101–5

Indian Act, 88
infectious salmon anaemia (ISA), 26, 147, 157, 225
Ingdal Bay, 13
Institute of Aquaculture, Stirling, 165
Institute of Marine Research (IMR), Bergen, 10, 34, 136
Intergovernmental Panel on Climate Change, 250, 257
International Atlantic Salmon Foundation, 87, 89
International Co-operative Research Board, 286, 300
International Council for the Exploration of the Sea (ICES), 8, 90
Inverness Firth, 49
Ireland, 47, 119–33, 154
Irish drift-net fishery, 83
Irish Marine Institute, 12, 124
Irving, J.D. Ltd., 229–30

Jan Mayen Island, 20, 50

Kelvin Valley Countryside Project, 182
Keogh Lake, 237
Kielder Reservoir, 190
Killary Harbour, 124, 129–30
killer whales, 52
KNAPK (Organization of Fishermen and Hunters in Greenland), 90–91
Kocik, John 228
Kyle of Durness, 111

Label Rouge mark, 163, 166
Labrador Sea, 29, 73, 256, 259, 261, 265
lake trout, *Salvelinus namaycush*, 45
lanternfishes, 31
Lea tags, 10
Listuguj First Nation, 228
Local Government, Scotland, Act 1887, 179
long-lines, 10
Lough Feeagh, 141
lumpfish, *Cyclopterus lumpus*, 91

Maastricht Treaty, 1992, 272
mackerel, *Scombrus scomber*, 19–20, 34, 63, 69, 71
mackerel fishery, 19–20, 34–5, 293
Maine Aquaculture Association, 152
Maine Atlantic Salmon Commission, 225, 228
Marine Harvest, 168
Marine Laboratory, Dunstaffnage, 165
microsatellites, 138, 140
micro tags, 10, 16–17, 286
mid-water trawl, 9

minisatellites, 138, 140
Miramichi Salmon Association, 228
Montrose, 10,15
Moray Firth, 50–51
Morley, Elliott, MP, 79
Muir Lake, 237

NASCO Rivers Database, 93, 99, 107
National Aquaculture Act, 1980, 149
National Centre for Environmental Prediction, 257
National Federation for Fishermen's Organizations, 80
National Fish and Wildlife Foundation, 91
National Marine Fisheries Service, 149, 154, 226
National Oceanic and Atmospheric Administration, 226
National Rivers Authority, 188
National Treatment Strategy for Sea Lice, 165–6, 168
Netarts Bay, 50
Net Limitation Order, 79–81
New Brunswick Salmon Growers Association, 225
New Brunswick Wildlife Trust Fund, 225–6, 230
Newfoundland, 200–220, 263–4, 284
nonylphenol (4-nonylphenol), 284
North Atlantic Current (NAC), 15
North Atlantic Oscillation (NAO), 73, 255–6, 266
North Atlantic Salmon Conservation Organization (NASCO), 84, 89–91, 93, 97, 148–50, 155, 162, 286, 300
North Atlantic Salmon Fund (NASF), 78–84, 90–91
north-east drift-nets, 78–83
North Sea, 16, 19, 274
Northumbrian Water, 2
Norway, 9–11, 13, 15–17, 34, 136–7
Norway pout, *Trisopterus esmarkii*, 32
Norwegian Sea, 8, 10, 15–16, 18–19, 21, 26, 30–31, 33, 35, 108, 259, 263, 265, 274, 284, 293

Ocean-Fish-Lift, 136
organophosphate pesticides, 196–7
O'Grady, Martin, 292
Oslo Resolution, 97, 110
osmotic imbalance, 27–8
otoliths, 48
otter, *Lutra lutra*, 48, 107

Pacific decadal oscillation, 235
Pacific lamprey, *Lampetra tridentatus*, 54
Pacific tuna fishery, 44

Passamaquoddy Bay, 146
pelagic research trawl, 10–11
pelagic trawl surveys, 28
pinger tags, 9, 228
plankton, 30–31, 255, 285–6
pollution, 175–85, 186–98
Porcupine Bank, 31
post-smolts
   cataracts, 284
   competition, 32
   diet, 28, 30–31
   distribution in relation to
      prey abundance, 283
      temperature, 17, 28, 283, 285
      salinity, 17, 283, 285
   effect of climate, 283
   effect of surface currents, 28, 283
   food of ranched post-smolts, 30
   growth, 33
   migration, 15, 283, 285
   mortality, 283
   overlap with other species, 19–20
   predators, 34, 48
   predation by gannets, 61–74
   salinity tolerance, 27, 283
   sea lice infestation, 136–7
   spatial distribution, 15, 17, 28–9, 283, 285
   speed of migration, 28
   surveys, 16
   swimming speed, 28–9
   temperature-related mortality, 35
   vertical distribution, 17, 285
precautionary principle, 93, 110
predation
   birds, 61–74
   marine mammals, 43–56

radio tags, 63, 110
radio tracking, 8, 156
Reason, Professor Jim, 1
redfish, *Sebastes* spp., 31
River Basin Management Plan, 273–4
river catchments, 188, 191
River Invertebrate Prediction and Classification System, 277
rivers, Canada
   Biscay Bay, 65–8
   Campbellton, 65–8
   Conne, 65–8, 208
   Exploits, 65–9, 200–201, 209
   Gander, 65–8, 202–3, 213, 215
   Humber, 65–8, 206
   Great Rattling Brook, 200–219
   Keogh, 235–52
   La Have, 69
   Little Salmon, 65, 67–8, 206
   Lomond, 65–9, 201–2, 213, 215
   Lower Clyde, 69
   Lower Terra Nova, 65–8
   Magaguadavic, 147, 225, 228
   Middle, 29, 200, 203
   Middle Brook, 65, 67–8, 213, 215
   Middle Exploits, 201–6, 208, 210–20
   Miramichi, 65, 69, 227–30
   Nashwaak, 228
   Northeast Placentia, 65–8, 201, 203
   Placentia, 213, 215
   Restigouche, 227–30
   Rocky, 65–8, 200–215, 218–19
   St. John, 69
   Salmon Brook, 65, 67–8
   Terra Nova, 200–205, 208–19
   Thompson, 236
   Tobique, 228
   Torrent, 65–8, 200–205, 207, 210–19
   Trepassey, 65–8, 201–3, 213, 215
   Trinité, 228
   Waukwaas, 235–7, 239–52
   Western Arm Brook, 29, 65, 67–8, 201–3, 213, 215
rivers, England
   Aire, 191
   Avon, 9, 12–13
   Bristol Avon, 187
   Calder, 187
   Camel, 195–6
   Don, 191
   Dove, 193
   Frome, 197
   Hampshire Avon, 195–7
   Itchen, 195–7
   Kennet, 193
   Medway, 187, 193
   Mersey, 186–7, 194–5
   Piddle, 197
   Ribble, 187
   Severn, 194
   Stour, 194
   Taw, 80
   Tees, 2, 9, 12, 187, 190, 192–3, 291
   Test, 9, 12, 197
   Thames, 186–7, 192, 195
   Torridge, 80
   Trent, 187, 193, 195
   Tyne, 2, 186, 189–90, 192–3, 291
   Ure, 191
   Wear, 190, 192
   Weaver, 187
rivers, Europe
   Danube, 274

Rhine, 274
rivers, Ireland
 Belclare, 124
 Bunowen, 124
 Burrishoole, 124, 132, 138
 Bush, 33, 196
 Delphi, 124
 Erriff, 123
 Gowla, 121, 123
 Invermore, 121, 123
 Newport, 124
 Owengarve, 124
 Owenmore, 138, 140–41, 143
rivers, Norway
 Figgio, 33, 35
 Gaula, 27
 Ims, 9–10, 12–13, 18
 Ingdal, 13–14
 Orkla, 9, 13, 34
 Surna, 27, 34
rivers, Scotland
 Aline, 98–9, 101
 Allan Water, 180, 183
 Allander Water, 182–3
 Almond, 177, 181, 183–4
 Attadale, 98, 101, 103, 106–7
 Ault Bea, 98–9, 101
 Avon Water, 176, 181, 184
 Awe, 111, 113
 Badachro, 98–9, 103
 Balgy, 98–9, 101
 Black Cart, 176–7, 180–81
 Broom, 98–9, 101
 Carnoch, 98–9, 101
 Carron, 177, 181, 183
 Clyde, 112–13, 175–8, 180–82, 184, 299
 Coe, 98–9, 101, 106
 Corrie, 98–9, 101
 Dee, 49, 53, 79
 Devon, 177, 180
 Dighty Burn, 179
 Dionard, 113
 Don, 49, 53
 Dubh Lighe, 98–9, 101
 Dundonnell, 98–9, 101
 Esk, 184
 Ewe, 98–101
 Fionn Lighe, 99, 101, 106
 Forth, 175, 177, 183
 Fyne, 112–13
 Gala Water, 178
 Garvan, 98–9, 101, 103, 106–7
 Glazert Water, 183
 Grimersta, 110
 Gruinard, 98–9, 101, 113
 Gryfe, 176–7, 180, 183
 Guiserein, 98–9, 101
 Inver, 113
 Inveranvie, 98–9, 101
 Inverie, 98–9, 101
 Irvine, 178
 Kanaird, 98–9, 101
 Kelvin, 176–8, 180–84
 Kerry, 98–9, 101
 Kilmarnock Water, 178
 Kishorn, 98, 101, 106
 Lael, 98–9, 101, 106
 Laxford, 113
 Leith, 177–8, 181
 Leven, 98–9, 101, 176–8, 181, 184
 Ling, 98–9, 103, 112
 Linhouse Water, 181
 Little Gruinard, 98–9, 101, 110–11, 113
 Locher Water, 176
 Lochy, 113
 Lussa, 49
 Moidart, 98–9, 101
 Morar, 98–100
 Murieston Water, 181
 Musselburgh North Esk, 177–8, 183
 Nethan Water, 176–7
 North Calder, 176, 180
 North Esk, 33, 35, 107–8
 Polla, 109
 Scaddle, 98–9, 101, 106
 Sguod, 98–9, 101, 103, 106
 Shiel, 98–101, 103
 Shieldaig, 98–9, 101–3, 105–7
 South Calder, 176, 180, 182
 Stroncreggan, 98–9, 101, 103, 106–7
 Strontian, 98–9, 101, 106
 Suileag, 98–9, 101, 106
 Tay, 81
 Teith, 180, 183
 Tweed, 81, 178–9
 Torridon, 98–9, 101
 Tournaig, 98–9, 101
 Tyne, 183
 Ullapool, 98–9, 101, 113
 White Cart, 176, 180–81, 183
 Ythan, 79
rivers, USA
 Connecticut, 292
 Dennys, 226
 Machias, 226
 Penobscot, 29, 69, 86–7
 Rogue, 54
 St. Croix, 226
rivers, Wales
 Afan, 192, 194

Conwy, 9, 12
Dee, 195–7, 274
Ebbw, 187, 291
Ely, 192, 194
Loughor, 187, 192
Neath, 187, 192, 194, 291
Ogmore, 187, 192, 194, 291
Rheidol, 192
Rhymney, 192
Taff, 186–7, 192–4, 291
Tawe, 9, 12, 192, 194, 291
Teifi, 195–6
Tywi, 195–6
Usk, 195–6
Wye, 80, 195–6
Ystwyth, 192
Rivers Pollution Act 1876, 179
Rivers Pollution Commission, 178–9
Rivers (Prevention of Pollution) (Scotland) Act 1951, 179
Rivers (Prevention of Pollution) (Scotland) Act 1965, 180
rotary screw traps, 237
Royal Society for the Protection from Cruelty to Animals (RSPCA), 166
Russia, 47

St. Croix Watershed Commission, 226
St. Pierre and Miquelon, 88–9, 288
saithe, *Pollachius virens*, 63
salmon escapes, 103–4, 106, 110, 139, 146–7, 181
salmon farming, 26, 93–113, 125–8, 130–33, 145–57, 159–71, 289–90
salmon fishery closures, 78–83, 86–7, 90–91, 151
Salmon and Freshwater Fisheries Review Group, 78
salmon nets, 48, 50, 78–83
salmon rod fisheries, 182–3, 190–94, 196
Salmonid Council of Newfoundland and Labrador, 88–9
Salmonid Fisheries Forum, 164
satellite rearing, 229
scale reading, 17
Scotland, 32, 34, 47–8, 53, 93–113, 159–71
Scottish Anglers' National Association, 167
Scottish Environment Protection Agency (SEPA), 97, 107, 165–6, 180
Scottish Executive Environment and Rural Affairs Department (SEERAD), 166
Scottish Fisheries Coordination Centre, 100
Scottish Landowners' Federation, 182
Scottish lochs
  Alsh, 112–13
  Awe, 112
  Broom, 112–13

Duich, 113
East Loch Roag, 112–13
Eil, 112–13
Etive, 112–13
Fyne, 112–13
Gare, 111–13
Laxford, 112–13
Leven, 112–13
Linnhe, 112–13
Lochy, 112
Lomond, 176–7, 180–81
Long, 111–13
Moidart, 111–13
Scottish Natural Heritage (SNH), 166
Scottish Office, Agriculture, Environmental and Fisheries Department (SOAEFD), 160
Scottish Quality Salmon (SQS), 159, 161, 163–71
Scottish Society for the Protection from Cruelty to Animals (SSPCA), 166
sea lice, *Lepeophtheirus salmonis*, 15, 26, 93, 95–6, 108–10, 119–133, 136–7, 165, 168, 289
sea surface temperature, 257–8, 260–66, 283–4
sea trout, *Salmo trutta*, 43, 45–7, 52, 55–6, 82, 93, 96, 119–33, 194
Sea Trout Task Force, 120
Sea Trout Working Group, 120, 133
seals, 34, 43–56, 110, 287
shag, *Phalacrocorax aristotelis*, 96, 110
shallow water hake, *Merluccius capensis*, 54
sharks, 34
sheep farming, 196–7
Sheffield, 191
Shetland Islands, 15, 18
Sir Alister Hardy Foundation for Ocean Science (SAHFOS), 285
skates, 34
slob trout, 47
smelt, 176
smolts
  effect of acid conditions, 27
  exposure to aluminium, 27
  migration, 27
  return rates, 27–8
  salinity tolerance, 27
  vulnerability to predators, 27, 52
smolt wheel, 222, 227–9
snow crab, *Chionoecetes opilio*, 91
Spain, 47
Special Area of Conservation (SAC), 97, 110
squid, 25, 71
  long-finned, *Illex loligo*, 63
  short-finned, *I. Illeceborus*, 63, 69
steelhead trout, *Oncorhynchus mykiss*, 44, 54–5, 236–6, 240–49, 251
Stolte Sea farm, 152

stomach thermal sensors, 73
Stonebyres Falls, 176
surface trawl, 10, 18, 70
Sweden, 32, 47
synthetic pyrethroids, 186, 197

tagging experiments, 10, 16, 286
temperature loggers, 73
Thames Salmon Trust, 193
transfer of adult salmon, 204–7, 210–12, 214
transgenic salmon, 148
Trent Salmon Trust, 193
Tripartite Working Group (TWG), 110, 159, 168–9
Trondheim fjord, 13–14, 30–31
Trout Unlimited, 152
Tweed Foundation, 183

UK Accreditation Service, 169–70
United Kingdom Register of Organic Farming Standards, 166
University of New Brunswick, 228
US Department of Agriculture, 149
US Environment Protection Agency, 150
US Fish and Wildlife Service, 149

US Food and Drugs Agency, 148
US National Fish and Wildlife Service, 90
US National Marine Fisheries Service, 228
US National Oceanic and Atmospheric Administration, 257

Vancouver Island, 235, 249, 252
Veterinary Medicines Directive, 166
Vigfusson, Orri, 80
Vøring Plateau, 34

water quality, 175–98
Water Quality Objectives, 273
Watershed Restoration Programme, 248
water temperature recorders, 226–7
West Highland Sea Trout and Salmon Group, 110
West Regional Fisheries Board, 124
whales, 34
Whitby, 83
White Sea, 50
whiting, *Merlangius merlangus*, 32
Wilson, Alan, MSP, 162
World Wildlife Fund, 144